41.50

D0561324

YOUR GENETIC DESTINY

Books by Aubrey Milunsky, M.D., D.Sc.

Know Your Genes
Choices, Not Chances
How to Have the Healthiest Baby You Can
Heredity and Your Family's Health

And Medical Texts (as Author, Co-Author, or Editor)

The Prenatal Diagnosis of Hereditary Disorders
The Prevention of Genetic Disease and Mental Retardation
Coping with Crisis and Handicap
Genetics and the Law (3 editions)
*Genetic Disorders and the Fetus: Diagnosis, Prevention and Treatment
  (4 editions)*

YOUR GENETIC DESTINY ·····················

Know Your Genes, Secure Your Health, and Save Your Life

Aubrey Milunsky, M.D. D.Sc.

PERSEUS PUBLISHING
Cambridge, Massachusetts

Perseus Publishing is a member of the Perseus Books Group

Cataloguing-In-Publication data is available through the Library of Congress.

Text design by Jeff Williams
Set in 10 pt Minion by Perseus Book Group
First printing, 2001

Visit us on the World Wide Web at http://www.perseusbooks.com

Perseus Books are available at special discounts for bulk purchases in the U.S. by corporations, institutions, and other organizations. For more information, please contact the Special Markets Department at the Perseus Book Group, 11 Cambridge Center, Cambridge, MA 02142; or call 617-252-5298

To my granddaughters
Julie and Miranda

In the beginning, you have your genes.
In the end, it's what you did with them
that makes the difference.

*"We used to think our fate was in the stars;
now we know in large measure,
our fate is on our genes."*

—JAMES WATSON, nobel laureate

CONTENTS

Genes and Infection • Genes and Behavior • Susceptibility Genes and Brain Injury or Hemorrhage • "Athletic" Genes

HARMFUL GENES

Gestational Diabetes • A Predisposition to Diabetic Complications • Heart Disease • Eye Complications • Kidney Complications • Suseptability to Infection • Nerve Damage • Hypertension • Obesity and Diabetes • Dietary Guidelines

15 Obesity: Genes or Environment?   193

The Obesity Epidemic • Obesity and Associated Disorders • Why and How People Get Fat • Signals and More Signals • Genes and Obesity • Genetic Disorders and Obesity • Risks of Obesity • Maternal Obesity • Maternal Obesity and Birth Defects • Treatment and Prevention

16 Genes and Cancer   203

What Is Cancer? • How Widespreas Is It ? • How Does a Normal Cell Become a Cancer Cell? • Genes at Checkpoints •Cells with Brakes • Gene Silencing • Cell Suicide • DNA Repair Systems • Means to an End • Imprinted Genes • Broken Threads • Causes of Cancer • Genetic Cancer

17 Is There a History of Cancer in Your Family?   211

Genetic Causes • Prenatal Risk Factors for Cancer • Cancer and Ethnic Origin • Birth Defects and Cancer • Cancers of the Breast, Ovary, Colon, and Prostate • Breast Cancer • Ovarian Cancer • Gene Tests Treatment Options • Colon Cancer • Prostate Cancer • Family Cancer Syndromes • Other Cancers • Cancer and Childbearing • Should Children Be Tested? • Privacy and Confidentiality • What You Should Know Before Having a Cancer Gene Test • Gene Therapy for Cancer

18 Mental Illness in the Family: Schizophrenia   240

Is Schizophrenia Inherited? • The Diagnosis • Family Studies • Twins and Schizophrenia • Adoption and Schizophrenia • Early Signs • Eye Movements and Susceptibility to schizophrenia • Genes and Schizophrenia • Treatment • Predicting the Outcome • What Happens Over Time? • Costs

19 Genes, Depression and other Mood Disorders   250

Mood Disorders • Depression (Unipolar Disorder) • Manic Depression (Bipolar Disorder) • How Common a Problem • Suicide • ADHD and Bipolar Disorder • Genes or Environment • Family Studies • Patterns of Inheritance • Genes and Anxiety Disorders • Alcoholism

AVOIDANCE AND PREVENTION OF GENETIC DISORDERS

PREFACE

Contrary to what you might think, your genetic destiny is not preordained. True, the structure and functioning of our bodies result from the genetic blueprints with which we were conceived and born. Indeed, every disease and every susceptibility is caused, influenced, regulated, or modulated by our genes. But despite our genetic blueprints, there is much we can do to secure our health—and even to save our lives or those of loved ones. This book is dedicated to informing you accurately of what you absolutely must know about your genes in order to recognize life-threatening and health-limiting factors and to exercise lifesaving and health-preserving options. If you choose not to know, you might not be altering your chances, but you most certainly will be limiting your choices.

Genes are miraculous in their orchestration of the form and function of our bodies. Given the complexity of our genetic blueprints, it is comforting how many of us continue to lead normal, healthy lives; we should keep this broader perspective in mind as we explore the consequences of errant genes and chromosomes. Yet even if you are presently enjoying good health, the wealth of information in this book is of immediate and compelling value and could prove crucial as you make decisions that affect your future well-being. It could be kept with advantage near at hand, as a critical repository of key genetic facts. As time passes and life unfolds, you and your family might well be confronted with one or more of the conditions described in this book.

You need no prior knowledge of genetics or biology in order to understand the facts outlined here. You do, however, need to know about serious illnesses, mental retardation, and birth defects in your extended family. The genetic aspects of many common diseases that affect virtually all families—including heart disease, cancer, hypertension, obesity, diabetes, mental illness, and Alzheimer's disease— are fully surveyed in this book. In addition, a number of genetic disorders are explored that are specific to particular ethnic groups. In the context of these common and less-common diseases, you will discover which genetic tests may be important for you or a close family member. As you read, you may even recognize the signs of a genetic disorder and susceptibilities in your family of which you were previously unaware. The book also provides explanations about the causes and risks of mental retardation and birth defects, together with information about newly available genetic tests and options for those planning to have children. The first ten chapters provide a foundation and framework, thereby making the chapters that follow even more meaningful. There is in fact much you may not have realized you need to know. Indeed, to know is to care, and knowing takes time and effort.

No one among the reading and listening public could have missed the continuing barrage of information about the Human Genome Project. Gene discovery has dominated science and has flooded the media. Important new DNA tests have emerged and have constituted critical advances in the diagnosis of genetic disease and the identification of gene carriers. New, predictive DNA tests can reveal the presence of a genetic disease decades before its physical onset, making it possible for us to seek early, lifesaving intervention. Because we are all potential carriers of a number of harmful genes (as further elucidated in these pages), knowledge of these advances is critical.

The Human Genome Project concentrated on mapping the three billion bits of information that constitute the human genetic code. Ironically, through the clarity born of discovery, one unintended result of this project has sharply refocused our attention on and heightened our appreciation of the environmental causes of the common diseases to which we are all subject, due to the susceptibility genes we have inherited. Some fortunate individuals whose long-lived families have few genetic disorders of consequence inevitably have some genes that convey one or more of the many susceptibilities to which we are predisposed and which are discussed extensively. This observation suggests that we could benefit from giving closer attention to the many factors of environment and of lifestyle that are known to increase our vulnerability to disease—factors that we might easily limit, manage, or avoid.

To be sure, there are more than 8,500 catalogued single gene diseases or traits. Much of this book concentrates on the quintessential information you need in order to prevent, avoid, and treat these disorders. The recent explosion in genetic knowledge, however, can be overwhelming to practicing physicians, and much of the information in this book has not yet penetrated general medical practice. Consequently, those seeking the most up-to-date information about a genetic disorder or wishing to take advantage of the latest advances in evaluation and intervention may need to consult a clinical geneticist. Moreover, those who have had a previously affected child or a genetic disorder in the family may need to contact or recontact a clinical geneticist, given the dramatic advances that have occurred in human genetics. Much personally valuable information is packed within these pages, and where possible, repetition is avoided. Readers are therefore encouraged to use the extensive index and appendices to assist in their search for all the information they seek.

Science invariably outstrips law, and the remarkable recent discoveries in genetics are no exception. Early instances of genetic discrimination occurred before legislative action caught up. Fears about the possible loss of health insurance or denial of life insurance, or about social stigmatization, have led some individuals at risk to forgo necessary genetic tests. Fortunately, most U.S. states have enacted legislation barring genetic discrimination, including the use of private genetic information in considering applications for employment as well as those for health and life insurance; the Congress is still wrestling with similar legislation at the federal level. Meanwhile, former president Clinton signed an executive order to guarantee the privacy of personal medical information, including genetic data. These developments are explored in greater detail in Chapter 26.

The deciphering of all our genes—the human blueprint—is a scientific and technical feat of enormous magnitude and importance to humankind. From these discoveries we have begun to understand fundamental flaws in ourselves that lead to common serious diseases, including cancer, heart disease, and brain and muscle disorders. These new insights will fuel the discovery of new therapeutic approaches to remedy heretofore irremediable diseases. Through its early, halting steps—and some notable successes—gene therapy will continue to evolve. Major initiatives in treatment are likely as we learn about the functions of the protein products of genes. However, our expectations must be moderated to fit the realities of scientific endeavor and the stringent standards of medical proof.

While rapid progress continues, there is much you can do now for yourself and your loved ones. Know your family history, be cognizant of your ethnic origin, determine your genetic susceptibilities, opt for necessary gene tests, take preventive actions, establish appropriate surveillance, and seek preemptive treatment where applicable. In this way, you can exercise control over your genetic destiny, secure your health, and—in more ways than you yet realize—save your life.

—*Aubrey Milunsky*

ACKNOWLEDGMENTS

The phenomenal recent developments in human genetics have generated a prodigious volume of scientific data. Thousands of scientists have contributed to this veritable explosion of knowledge about our genes and those of other species important to our health and salvation. I acknowledge with gratitude and humility their enormous contributions. The work on which this book is based includes many hundreds of specialist medical books and technical journals, not all of which could reasonably be credited or listed within its covers. For readers seeking more technical information, a list of selected reference books used is provided in Appendix D. From this massive body of information, as well as decades of experience, I have synthesized the essential facts and lessons critical to our genetic health. All of the information provided here has a scientific basis and a recognized source.

I acknowledge with gratitude the outstanding work and support provided by my executive secretary, Marilyn McPhail, whose standards of perfection and eye for detail have made a challenging task more pleasurable.

It is with special pride and joy that I acknowledge the critical advice and assistance provided by my son, Jeff, who is now a triple Board-certified pediatric geneticist and director of clinical genetics and associate director of the Molecular Genetics Diagnostic Laboratories at Boston University School of Medicine. If genes have meaning, I am basking in the warmth and joy of their reflected function!

The manuscript was also carefully and critically read by my wife, Laura Becker, Ph.D. I am most grateful for her outstanding linguistic skills and insightful comments, as well as for her love and companionship.

KNOWLEDGE IS LIFESAVING ···············

chapter 1

What You Should Know, and Why

....

Contrary to popular belief, your genetic destiny is not preordained. True, you are born with a genetic blueprint that dictates the structure and functioning of your body. Notwithstanding this irrevocable fact of nature, you are not a hapless victim, inevitably bound to suffer the inexorable consequences of any flawed genes you may have inherited. Although every aspect of health and all disease is controlled, regulated, modulated, or influenced by your genes, genes are not the last word; there is much more you should know and much you can do to influence the outcome. The purpose of this book is to empower and encourage you to know your genes so that you can secure your own health and that of your loved ones.

The first step is to develop and maintain as thorough a knowledge of your family medical history as possible (see Chapter 2). Second, remain attentive to the remarkable and rapidly escalating advances in genetics reflected daily in the mass media and on the Internet. Third, select a physician who understands the importance of genes and is willing to assist you in genetic health maintenance. Until recently, our genetic fate was like a big black box—sealed and impenetrable, unfathomable and irrevocable. This is no longer the case. You might well be surprised at how much new information has come to light on human genetics, and how relevant it is to you and your family. There is much in this book that you need to know, that could save your life and secure your health and those of all your loved ones. You do need to care, to know.

To Know Is to Care

Ignorance is not bliss, especially when it comes to health. Only a few decades ago, the best that genetics could offer was a basis for calculating our statistical chances of being afflicted by a disease or disorder due to a defective gene, of carrying such a gene, or of passing the gene along to our offspring. More recently, paralleling the discovery of genes that cause specific genetic diseases, precise diagnostic methods have emerged. These advances have created opportunities for highly effective detection and diagnosis of genetic disorders—even prenatally—as well as for definitive predictive tests. Particularly important are new procedures enabling us to identify genetic defects that spell potentially life-threatening, future complications (such as cancer or a heart disorder). Predictive or presymptomatic testing is discussed fully in Chapter 24. The point I wish to emphasize here is that knowledge of your family history could prompt a genetic analysis that might enable you not only to save your own life but also the lives of those dear to you.

Imagine, as you cross a street, seeing a truck barreling down a hill and recognizing that you have no escape. If you had been forewarned and spotted the truck before you entered the crossing, you could have taken evasive action and avoided being hit by the truck. Similarly, if you were standing on a street corner watching a loved one cross the street and you saw the truck bearing down, would you not automatically call out a warning? Lacking any such warning, you would share the desperate sadness I feel when I see patients who have ignored their family history and discovered too late that tests were available that might have been lifesaving. I have repeatedly seen siblings and cousins, uncles and aunts, and even parents whose lives were lost or placed at serious risk because of their failure to pay attention to genetic disorders in their family. It is worse still to communicate the diagnosis of a serious genetic disorder in a child and witness the devastating realization by the parents (and other family members) that one of their own was affected but had failed to warn close relatives about their risks and available tests.

It may take a great deal of effort to unearth the facts of your family's health history, especially given the dehiscence of modern families. Discord is common in today's families, and full communication—especially about private health matters—is frequently limited or absent. Yet these barriers must be overcome if you are concerned about yourself and your loved ones. You must also overcome the natural fear of finding out that you are affected by or are carrying a culprit gene. Although denial may be helpful as a psychological defense mechanism, it can be a fatal flaw where genetic risks are concerned. Obsessing about potential risk factors is similarly unhelpful because it tends to paralyze us and prevent timely action. The all-important first step is to care enough to become as well informed as possible about your family medical history. This is an ongoing process: You should update the information you have on a regular basis, at least annually (see Chapter 2).

Why You Should Know

Never before in the annals of human history was it possible literally to "know your genes." Now, as a consequence of the Human Genome Project—an international effort to map the structure of all human genes—the location and identification of every human gene is essentially complete. These advances are comparable in significance to the splitting of the atom and walking on the moon. They will have a huge impact, because genes play a major role in all matters of health and illness—not only in inherited disorders. Genetic disorders affect at least one person out of every ten living in the Western world. In the United States, over 26 million people are affected by an inherited disorder. Of these, most suffer from diseases caused by the interaction of multiple genetic and environmental factors; many others are directly afflicted by a disease caused by a single gene (such as cystic fibrosis, Huntington's disease, sickle cell disease, and hemophilia).

Thus far, more than 8,500 genetic diseases or inherited physical or biochemical traits caused by single defective genes have been catalogued. These diseases and conditions account for 25 to 30 percent of admissions to the major children's hospitals in the United States and Canada. Their preponderance in proportion to other causes of hospitalization is partly due to the waning of serious infectious diseases over the past few decades as well as to improved early recognition of genetic disorders and to the increased life span of some of those affected. Moreover, three or four of every 100 babies born have a recognizable major birth defect that may or may not be genetic. By age 7, an additional 3 to 4 percent of children are found to have major birth defects that were not evident when they were newborn. Disorders involving the heart, muscles and ligaments, kidneys, eyes, brain, and skin were among those most often diagnosed at this later stage. Overall, then, between 6 and 8 percent of all children are born with or later manifest major birth defects. In addition, more than 6 million individuals in the United States are mentally retarded due either to hereditary or acquired disorders.

Clearly, genetic disorders may threaten life or cause all manner of health problems. The key reason to find out as much as you can, as soon as you can, is to obtain timely diagnosis and treatment that may preempt or ameliorate the condition. By knowing your genes, you can save your life: Once you recognize a condition as genetic, you can determine your specific risk, elect precise diagnostic tests, establish careful surveillance by regular testing, and take other preemptive or therapeutic measures (such as elective surgery). The same considerations apply to your children and close family members.

We are all carriers of some helpful and some harmful genes (see Chapters 6 and 7). Increasingly, it is becoming possible to determine through laboratory testing which defective genes we carry that make us susceptible to common genetic disorders and that increase our risks of having affected children. In this context, it is important to recognize that we all have genes that make us susceptible to specific diseases and to certain environmental agents. These genetic susceptibilities or predispositions may result in illness, allergic reactions , adverse effects of medications, stomach complaints from dietary items (e.g., milk), and even death. The identification of our genetic susceptibilities may provide us opportunities to avoid illness or worse catastrophes (see Chapter 9).

What You Should Know

All aspects of health and illness are both governed by genes and influenced by environmental factors. The vast majority of disorders that affect mankind are due to interactions between genes and environmental agents such as bacteria and viruses, dietary components, and toxins. Topping the list of what you should know are details about your family medical history (see Chapter 2). Ethnic background is also extremely important, as many genetic traits are linked with ethnicity. Indeed, knowledge of your ethnic origin is second in importance only to knowledge of your family medical history. The determination of your genetic risks based on your ethnic origin could prompt specific gene tests enabling you and your family members to avoid a number of disorders (see Chapter 8). You should maintain a family tree (pedigree) chart, updated annually, and request that your physician place a copy in the front of your medical record (which invariably is not the case at present). Unfortunately, a number of laws still exist that impede the access of adoptees to information about their biological parents (see Chapter 26). Such laws must be changed.

It is important to realize that there is no family without some flawed genes. Recognition of a clearly inherited disorder should lead you first to seek a referral for genetic counseling. This would apply not only to yourself but to all of your family members who could be at risk. Many are unaware that there are Board-certified specialists in clinical genetics who provide genetic diagnoses and counseling. These physician specialists, who are certified by the American Board of Medical Genetics and most of whom are members of the American College of Medical Genetics, are likely to be found in every major university medical center. A consultation with a clinical geneticist could prove the most important step you take in your life.

If faced with evidence of a genetic disorder in your family, you should make every effort to obtain an accurate diagnosis. Increasingly, the diagnostic process involves analysis of a specific gene. A precise genetic analysis will facilitate an accurate, early diagnosis and give you time to take preemptive steps. It also will permit you and your physician to identify the best means available for monitoring and maintaining your health.

A Responsibility to Tell

The discovery that you harbor a certain genetic defect may have important—even momentous—implications for your personal future plans as well as your present health. But remember also that genes travel: Your children, your parents, grandparents, grandchildren, uncles, aunts, and cousins may also unwittingly be at risk. Each of us has the ethical responsibility to communicate vital information to our relatives, although the choice of whether to act on such knowledge is theirs alone. Sadly, I have repeatedly seen unnecessary illness, genetic defects, and death caused by a lack of communication about such matters within families. Hiding the truth behind a cover of concern for the feelings of individual family members is unacceptable when the wellness and the very life of another is at stake. We have a moral imperative to tell.

The death of a child due to a grave birth defect or genetic disorder is a painful tragedy that lives on in eternal memory. Years later, when the siblings of such a child are planning their own families, the facts about their deceased sibling become important because of possibly increased risks to their own offspring. Yet well-meaning parents too often fail to tell their children about such tragedies. We prefer to avoid talking about severely retarded, institutionalized relatives, stillbirths, and other family defects that signal increased risks to us. Protectiveness, shame, culpability, superstition, unwillingness to relive the pain in telling, and ignorance of the importance to others of knowing the facts are the main motivating factors I have observed. Families simply must realize their responsibility to tell. Equally important is the need for couples who have had a defective child to inform their siblings, uncles, aunts, and cousins who are still in their childbearing years. Families that care will share their knowledge.

Types of Genetic Disorders

A clear understanding of the different ways harmful genes are transmitted is important (see Chapter 6). Some genetic diseases involve only a single gene, as is the case for Huntington's disease, a progressive brain degeneration that causes dementia, speech defects, and purposeless muscle movements. A single-gene disease, inherited from one parent, is called *dominant*. Another type involves defects in a pair of genes, as in cystic fibrosis, a disorder that causes chronic lung infection and malabsorption of food. This *recessive* disorder is transmitted equally by two parents who are completely healthy and who usually are not even aware that they carry this condition in their genes. *Sex-linked* disorders involve single genes that cause diseases that mainly (but not exclusively) affect males. Hemophilia, a bleeding disorder caused by a missing blood clotting factor, is linked to a defective gene transmitted by females but occurring almost exclusively in males. The remarkable *mitochondrial* inheritance is due to defects in genes that congregate in tiny structures called mitochondria, which are dispersed in the cell fluid around the nucleus. The entire set of genes in mitochondria originates from the mother. A few, rare genetic conditions are due to mitochondrial malfunction. In such cases, a mother would transmit the disease to all of her children.

Genetic disorders that result from the interaction of multiple genes with one or more environmental factors are called *multifactorial* or *polygenic*. Examples include high blood pressure, coronary heart disease, schizophrenia, and many other conditions discussed in this book.

Major birth defects, including mental retardation, may arise as a consequence of defects in the development of cells in the early embryo. Such defects include heart malformations (such as a hole in the heart, called *ventricular septal defect*), an open spine defect (called *spina bifida*), a cleft lip and palate, and hundreds of others. Defects such as these may occur as the result of the failure of cells to migrate, to organize into defined tissues, to differentiate into specialized cell functions, or even to die when programmed to do so. A whole panoply of environmental factors (including diet, infectious agents, toxins, and medications) may by themselves, or through interaction with our genes, result in such defects (see Chapter 9). These disorders may be present at birth (called *congenital*) and could be due to environmental factors alone or to the interaction of such factors with our genes. You will obtain a clearer understanding of the various ways in which genetic disorders and birth defects occur from the chapters that follow.

Good Genes, Bad Genes

The oft-heard compliment regarding "good genes" invariably refers to admirable qualities and characteristics that a person might have inherited from a parent or grandparent. While it might be wishful thinking to have a category of "good" genes, we would all like to think that these genes are "healthful," enabling, and protective. "Healthful" genes might be those that help us maintain and achieve good physical and mental health with at least average intellectual capacity. Such genes probably do exist but almost invariably can be expected to function through close interactions with environmental influences, which include family and the physical environment. "Good" genes may also enable superior performance—physical, artistic, musical, and intellectual. Remember the Swiss Bernoulli family of mathematicians, in which there were three generations with eight mathematicians of unusual ability and more than 120 descendants, the majority of whom achieved distinction. "Good" genes might also be expected to be protective—for example, against infections or even cancers. Genes that govern the veritable army of immune mechanisms warding off environmental insults to our bodies are in this group: the genes that orchestrate antibodies, white blood cell functions, tissue repair systems, and many other protective mechanisms.

"Bad" or disadvantageous genes include those that are structurally defective and that lead directly to diseases as well as those that act in concert to produce only dysfunction. Much of this book is about the importance of recognizing and detecting genes that, make us susceptible to specific diseases or predispose us to react in ways that endanger our lives or health. Detection of a defective "mutated" gene could be the key tip-off that could save your life. For example, there is a single gene that, when defective, predisposes an individual to develop alarmingly high temperatures (for example, 108°F) while undergoing a routine surgical operation under anesthesia. Deaths on the operating table from malignant hyperthermia are well known but thankfully rare and treatable if detected immediately. Many culprit genes predispose our bodies to react adversely—sometimes even with fatal consequences—to medications we may take or environmental factors to which we may be exposed. Examples abound. Allergy to penicillin on rare occasions might be so overwhelming as to be fatal (anaphylaxis). When exposed to sulfa medications or certain other drugs, those born with an enzyme deficiency called glucose-6-phosphate dehydrogenase (G6PD) are particularly likely to develop serious hemolytic anemia (in which the life span of circulating red blood cells is dramatically shortened). Remarkably, disadvantageous genes may also confer benefits. Carriers of the sickle cell disease gene have been shown to be more resistant to malaria. The same can be said for female carriers of the sex-linked G6PD deficiency.

It's All in the Genes

Have you ever realized how unique you are among the more than 6 billion people that inhabit the earth? Unless you are an identical twin (or an extremely rare identical triplet), it is unlikely you will encounter anyone else with a genetic makeup identical to your own. Even more remarkable is the astounding realization that humans share 99.9 percent of their genes and at least 98.5 percent with chimpanzees. There must therefore only be a mere handful of genes that enable us to walk upright, compose music, reason, and make moral distinctions. We do not yet understand the precise differences in genes between humans and chimpanzees. But this question is of more than rhetorical interest, because apes are less susceptible than humans to diseases such as cancer and AIDS. Biochemical differences between humans and apes are being discovered and work is in progress to define the genetic origins of the two species. Early results indicate that a certain type of sugar (sialic acid) is missing from the surfaces of all human cells. This molecule may play a variety of roles. For example, it may influence the communication pathways used by genes to regulate an infinite number of functions.

The number of actual and potential interactions and communications within and between the cells of an organism is infinite. Common sense would suggest that similar cell functions must be necessary to maintain life in any species. Since we know that all bodily functions and structures are dictated by genes, it should come as no surprise that key genes responsible for fundamental life-maintenance functions in humans are also present in even the most primitive of species. Advances in biotechnology have shown that many genes common to species such as zebra fish, frogs, worms, and fruit flies have been conserved in humans and other mammals (e.g., mice and other rodents, pigs, and other species). We now know that highly conserved genes—for instance, those found even in the zebra fish or frog—when occurring in mutant form in humans can cause serious and even fatal disorders. This knowledge of gene conservation has been valuable in the detection of human genes. For example, once our team had identified a gene responsible for pigmentary and other anomalies in mice, we were able to locate a comparable (syntenic) region on human chromosome 2 and discover a gene that causes a disorder called Waardenburg syndrome (characterized by variable presence of deafness, a white forelock, eyes of different color, patchy hypopigmentation, and wide-set eyes).

The Gene Orchestra

Like the instruments in a large symphony orchestra, not all genes are in action simultaneously. Many genes function only for a few hours or days before switching off, never to function again. Such genes play critical roles—for example, in directing how the sperm fertilizes the egg, or in the early stages of the embryo, dictating the timing of cell replication, organization, migration into defined regions, and differentiation into specialized tissues and organs. Many genes participate in these many stages, but only briefly. Later in life, there are occasions (for example, when the body is threatened by certain types of cancer) when some embryonic genes turn on again and even begin manufacturing embryonic proteins.

Imagine for a moment the orchestra of genes necessary to create and run the human heart. Structural genes are responsible for influencing primitive embryonic cells into becoming the genetically predestined specialized cells of the heart muscle, capable of producing rhythmic beats for a lifetime. Other genes meanwhile simultaneously are forming and organizing the inner lining of the heart and the cement (fibrous collagen tissue) that holds all of these cells together. In another orchestral section, other genes are responsible for developing and laying out the coronary arteries and their branches as well as their inner linings and vessel wall structure. Synchronized, harmonious gene action is vital to ensure that the inner lining of the coronary arteries, the collagen and muscle of the arteries and veins, and the connections between their tributaries are formed and integrated. In simultaneous but separate action, another set of genes functions to develop and thread the key nerves that convey the electrical impulses controlling the heartbeat. The geography of the coronary arteries, and even the electrocardiogram, directly reflect the functions of genes. Defects in any one of these critical genes may result in fatal genetic diseases of the heart muscle (cardiomyopathy) or inherited disturbances of heart rhythm (arrhythmias) like those that have caused the deaths of various famous sports figures on the court or the playing field. Genes even dictate that the heart be positioned on the left side. When these so-called laterality genes are defective, the heart could fail to rotate, remaining instead on the right side of the chest. The same (and possibly other) laterality genes can reverse the positions of the spleen and liver.

This brief introduction serves as a broad outline of why and what you should know about your genes, and the importance of this knowledge to you and your loved ones. This early preview of the complex nature of gene function sets the stage for more detailed discussions. Remember, to know is to care, and caring takes effort.

chapter 2

Personal Considerations and Your Family History

··································

You are a carrier of a handful of defective genes. We all are. Which genes do we carry, and how important or harmful are they? They may have remained unknown until recently. Most of the defective genes we carry are cryptic—that is, they remain undetected throughout our lifetimes. These usually recessive genes rarely impair our own health, but when mated with a partner's similarly defective genes, they may produce serious disease or defects in our offspring. Recessive, dominant, sex-linked, and other types of genes (see Chapter 6) may lead to disorders that if detected early enough can be avoided, treated, or ameliorated. Information about such possibilities can be culled from your family history.

Your Family History

If you reached for a pencil and paper now and drew a diagram of your family tree (pedigree), how far would you be able to get, and how accurate would it be in accounting for all stillbirths, miscarriages, birth defects, genetic disorders, cancers, heart disease, hypertension, mental retardation, Alzheimer's disease, sudden deaths, diabetes, mental illness, and the sex of affected individuals? You would focus on your siblings, parents, uncles and aunts, nieces and nephews, first and second cousins, grandparents, and children. You would include your ethnic or ancestral origin, given that serious genetic disorders typically affect particular ethnic groups (see Chapter 7). Of particular importance is consanguinity— unions between direct blood relatives such as first cousins, or uncles and nieces. First cousins have a risk of between 6 and 8 percent of having a child with a birth defect, genetic disorder, or mental retardation—about double the risk borne by the general population. The presence of a particular recessive genetic disorder (e.g., thalassemia) in the related families may create an even greater risk of the same genetic defect occurring in their offspring—as much as 25 percent.

Particular attention should also be given to familial disorders of early onset, such as breast cancer or colon cancer that occurs before the age of 40. Serious disorders occurring at early ages are almost invariably genetic.

Whom to Ask

Frequently in a family there is a grand historian—often a grandparent, an aunt, or an uncle who has detailed information, including the health status of all relatives. Such oral historians are precious and their memories should be taxed and committed to paper. Accurate information is critical in the construction of a family pedigree but is often difficult to obtain. Death certificates are notoriously uninformative and many are even misleading. A stated cause of death such as "pneumonia" may obscure the fact that the person had a chronic genetic disorder of the nervous system and succumbed to pneumonia. Autopsy reports are usually reliable, but they also are not infallible. Medical records, which can be extremely valuable, can be obtained only by next-of-kin relatives. Photographs of children born with physical defects and/or mental retardation could in retrospect reveal an undiagnosed genetic syndrome. Genetic test reports—for example, about chromosomes or gene analyses—can provide definitive diagnostic and risk information. On occasion, families at high risk of having an affected child with a serious defect have requested the exhumation of a previously deceased child or of a close relative.

Constructing Your Family Pedigree

The simple exercise of drawing your family tree and including the necessary medical details could well prove to be one of the most important steps you take to secure your health and save your life or those of your loved ones. It is worth taking the time and effort to obtain precise and complete medical information. Remember, death certificate information is frequently misleading because it invariably focuses on the cause of death. The fact that the individual had a specific genetic disorder may not even be mentioned. Specific notes should be made and questions asked about miscarriages, stillbirths, birth defects, deaths in the first month or two of life, mental retardation, any genetic disorder, marriages within the family (first cousins, uncle-niece, and so on), and the ethnic origin of each grandparent. The family members it is most important to include in a basic family chart are grandparents, parents, uncles and aunts, first cousins, children, and grandchildren. Adopted individuals should have the right to obtain information about their pedigrees and to be informed through social service agencies about any newly discovered genetic diseases in their biological family that may raise their personal risks or those of their children. Once drawn, the pedigree should be kept with your important papers and updated whenever family members have children, with or without birth defects, or develop serious illnesses.

The symbols to be used in constructing a pedigree are shown in Figure 2.1. Using these symbols, you should be able to construct your own pedigree, basing it on the example shown in Figure 2.2. Notice that in Figure 2.2 each generation is labeled with a roman numeral on the left and that each person in each generation is numbered so that we can refer to the individuals more precisely. Although I use the numbers here for discussion purposes, you should write the names of the individuals below their symbols.

To give you some idea of how valuable pedigrees are, I describe on pages 17 and 18 the medical facts I gleaned and communicated to a woman (indicated by an arrow in Figure 2.2 as the "proband") who came to see me because she had given birth to a son with mental retardation.

Decisions to Transmit Harmful Genes

The "new genetics" facilitates the detection of harmful genes, which if transmitted could result in birth defects, disfigurement, mental retardation, physical and/or emotional pain and suffering, and chronic illness or death. Never before was it possible to test for a wide range of such genes, recognize risks for future offspring, and consider options that include not having children, employing new assisted reproductive techniques, or having prenatal diagnosis and selective abortion if the fetus is defective (Chapter 22).

Few potential parents would so disregard the health and welfare of a future child that they would choose not to learn about the risks and options available. The consequence of knowingly conceiving and bearing a child destined to suffer and die must be even greater emotional pain to parents than if no opportunity existed to avoid this tragedy.

Parents who experience the sudden grief of learning that their child has a major birth defect, might die, or is likely to be severely mentally retarded experience complex reactions and adaptive mechanisms. Depression and feelings of guilt are inevitable, even though no culpability can be ascribed to either parent. Bitterness is common and occasionally is accompanied by envy of close relatives

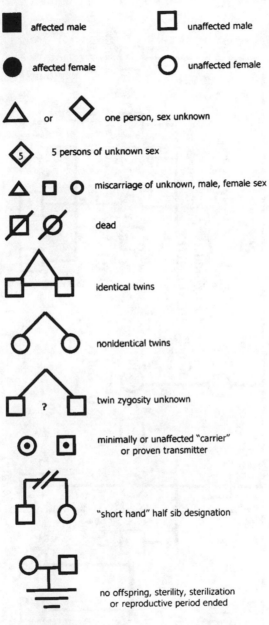

FIGURE 2.1   Pedigree Symbols

16

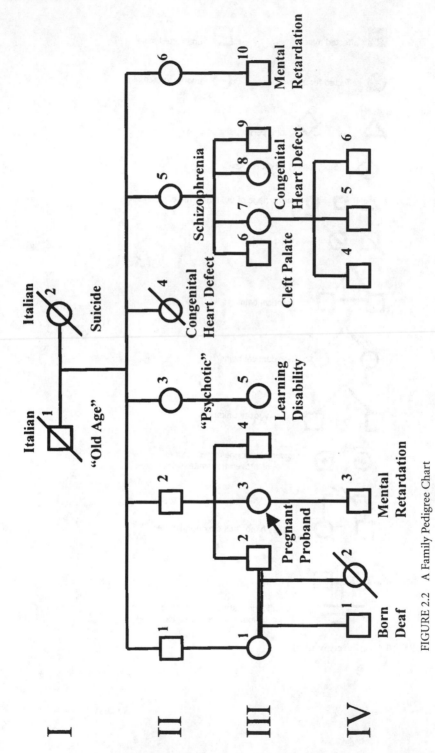

FIGURE 2.2   A Family Pedigree Chart

Allison's son (IV-3) had been fully evaluated and no cause had been determined for his mental retardation. He was then 12 years old, and Allison and her husband had decided, given his serious disability, to try one more time to have another child. She had elected to have prenatal genetic studies (see Chapter 22) due to her advanced age (39).

When I asked her specifically to tell me about any family member with a birth defect, mental retardation, genetic disorder, cancer, or serious chronic illness, she related the following details about her family, which she had obtained from her father prior to her consultation with me. She told of a psychotic aunt (II-3) on her father's side as well as another aunt (II-5) institutionalized for schizophrenia. Her father had had a sister (II-4) who had died in early childhood, having been born with a serious heart defect. Her grandmother (I-2) had committed suicide at a young age. Both grandparents were of Italian ancestry.

One first cousin (III-5) had a significant learning disability, but since this individual had moved to California, she had lost touch and was unaware if there were any other associated problems. Allison also had another first cousin (III-6), born with a cleft palate, whose sister (III-8) was also born with a serious heart defect, remedied by surgery during childhood. Another male first cousin (III-10) was diagnosed with moderate mental retardation in early childhood.

Allison's brother (III-2) had married a first cousin (III-1), and their first child had been born deaf. A second child (IV-2), who had appeared healthy at birth, had fallen ill only days later and died at the age of about two weeks. Allison was unaware of any diagnosis, and no autopsy had been done.

Through the relation of these facts, two evident patterns of abnormalities came into sharp focus. The first was the occurrences of mental retardation (IV-3 and III-10) associated with heart defects (II-4 and III-8), cleft palate (III-6), and mental illness in three individuals (I-2, II-3, and II-5). This first pattern included the learning disability in the first cousin (III-5).

This concatenation of features was directly reminiscent of a disorder called velocardiofacial syndrome. All of the features mentioned occur in this syndrome and

(continued)

with normal children. Either or both parents may indulge in self-pity or seek solace in alcohol. Initial denial of both the diagnosis and the prognosis is frequently followed by anger directed at physicians and other health care providers for not having prevented the tragedy. All of these feelings are frequently compounded by

(continues)

frequently are associated with other items not mentioned or noticed in the family, such as characteristic facial features, some difficulty in swallowing, a small cleft in palate, an undiagnosed heart defect, a learning disability, poor performance in school, mild mental illness, and many other signs, both major and minor.

Allison's son (IV-3) had not been evaluated for this condition because the necessary technology (permitting the splicing of a piece from a gene on chromosome 22, called a deletion) had not been available when his chromosomes were studied. Using a type of analysis called fluorescent in situ hybridization, or FISH (see Chapter 5), we were able to establish a diagnosis for her son and subsequently for a few of her other family members who were willing to be studied. Moreover, Allison and her father were noted to have similar facial features, and only later acknowledged learning problems and some incoordination of the palate during swallowing.

The second pattern of concern was the mental retardation in two males (IV-3 and III-10), who were the offspring of two apparently normal females. This signaled the possibility of fragile X syndrome (see Chapter 9), which was rapidly excluded by precise DNA diagnostic tests that had not been available when Allison's son was first evaluated.

Later we were able to evaluate the child (IV-1) who was born deaf. Since this child was the product of a marriage of first cousins, there was a high probability that the deafness was due to a specific mutation in the gene that most often causes recessive deafness (Connexin-26). This did indeed prove to be the case. The implication was that this couple (III-1 and III-2) had a 25 percent risk in any future pregnancy of having another deaf child. Although the child who died (IV-2) was undiagnosed, this early death was most likely due to a biochemical genetic disorder related to the consanguinity (blood relationship) of the parents. If that were true, they would also have a 25 percent risk in a future pregnancy of having a similarly affected child. No tests, however, would have been recommended, since no diagnosis of the deceased child had been made.

When we had completed our laboratory analyses, I informed Allison that she had a 50 percent risk of having another child with the same syndrome. Fortunately, the prenatal genetic studies confirmed that the fetus she was carrying did not have this syndrome or any other chromosome abnormality.

increasing frustration over time, as the lack of a cure and the absence of any progress dawn with stinging reality.

Most often the presence of a child or adult with a serious birth defect or genetic disorder in the home becomes a chronic emotional, physical, and economic drain on the parents (or family), often leading to a severe state of exhaustion that affects every facet of life. These stresses and strains invariably increase marital conflict. The sex life of the couple becomes a major casualty, further feeding the fires of anger and frustration. Separation and divorce are only too well known in families in which such tragedies have occurred. The enormous drain on the energies of the parents frequently leads to an unwitting relative neglect of their unaffected children. Except for those parents with outstanding economic resources, the sheer physical and emotional exhaustion deprive parents of the necessary energy and time to still attend to their healthy children, as they would otherwise have done. Not infrequently in these situations, emotional, behavioral, and psychological problems develop in the healthy siblings, which may remain unrecognized and unaddressed.

There are many families in which children with chronic defects or genetic disorders manage admirably. It is, however, a very rare family that truly benefits from such tragedies. Although the qualities of compassion, patience, and love can be learned by caring for a handicapped child or adult—a healthy adaptation—this scenario is unfortunately the exception rather than the rule. More numerous are the cases of siblings from affected families who seek counseling with clinical geneticists in an effort to avoid similar outcomes to their childbearing.

Regardless of your spiritual inclination or religious persuasion, you and your partner owe it to yourselves and your future offspring to be as fully informed as possible about your genetic risks and options when considering planned childbearing. The decision to knowingly transmit harmful genes to a future child has awesome implications not only for the parents but also for the affected child. Physical pain and suffering, chronic psychological distress, loneliness, depression, stigmatization, extreme anger and frustration because of absolute dependence, eventual institutionalization, and early death are the outcomes I have most frequently observed.

Decisions to transmit genes should be preceded by full knowledge of family history and by careful attention to the risks and options outlined in this book. Decisions to knowingly transmit disease-causing genes are extremely personal. They do, however, have lifelong consequences not only for the parents but also for the affected child and that child's siblings (who may eventually bear the burden) and grandparents. Such decisions should not be made without consulting a clinical geneticist and, if desired, a spiritual adviser.

THE THREADS OF YOUR LIFE ·············

chapter 3

Too Many or Too Few Chromosomes

· ·

These tiny threads—the chromosomes that house our destinies, our genes—control our well-being and that of our families. Yet couples often have told me that they have no idea what chromosomes are and how they differ from genes. Because an understanding of chromosomes is vital to our comprehension of diseases and their origins, we will explore the nature and functioning of chromosomes before turning to an in-depth discussion of particular abnormalities. You will need no prior knowledge of biology to understand the straightforward explanations that follow.

Cells

Our bodies are made up of billions of cells. Many have special functions: We have brain cells for memory and intelligence, heart cells for rhythmic contraction, inner-ear hair cells for hearing, and so on. The lifespan of cells is variable and dependent upon the organs in which they originate. Although we cannot grow new brain cells (in fact, we steadily lose brain cells as we get older), the cells lining our intestines are lost and replaced by new ones every 24 hours or so. Millions of our cells are shed every day and almost all are rapidly replaced. Others live longer: Sperm cells in the testes may live about 2 months, and ova (eggs) in the ovaries may live longer than 50 years. Due to their comparatively long life, ova are exposed to more environmental influences than are other cells—influences such as gene-damaging toxins, drugs, and X-rays.

Every cell, except mature red blood cells, contains a tiny center of operations called the nucleus. The nucleus is the control center of the cell as well as the receptor of the messages, or blueprints, that we inherit from our parents and grandparents, which determine cell function and the characteristics we pass to each of

There was a clumsy awkwardness about the tall, gangly, overweight man as I ush-ered him and his diminutive wife into my office on a cold, gray day. They had been referred to me for consultation because of a chromosome abnormality we had detect-ed in the fetal tissues after his wife's recent miscarriage. Before launching into an explanation about the abnormality in number of chromosomes that we had found in the fetus, I obtained a detailed medical history as well as a family pedigree. No remarkable medical points emerged about his wife or her family; however, he recounted a childhood complicated by learning difficulties and behavioral problems in school, throughout which he reported he had "no friends." He stated that he had been a below-average student and that he had disliked sports intensely, mainly because of his poor coordination. On direct questioning, he recalled that at puberty his breasts became so prominent that even in summer he would not wear a swimsuit. From puberty onward he steadily gained weight, which he thought made the prominence of his breasts less obvious. He developed a normal-sized penis and normal pubic hair.

He graduated from high school with considerable tutoring and joined a compa-ny as a sales representative. He married early and his wife had had two preg-nancies thus far, both of which had miscarried. It was the second pregnancy that had precipitated the consultation.

Again on direct questioning, he admitted to a concern about his right breast, in which he thought there was "a lump." My examination confirmed that this tall man also had very long arms. In addition, he had a few minor abnormalities which includ-ed incurved fifth fingers, and a single transverse crease across his left palm. There was indeed a discrete, half-inch-diameter lump in his right breast, both breasts being very prominent. In addition, both of his testes were particularly small.

our children. Each nucleus contains tiny threads of chemical compounds called *chromosomes*, the most important component of which is DNA (deoxyribonu-cleic acid)—the substance of which *genes* are made.

Normal Chromosomes

It was known almost a century ago that when a special dye was added to a cell at a critical point in its formation, threadlike structures of protein would take up the stain, becoming easier to see. They were therefore called chromosomes, from the Greek words *chroma*, meaning "color," and *soma*, for "body." Late in the nine-

I concluded that he almost certainly had a chromosomal condition known as *Klinefelter syndrome,* in which a male has an extra X chromosome. I also feared that he had breast cancer, and I immediately arranged for a biopsy of the lump. At the same time, I obtained a blood sample for chromosome analysis to confirm my clinical diagnosis. He did indeed have breast cancer, which occurs with increased frequency in men affected by this condition. My diagnosis later was confirmed by the routine chromosome study.

During this first consultation and a follow-up visit, I explained the nature of chromosome abnormalities generally and the mechanism by which this particular abnormality had occurred, and informed the couple that sterility was a predictable characteristic of Klinefelter syndrome. Only then did he open up with the truth that his wife had become pregnant twice through artificial insemination by a different donor and that his personal embarrassment had prevented him from disclosing this fact earlier. Because the fetal tissue had exhibited an unrelated abnormality in the number of chromosomes, we also discussed that disorder, the mechanism of its origin, and the risk of its recurring, as well as the couple's options for future pregnancies.

Fortunately, surgical removal of the breast lump was performed without incident, and no spread of the cancer was found. During their second visit, after the diagnosis of the breast cancer but before surgery, on a bright and sunny day, I recalled our earlier consultation on that cool and gray day and reflected that this timely diagnosis of breast cancer had come about, ironically, because of a fetal loss due to an unrelated chromosomal abnormality—a defect for which he was not culpable but without which he might have lost his life.

teenth century, chromosomes were already considered the likely carriers of hereditary factors.

All of the information necessary to direct the formation and function of a human being—or of any other living thing, from bacteria, to plants, to elephants—is contained in these complex, thin threads. The chromosomes are, in turn, composed of *genes,* which are the units of heredity. Single genes are so small that they remain invisible even when viewed through the most modern instruments, including the electron microscope.

You receive half of your chromosome complement from your mother and half from your father. In other words, the genes constituting your chromosomes are equally contributed by both parents. In turn, you will pass along half of your chromosomes and genes to each of your own children. A look at what normally happens to chromosomes as they pass from parent to child helps us understand what can happen to chromosomes when things go wrong.

Number, Size, Sex, and Identity

The number of chromosomes and their structure vary greatly among living organisms, ranging from 4 to 500 in each cell. Chimpanzees and gorillas have 48 chromosomes per cell. In 1956, it was discovered that humans have 46 chromosomes in every cell (except sperm and ova, which have 23 each), and not 48 as was previously thought. The chromosomes in a cell can be viewed through a microscope and photographed; after special preparation, they appear as shown in Figure 3.1. Each chromosome in the photograph can be cut out, paired with a corresponding chromosome, arranged in order from largest to smallest, and pasted on cardboard; or photographed, arranged, and printed out by computer—as shown in Figure 3.2. There are 23 pairs (a total of 46 in all). One chromosome of each pair is from the father, and the other is from the mother. The first 22 pairs are numbered as shown.

The remaining two chromosomes in every cell are called *sex chromosomes*, since they carry the message that determines sex. Note that the two sex chromosomes are arranged separately from the other 22 pairs in Figure 3.2. Each parent normally passes along one sex chromosome to his or her offspring. The two sex chromosomes in females are denoted as XX and in males as XY. The mother can contribute an X chromosome only, while the father can contribute an X or a Y. When the mother contributes a single X and the father also contributes an X, then the offspring will be female (XX). The presence of the Y chromosome always dictates that the offspring will be male (even in abnormal situations when there are two or even more X chromosomes together with the Y chromosome).

Each chromosome can be distinguished from another in ways more accurate than by size alone. Using various staining techniques, it is possible to distinguish horizontal bands, or cross-striations, on every chromosome (see Figure 3.2). As with our fingerprints, the total banding pattern along each chromosome is unique for each individual. About 550 bands are usually examined in routine chromosome analysis. It is now possible, however, to "catch" chromosomes at a slightly earlier phase, during cell division, at which time they appear longer under the microscope. When these chromosomes are stained, at least 800 to 1,400 bands (some say even as many as 5,000) can be visualized, thereby making it possible to detect extremely small defects.

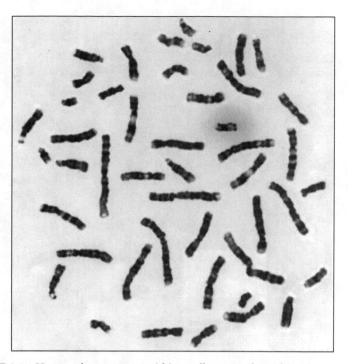

FIGURE 3.1   Human chromosomes within a cell, as seen through a microscope

These high-resolution chromosome-banding techniques (as they are called), and even the more traditional staining techniques, allow minor differences between individuals to be recognized. Experts can therefore distinguish between them and tell, especially for certain chromosomes, whether they come from one person or another—just like a fingerprint. For example, particular structural differences are common between individuals for chromosomes 1, 9, 16, and the male Y chromosome. These chromosome differences, called *polymorphisms,* can be used to trace an individual chromosome through a family. I recall a prenatal diagnosis I made some years ago, in which a rather odd-looking Y chromosome was found in the fetus. To exclude the possibility of a potential structural abnormality, I requested a blood sample from the husband to check his Y chromosome. Study of his chromosomes showed a completely different Y chromosome. This gentleman, who was a sailor and who spent much time away from home, was clearly not the father.

Chromosomal polymorphisms also have been used to determine the origin of the extra chromosome that appears in conditions such as Down syndrome. From

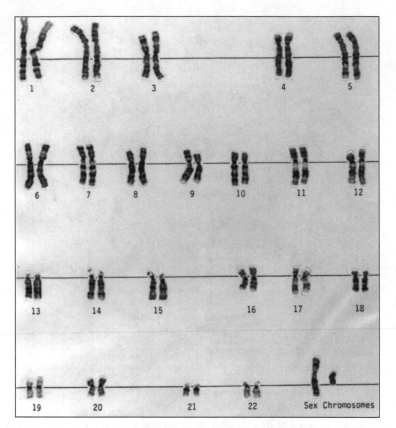

FIGURE 3.2   Normal human chromosomes from a single cell, arranged in descending order of size (except, by convention, for pair number 22 and the sex chromosomes). Staining reveals the cross-striations, or bands. Note the one female (X) and the smaller male (Y) chromosome at the end.

such studies we now know that the extra number 21 chromosome in Down syndrome originates from the mother in about 95 percent of cases and from the father in about 5 percent.

Sperm and Eggs

In the formation of both the sperm and the ovum, or egg, the number of chromosomes per cell is only half that present in other human cells—23. When the sperm and ovum fuse at fertilization, they form one cell that again contains 46 chromosomes. How do the sperm and egg form with only 23 chromosomes each?

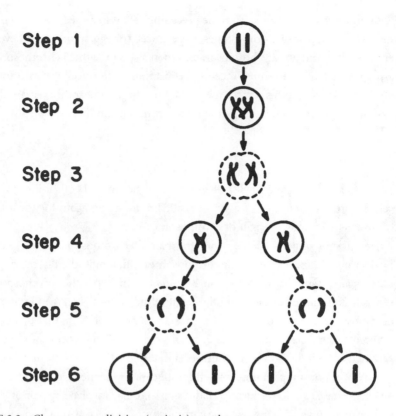

FIGURE 3.3   Chromosome division (meiosis), step by step

And once fertilization occurs, as we grow from a single cell into a whole person, how do the chromosomes divide? The sequence of events is best followed by observing Figure 3.3 carefully. Let us start with one cell nucleus, in the testis. (The same process happens in the ovary.) To make it easier, we will follow only one of the 23 pairs of chromosomes in that cell nucleus. The same process, called *meiosis,* occurs for each of the 23 chromosome pairs in every cell from which a sperm or ovum originates.

Step 1 shows the single cell containing a pair of chromosomes. In Step 2, the chromosomes split longitudinally and begin to pair off. In Step 3, the cell nucleus begins to divide. Step 4 shows the cell nucleus (and the cell that it occupies) that has divided into two new nuclei, each containing a pair of chromosomes. In Step 5, the two chromosomes in each new nucleus begin to move apart as the cells and their nuclei divide. In Step 6, new cells and nuclei are formed, each beginning with only one chromosome from the preceding cell. We can see that from the original cell with a pair of chromosomes, a sperm cell (or egg, in the ovary) eventually emerges that contains 23 chromosomes. When a sperm with 23 chromosomes and an egg with 23 chromosomes meet in fertilization, a single cell is constituted with 46 chromosomes. In this way we receive half of our chromosomes (and with them, half of our genes) from our father and half from our mother.

Multiplication by Division

We all start like this, as a single cell with 46 chromosomes. Let us follow that single cell (Figure 3.4) as it divides, a process called *mitosis,* and again focus for the sake of simplicity on only one pair of chromosomes.

Step 1 shows the single cell with the pair of chromosomes enclosed by the nuclear membrane. In Step 2, the chromosomes split longitudinally, making a total of two pairs, beginning the stage called *metaphase.* In Step 3, the chromosome pairs separate and the cell nucleus and the cell itself begin to divide. And in Step 4, one chromosome member of each pair is found in a new cell. Note that there are now two cells, each generated from the original single cell. This process of cell division, repeated an infinite number of times, eventually constitutes the human body.

A very important phenomenon with implications for the risks of inheriting any genetic disease can occur during cell division and may involve any pair of chromosomes. A variable-sized piece of one chromosome—a piece as small as a single gene or lengthy enough to hold thousands of genes—can cross over, or interchange, with a segment on the other member of the pair. Like the game of musical chairs, this mechanism results in a chromosome bearing "unexpected" genes, which when transmitted to an offspring could cause a specific genetic disorder. This process, also known as *recombination,* can occur both in meiosis and in mitosis. The crossing-over of genes during meiosis yields about 8 million different possible genetic combinations in the egg or sperm. Recombination thus contributes heavily to genetic variation.

Chromosomal Disorders

Numerical abnormalities also may occur (too many or too few chromosomes); and structural defects, or *rearrangements,* may appear in one or more chromosomes. Because these two types of abnormalities give rise to different outcomes, they are best discussed separately.

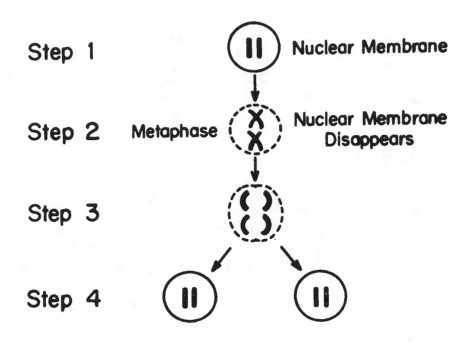

Step 1   Nuclear Membrane

Step 2   Metaphase   Nuclear Membrane
                              Disappears

Step 3

Step 4

FIGURE 3.4   The process of cell division (mitosis)

Numerical Chromosome Abnormalities

Too many or too few chromosomes may result during the process of cell division. This phenomenon is shown in Figure 3.5. It is the same basic diagram as Figure 3.3; Steps 1, 2, 3, and 4 are identical. But a crucial difference occurs in Steps 5 and 6: In Step 5, the cell nucleus and the cell that it occupies divide into two, but this time the chromosome pair fails to separate. In Step 6, both remain in one cell, which is now, say, the egg; the other cell ends up with all of the other chromosomes but lacks this particular one. Most commonly in studies revealing this type of abnormality at birth, this is the number 21 chromosome. When the egg with the extra chromosome is fertilized by a normal sperm, the resulting single cell has an extra chromosome.

The chromosome separation also could go awry earlier in the process—for example, between Steps 3 and 4. The sperm or eggs formed would again end up with either one too many or one too few chromosomes. Any person born with an extra number 21 chromosome in all or many cells will have the features of Down

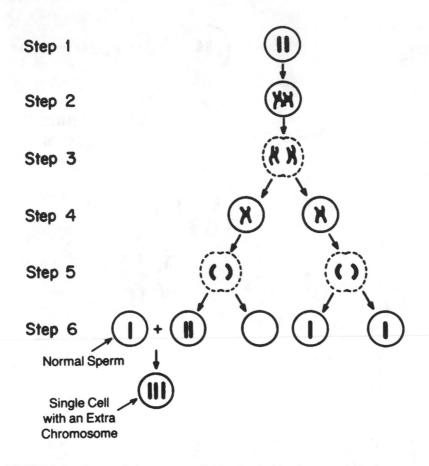

Step 1

Step 2

Step 3

Step 4

Step 5

Step 6

Normal Sperm

Single Cell
with an Extra
Chromosome

FIGURE 3.5   Abnormal chromosome division (meiosis)—for example, in the ovary—
can result in one cell with an extra chromosome.

syndrome, or trisomy 21 (once known as mongolism).* The commonest type of
Down syndrome involves trisomy 21, which accounts for about 94 percent of
cases diagnosed. In the remaining 6 percent, the same syndrome is caused by
abnormal rearrangements of the chromosomes or by mixtures of normal and trisomy 21 cells (see the discussion of *mosaicism,* in Chapter 4).

    The phenomenon of one chromosome sticking with another during cell divi-
sion (called *nondisjunction*) is not confined to the extra number 21 chromosome

---

*A trisomy is a disorder caused by the existence of an extra, unpaired chromosome in a cell.

in Down syndrome but may also occur with number 13, number 18, or any other chromosome, almost invariably causing miscarriage or serious defects in the child. This tendency to have "sticky" chromosomes increases with a woman's age, as do the risks of bearing a child with a trisomy such as Down syndrome. In addition to the mother's advanced age, there are other unknown influences that make the extra chromosomes stick in these trisomic disorders. Many factors have been suggested as causes, including a "stickiness" gene, X-ray exposure before pregnancy, viral infections, and diabetes or thyroid disease in the mother. In support of a hereditary factor is the fact that individuals with a near relative (sibling, aunt, uncle, or first cousin) with Down syndrome or another trisomy have a slightly increased risk (an increase by less than 1 percent) themselves of having a trisomic child.

Down Syndrome

Trisomy 21 or Down syndrome is evidenced by mental retardation and typical facial and other features (see Table 3.1). More than 85 percent of children born with this syndrome survive at least one year, and more than 50 percent live more than fifty years. Premature aging, with an increased frequency of early Alzheimer's disease, is common. Nevertheless, the average life expectancy of a person affected by Down syndrome now exceeds 60 years, given advances in the surgical correction of heart defects and in the treatment of common infections. The relatively long life expectancy has meant that brothers and sisters are almost invariably left with the burden of caring for those with Down syndrome. If there are no siblings, or none willing to take on their care, the State almost always ends up with this responsibility. Complications such as leukemia (20 times more common in Down syndrome), diabetes, and common illnesses most often claim the lives of those with Down syndrome. Additional, longer-term complications include hearing loss (affecting 40–75 percent of children and adults), mainly due to the high frequency of ear infections; eye abnormalities such as cataracts, glaucoma, cross-eyes, and major refractive errors (found in about 61 percent); and greater susceptibility to infection, which remains a major cause of morbidity and mortality among those afflicted with Down syndrome. Thyroid disorders also are common, making annual thyroid tests a reasonable recommendation. Epileptic seizures occur in 5–10 percent.

Between 2 and 4 percent of those with Down syndrome have a mixture of normal cells and cells with an extra chromosome 21. This mosaicism is still reflected in the typical features of Down syndrome, but these features are in some cases less striking or severe. The structural rearrangement (translocation) that causes Down syndrome is discussed in greater detail in Chapter 5.

Other Trisomies

Virtually all chromosomes have been found in triplicate, mostly in the very early, spontaneously miscarried, abnormal embryo/fetus, or (much less frequently) in the stillborn or liveborn infant. Two important trisomies involve chromosomes 18 and 13, the features of which are shown in Table 3.1. Since the risks of trisomies 21, 18, and 13 increase with maternal age, they remain important considerations for prenatal diagnosis (see Chapter 22). Extremely few fetuses with trisomies 13 or 18 survive during pregnancy. For the few that do, the average survival for trisomy 13 is about three days and for trisomy 18 about a week. Although fewer than 5 percent of these infants survive to one year of age, some are very long-term survivors with profound retardation and other defects. Generally, the risk of recurrence following conception of offspring with trisomies 21, 18, or 13 is about 1 percent. Prenatal diagnosis is recommended in such cases (see Chapter 22).

Miscarriage and Stillbirth

You may be surprised to learn that as many as one-third of all recognized pregnancies end in miscarriage. Even more startling, an estimated 78 percent of all conceptions do not result in a live birth. About 22 percent of miscarriages occur at the time of the first menstrual period after conception, even before pregnancy has been detected.

Among the most common causes of miscarriage are chromosome disorders in the developing embryo or fetus. About 1 in 10 sperm and 1 in 4 mature eggs have a chromosome abnormality; and up to half of the ova of women with infertility have chromosome defects. Remarkably, chromosome defects occur in some 8 percent of all pregnancies (about 1 in every 13 conceptions). Miraculously, the body is able to detect and reject more than 99 percent of them. About 50 percent of the embryos or fetuses miscarried in the first three months of pregnancy have some type of chromosome disorder.

In couples who experience miscarriage repeatedly, there is an increased likelihood that one of the partners carries (but in most cases is not affected by) a chromosome defect. After two miscarriages, there is a 1 to 3 percent likelihood that one member of the couple carries a chromosome defect that is being passed on

TABLE 3.1   The Most Common Features in Three Disorders Caused by an Extra
Chromosome and Associated with Advanced Maternal Age*

Trisomy 21 (Down syndrome) 1:800 live births	Trisomy 13 (Patau syndrome) 1:22,000 live births	Trisomy 18 (Edward syndrome) 1:7,500 live births
Mental and physical retardation	Mental and physical retardation	Mental and physical retardation
Short stature	Undescended testes	Failure to thrive
Flat face	Widely spaced eyes	Undescended testes
Large tongue	Low-set and malformed ears	Difficulty feeding with poor suck
Oblique, upslanting eyelids	Jitteriness and spells of no breath (apnea)	Heart defects
Cross-eyes	Small head (microcephaly)	Low-set, malformed ears
Flat back of head	Small, abnormal eyes	Elongated skull
Short limbs	Cleft lip or palate	Short sternum
Short, broad hands; short fingers, especially the fifth	Six fingers and toes	Single umbilical artery
High-arched, narrow palate	Heart defects	Contracted, overlapping fingers
Poor muscle tone	Presumptive deafness	Poor or excessive muscle tone
Flat bridge of nose	Small chin	Short neck
Abnormally shaped, small, and low-set ears	Excess skin at nape of short neck	
Incurved fifth finger	Strawberrylike birthmarks	Groin or umbilical hernia
Heart defects (in about 50 percent)	Long, curved nails	Small chin
Single crease across palm	Flexible thumbs	
Excess skin at nape of neck	Contraction deformity of fingers	
Intestinal (duodenal) obstruction	Single crease across palm	
Premature aging	Prominent heel	
	Poor or excessive muscle tone	

*An affected individual might not have every listed feature.

to the embryo. After three miscarriages, that likelihood increases to 3 to 8 percent. At this point it is standard medical practice to recommend blood chromosome analysis of both partners. If a chromosome abnormality is in fact detected in one partner, the couple's risk of bearing a chromosomally abnormal child can be determined. This risk is rarely more than 20 percent and is usually only about 1 percent. Fortunately, prenatal diagnostic tests are available (Chapter 22), and they are recommended for all pregnancies when one partner carries a chromosome defect.

Given the body's remarkable ability to detect and reject chromosomally defective embryos and fetuses, the frequency of such defects drops from a high of about 78 percent at about two weeks of pregnancy, to about 3 percent at sixteen weeks, and to 0.5 percent in liveborn infants. Stillborns and babies who die in the first few weeks of life have a 6 to 11 percent frequency of chromosome defects. It is critically important to check the chromosomes of every stillborn baby (even if no physical defects are found), not only to determine the cause of stillbirth but also to detect any potential risks to future pregnancies and identify appropriate options, including prenatal testing.

Birth Defects Due to Chromosome Abnormalities

Fortunately, the remarkable efficiency of the body in recognizing and rejecting defective embryos or fetuses results in very few babies actually born with these disorders. For example, about 75 percent of pregnancies with Down syndrome do not result in a live birth. Notwithstanding the efficiency of the body, chromosome defects still occur with a frequency of about 1 in every 156 live births (0.65 percent). About 0.4 percent of liveborn babies have a serious chromosome defect, and an additional 0.2 percent carry a chromosome defect that may later have repercussions when they begin childbearing. Fortunately, all chromosome defects can be detected in early pregnancy so that parents can consider the option of elective abortion (see Chapter 22). Moreover, being aware of your family history may prompt chromosome tests before pregnancy, which could provide early warning of significant risk. Early warning permits the issue of elective abortion to be circumvented in favor of other options outlined in Chapters 22 and 25.

Although too many or too few non–sex chromosomes frequently cause serious or fatal disorders, the health implications generally are not as ominous in the case of sex chromosome disorders, which are considered in the next chapter.

chapter 4

Sex Chromosome Disorders

....

Super females? Super males? On the contrary, having more than the normal complement of sex chromosomes is no advantage but in almost every case is associated with features that may include reproductive problems; speech, language, and
learning difficulties; intellectual deficits; and mental illness. Sex chromosome disorders are fairly common, occurring in about 1 in 300–400 births. About 10,000
children are born with sex chromosome disorders each year in the United States.

Numerical disorders of the sex chromosomes (too many or too few) are discussed below, and abnormalities in chromosome structure are addressed in the
next chapter.

Males with an Extra X

As one might expect, the addition of an extra female chromosome to a male has
a feminizing effect. The presence of two X chromosomes (along with one Y chromosome) instead of one X in every cell of a male results in a condition called
Klinefelter syndrome. Between 1 in 500 and 1 in 1,000 males are born with this
disorder every year in the United States (a total of at least 4,000 per year). Based
on the number of live male births recorded, we can estimate that there are more
than 150,000 males in the country today with this condition. Most of them probably have not yet been diagnosed: This syndrome is most commonly diagnosed
at or after puberty—when it is most easily detected. The characteristics that may
be observable in childhood include tall stature; speech, language, and learning
disorders; a tendency to score lower on IQ tests; and personality and behavioral
problems. (See Table 4.1 and the case described on pp. 24–25.)

In adulthood, males with Klinefelter syndrome are often remarkably indolent
about their own care and future, and are prone to alcoholism and antisocial
behavior. Breast development, which normally becomes obvious at puberty, is
very pronounced in about 15 percent of cases and causes extreme embarrassment. In many cases, surgical correction by mastectomy is obtained. The male

TABLE 4.1   The Most Common Features and Disorders Associated with Klinefelter
Syndrome*

Common Features	Associated Disorders
Small testes, normal penis	Obesity in adulthood
Sterility/infertility	Osteoporosis (thin bones)
Mental retardation	Diabetes mellitus
Feminine build	Varicose veins and leg ulcers
Excessive breast development in 50 percent	Rheumatoid arthritis
Tall stature (especially long legs)	Lupus erythematosus
Personality/character trait disturbances	Germ-cell tumors
Emotional/behavioral problems	Early tooth decay
Psychiatric/sexual problems	Erectile dysfunction
Various nervous system abnormalities	Breast cancer
Bone anomalies	
Delayed speech	
Attention deficit disorder	
Learning disorders	

*Klinefelter syndrome occurs in 1 in 500 to 1 in 1,000 males. An affected individual
might not have every listed feature.

hormone testosterone (by skin patch) has been used in some cases to deepen the
voice, stimulate the growth of facial hair, and improve libido (sexual drive), bone
density, and overall self-esteem. Generally, men affected by this syndrome tend to
be reticent, passive, and low-key. Mental illnesses, both neuroses and psychoses,
are more common among them, as are periods of depression and of mania. Other
common disorders (noted in Table 4.1) also appear somewhat more frequently in
men with this syndrome. Sexual behavior is normal and homosexuality is not a
consequence, but the libido tends to be depressed.* Breast cancer is at least 20
times more common among men with this disorder (a Swedish study has indicated that it is 50 times more common), accounting for some 4 percent of breast
cancers in men. Their life expectancy appears to be somewhat shorter than that
of the general population. Stroke, brain hemorrhage, lung infection, disease of

*Homosexuality, transvestism, and similar phenomena are not associated with abnormalities in
the sex chromosomes.

the aortic heart valve, and cancer of the breast account almost entirely for their increased mortality rate.

Characteristically, the testicles of males with Klinefelter syndrome are smaller than normal, and no sperm are found in the semen. Fertilization is almost never achieved without medical intervention. A few instances of successful pregnancy have been reported following aspiration of a single sperm through a needle introduced directly into the testis, which is subsequently used to fertilize an egg (this procedure is called *intracytoplasmic sperm injection*). The potential hazard in these rare cases is that the recovered sperm may have two sex chromosomes instead of one. In such a case, an offspring could be born with either three X (female) chromosomes or with Klinefelter syndrome.

Men born with a mixture of XXY and normal XY cells are described as *mosaic for Klinefelter syndrome*. They may be extremely difficult to diagnose and their condition may not come to light until they have a child who is diagnosed with a sex chromosome disorder.

Males with an Extra Y

Men with one female and two male sex chromosomes (XYY) in every cell are usually tall, have a tendency to acne, are likely to have lower IQs than their siblings, and in many cases have speech, language, or learning disorders (see Table 4.2), although their recorded IQ scores range from a low of 70 to a high of 145. XYY males tend to be impulsive, react poorly to frustration and adverse circumstances, and exhibit wild tempers. They are not, however, more violent or aggressive than other males, and their sexual behavior is not abnormal. Antisocial behavior in XYY males with a lower IQ and emotional lability lands this group in trouble with the police at least ten times more often than males with normal chromosomes.

Boys with XYY chromosomes often go undetected: About 1 in 1,000 males born have the XYY complement, and most are never diagnosed. Many diagnosed XYY males retrospectively describe themselves as children with extremely defiant natures, destructiveness, terrible tempers, and inclinations to climb to dangerous places—all evident by the age of four years. However, many boys with normal chromosomes also exhibit some of these features. Later in childhood, speech, language, and learning difficulties as well as behavioral problems tended to hamper their educational achievement. Nevertheless, a majority of XYY males have perfectly normal IQs, and at least one has been reported with a genius-level IQ of 145.

TABLE 4.2   The Most Common Features of Males with an Extra Y Chromosome*

XYY Males

Tall stature	Large teeth and long ears
Normal overall appearance	Severe acne in adolescence
Personality/behavior disorders	Occasional undescended testes or small penis
Antisocial/criminal behavior in some	Learning disorders
Tendency to low IQ	Tremor

* About 1 in 1,000 males have the extra Y chromosome. An affected individual might not have every listed feature.

XYY males are fertile. Their sperm, however, may contain one X, only one Y, an X and a Y, or two Y chromosomes. Consequently, when one of their sperm fertilizes a normal ovum containing one X chromosome, the product may be a normal boy, a normal girl, an XYY male, or a male with Klinefelter syndrome (XXY).

Females with an Extra X

No diagnostic physical features characterize the triple X female (Table 4.3). Minor variations occurring more frequently (and not affecting health) include a relatively small head in relation to height, incurved fifth fingers and toes, low-set ears, and poor coordination. About 1 in 1,200 females are born with this disorder, but most have never been diagnosed. In childhood this disorder might be detected only as part of an evaluation for disorders of speech, language, and learning. Although mental retardation is not a primary feature, the average IQ is about 85 (with a range of 64–120). The majority require special education classes. In adulthood, mental illness (psychosis or schizoid personality) appears to occur more frequently. As a group, triple X females tend to be passive and immature, have difficulty in forming interpersonal relationships, and frequently have psychological problems. There also appears to be a somewhat higher frequency of epilepsy among them.

Menstrual difficulties are relatively common and include late onset of periods, scanty or skipped periods, infertility or sterility, and early onset of menopause. Triple X women have a normal sex life and may bear male or female offspring, who may be born with an extra X chromosome.

TABLE 4.3   The Most Common Features of Females with an Extra X Chromosome*

XXX Females
Normal appearance
Menstrual irregularities
Mental illness
Relative infertility
Early menopause
Learning disorders

* About 1 in 1,200 females have an extra X chromosome. An affected individual might not have every listed feature.

Rarer Additions

Rarely, persons are born with four or five X chromosomes or three or even four Y chromosomes. Severe or profound mental retardation can be expected when these additional X or Y chromosomes are present.

Too Few Sex Chromosomes

To survive, we all need at least one X chromosome. We can manage without a Y chromosome, but not without the critical genes present on the X chromosome. Nevertheless, a deficiency of one sex chromosome can cause serious problems.

Turner Syndrome

The most common chromosome abnormality in miscarriages occurring during the first three months of pregnancy is a missing sex chromosome. This condition, known as *Turner syndrome,* occurs in about 20 percent of all miscarriages. Remarkably, fewer than 1 percent of Turner syndrome conceptions are carried to term and result in a live birth. Curiously, there also appears to be a five- to tenfold increase in twinning among the brothers and sisters of affected individuals. Strangely, too, most of these twins are identical.

A child born with only a single X chromosome will appear female in about 96 percent of cases, male in about 3 percent, and of ambiguous (uncertain) sex in less than 1 percent (see the discussion on mosaicism, below). In the vast majority of cases, the missing sex chromosome is of paternal origin. At birth such an

TABLE 4.4   The Most Common Features and Disorders Associated with Turner Syndrome*

Common Features	
Short stature	Kidney or urinary-tract anomaly (or both)
Absent menstrual periods	Underdeveloped, curved nails
Sterility	Short fingers and toes
Infantile sex organs	Pigmented moles (nevi)
Lack of breast development	High-arched palate
Transient puffiness of hands and feet	Abnormal teeth
Webbed neck	Short neck
Increased carrying angle at elbow	Cross-eyes
Low hairline in neck	Hearing impairment
Shield-shaped chest	Small chin
Cardiovascular anomaly	Bony defect of chest wall
Juvenile rheumatoid arthritis	Learning and language disorders
Schizophrenia	

Associated Disorders

Thyroid disorders

Diabetes mellitus

Osteoporosis (thin bones)

Hypertension (high blood pressure)

Obesity

Celiac disease

Tiny spider-like blood vessels in the bowel (may bleed)

Deafness

Orthodontic problems

* Turner syndrome (XO syndrome) occurs in 1 female out of 2,500. An affected individual might not have every listed feature.

infant is frequently noted to have striking puffiness of the back of the hands and feet. This puffiness disappears slowly in the first year of life. Birth defects of the heart, aorta, and other arteries occur in at least 50 percent of cases. Webbing of the neck, and typical facial and other features, are noted in Table 4.4. IQs are usually within the normal range, but frequently lower than in siblings. There is an increased frequency of learning difficulties and of problems with spatial perception and mathematics. Some reports have characterized the personalities of those with this syndrome as immature, with a greater occurrence of psychological problems. The ovaries fail to develop and are replaced by a streak of connective tissue. Consequently, without treatment, no breast development occurs and the womb remains infantile in size. In adulthood, without treatment, women with a single X chromosome are sterile, have no menstrual periods, lack breast development, and have very short stature.

Although these traits are characteristic of the majority, spontaneous puberty occurs in 5 to 10 percent of women with Turner syndrome, and between 2 and 5 percent achieve pregnancy. These exceptions might be due to chromosomal mosaicism (which will be explored in more detail later). This mixture or mosaic of chromosomes may endow these women with hormonal and reproductive function. Moreover, egg donation with fertilization achieved in the laboratory, and with transfer and implantation of the embryo into the hormonally prepared uterus, also has been successful in as many as 46 percent of transfers attempted. These pregnancies have higher risks and need specialized care throughout.

Children born with ambiguous genitals (a characteristic that might be due to a variety of genetic disorders besides Turner syndrome) need prompt specialist attention. Medical, surgical, hormonal, and psychological treatment may be necessary. In the rare instances in which a remnant of a Y chromosome is embedded in the solitary remaining X, the genitals will appear male. More importantly, these males will invariably be mentally retarded. If the external genitals are female and an errant fragment of the Y chromosome is embedded in the solitary remaining X, a malignant tumor (called *gonadoblastoma*) may develop in the tissues that were to have become the ovaries. As soon as such a diagnosis is made, surgical removal of these tissues is necessary, given the 26–30 percent likelihood of cancer developing in them.

Cyclical hormone treatment is used from puberty until about 45 years of age to foster breast development, menstruation, and other secondary sex characteristics. Human growth hormone use by injection beginning at 4–6 years of age has enabled these children to grow taller than they would have without treatment.

An increased frequency of various other disorders (listed in Table 4.4) has been reported in adults with Turner syndrome.

The single X chromosome complement in Turner syndrome accounts for only about half of those affected; a wide range of other abnormalities involving the X (and occasionally the Y) chromosome are responsible for the other half. Since the implications, although more variable, are very similar, suffice it to say here that the other chromosomal constitutions include mixtures of normal cells with cells containing a single X (mosaics), or all cells with variable size deletions of portions of an X chromosome, as well as other more complex structural rearrangements. In less than 3 percent of cases the Y chromosome may be involved, in which case the outcome could be considerably more serious, including mental retardation and significant risk of tumors developing in the tissue that failed to become a normal ovary. All of these complex situations call for consultation with the appropriate specialists.

Mosaic Chromosome Patterns

In the process of cell division that follows the fertilization of the egg, a mixture of normal and abnormal cells with two different sets of chromosomes may emerge: one with a normal set and one with an extra (missing, abnormal, or rearranged) chromosome. The resulting individual may be made up of a thorough mixture of cells, or a predominance in certain organs or tissues of normal or abnormal cells. For instance, the brain, the genital organs, the blood, and the skin may have cells with too many chromosomes, with all the other organs having normal cells. Individuals with a mixture of chromosomally normal and abnormal cells are described as having *chromosomal mosaicism.*

If, for example, 40 percent of the cells have the normal 46 chromosomes and 60 percent have the extra number-21 chromosome, then the features of Down syndrome will be present but are likely to be milder than usual, depending upon which organs have normal cells. As few as 1 to 4 percent of the cells of a person could have too many or too few chromosomes. Such individuals will usually be healthy, but they might have infertility or miscarriage problems. The difficulty in these situations is determining whether the low degree of mosaicism is actually causing the infertility or recurrent miscarriages. Mosaicism of the regular chromosomes is much less common than that of the sex chromosomes.

Generally, if no mosaic pattern is found in either blood or skin, mosaicism is very unlikely. It cannot, however, be ruled out on this basis, since it is conceivable that all cells in blood, bone marrow, and skin may have a normal constitution at the same time that those in the brain, ovaries, or testes have an abnormal set of chromosomes.

In sum, numerical disorders of the sex chromosomes are not as problematic as are those of the non–sex chromosomes. Structural rearrangements of both types of chromosomes, which we consider next, similarly vary in their implications from irrelevant to grave. Knowledge about any such rearrangement in the family history must not be ignored.

chapter 5

Chromosome Rearrangements

••

It is unnerving to realize that the protein scaffold and linked genes forming our chromosomes in certain situations may break. Breakages occurring in one or two (rarely more) chromosomes during formation of the egg or sperm are the most significant. Breakages at that critical moment can result in transmission of the broken chromosome(s) to the new embryo. Serious health implications also follow from chromosome breakage in bone marrow cells or in cancerous tumors.

When Chromosomes Break

About 1 in 120 newborns have some type of chromosome breakage, usually called a *structural rearrangement*. If these breaks interfere with the function of one or more genes, or if genes are lost or duplicated in the process, the health consequences might include birth defects, a genetic disorder, mental retardation, and certain types of cancer. Various structural rearrangements may occur, and they differ in their consequences.

Translocations

The most common rearrangement is when two different chromosomes snap and the broken pieces of one chromosome reattach to pieces of the other. This process is called *translocation*. Half of all such cases occur spontaneously during the production of the sperm or egg. The other 50 percent are directly inherited from a parent and passed down through the generations. The health implications differ depending upon which two chromosomes are involved. For example, an inherited form of Down syndrome is caused by a translocation between a number 21 chromosome and another, most often a number 14.

When the chromosome pieces change places without any piece being lost in the exchange, the process is called a *balanced translocation*. This process occurs in about 1 in 500 individuals and is not usually associated with any negative health consequences. However, in balanced translocations that are not inherited there is a slightly greater than random risk of mental retardation and/or other birth defects. The reason for this is unknown; but the additional risk may reflect damage to a gene at the site of breakage, malalignment of the exchanged pieces, or interference from neighboring genes (see the discussion in Chapter 6). Moreover, there is also an increased risk of chromosome "stickiness" (nondisjunction) and hence a slightly greater than random risk of a child's having an extra chromosome (for example, trisomy 21, or Down syndrome).

The chromosome with the translocated piece is vulnerable, so during formation of the egg or sperm or during fertilization, a break may occur again at the site of the original break, this time of only one chromosome. The piece broken off may then literally disappear. This *deletion* virtually always results in serious birth defects and/or mental retardation. Not all deletions are visible under the microscope.

In yet other examples, the chromosome piece that had exchanged places with another during a translocation may duplicate, yielding three identical pieces, including the similar segment from the unbroken chromosome. These three segments (called *partial trisomy*) almost always result in visible birth defects and/or retardation. Structural rearrangements with translocations resulting in deletions or duplications are termed *unbalanced translocations*. A parent who carries a translocation is at risk of having a child with a translocation that includes a partial trisomy or a deletion.

Sex-Chromosome Translocations

Males normally have only a single X chromosome, whereas females have two. Much of the Y chromosome is inert and thus far appears to contain no genes that threaten survival. The second X chromosome in the female is largely (but not completely) silenced in a process that occurs around the second week following conception (see Chapter 6). This process, called *genetic inactivation,* is random (either X chromosome inactivated). In all descendant cells, this same inactivated X chromosome continues, each X "remembering" its permanent state of activation or inactivation. The inactivated segments are nonfunctional.

Translocations occurring between an X chromosome and any other chromosome are well known. Usually the normal X will be preferentially inactivated. If the translocation somehow disrupts or interferes with the expression of a gene on the X chromosome, or if the particular X chromosome involved in the transloca-tion has a disease-causing gene, that female will have a genetic disorder.*

Deletions

Small segments of chromosomes containing one or many genes may be deleted as the consequence of a new mutation or an inherited disorder. Deletions may occur anywhere along the length of any chromosome, and they may be visible under the microscope or be so tiny that they are beyond the resolution of the microscope.

In microscopically visible deletions, multiple contiguous genes have been delet-ed, possibly resulting in multiple genetic disorders in the same person. In one such case, a boy with a large deletion in the short arm of his X chromosome had Duchenne muscular dystrophy and three other genetic disorders simultaneously. Ironically, while living with this burden, he was killed in an automobile accident.

So-called microdeletion syndromes are discussed in Chapter 7. The techniques used to detect such chromosome deletions are outlined below in the discussion on "FISHing."

Tiny, hidden deletions may occur near the very tips of each chromosome. These gene-rich regions are capped by specialized DNA that seals the chromo-some and prevents its fusion with any other chromosome. Special techniques (such as FISH; see below) now permit the detection of tiny deletions tucked immediately below these chromosome caps, or telomeres. *Subtelomeric deletions*, as they are now called, rank as the second most important cause of unexplained mental retardation. Consequently, in situations where evaluation has not result-ed in a precise diagnosis of the type and origin of mental retardation in an indi-vidual, a study for subtelomeric deletions is indicated. Over 7 percent of such individuals have a hidden deletion, as I and my colleagues have also found in our research at Boston University's Center for Human Genetics.

Inversions

Any chromosome also may break in two places, freeing a segment that then turns upside down and reattaches. These *inversions* frequently have no health conse-

---

*¹The *expression* of a gene is the transmission of genetic information from the gene to a protein, which determines whether the gene has particular effects on the body. Such effects may include (but are not limited to) visible, heritable physical traits.

quences; but on occasion they may be associated with infertility, recurrent miscarriage, or increased risks of birth defects and mental retardation because of vulnerability at the break site. Again, about half of such rearrangements are inherited, and the remainder arise spontaneously.

Other forms and consequences of chromosome breakages also may occur. Because they are much rarer and have highly technical explanations, I have not addressed them here.

Clues Suggesting a Rearrangement

Most of us have no idea whether or not we have been born with a balanced chromosomal rearrangement, such as a translocation or an inversion. Each year in the United States alone, over 8,000 children are born with balanced and over 2,500 with unbalanced chromosomes. Important clues arise mostly in connection with reproduction. Women who experience repeated miscarriages may find that either they or their partners are balanced translocation carriers, or that one has a chromosome anomaly, or harbors an inversion. After two miscarriages occurring in the first three months of pregnancy, the likelihood is between 1 and 3 percent that either partner has or carries a structural chromosomal rearrangement. After three losses, the risk escalates to between 3 and 8 percent. At this stage, blood chromosome analysis is advised for both partners.

Couples experiencing problems in achieving a pregnancy also have an increased risk that either partner carries a chromosomal rearrangement. Evaluation for infertility should always include a blood chromosome analysis for both partners.

Other clues are the prenatal detection of a fetus with anatomical defects, or a child born with birth defects or one who subsequently manifests developmental delay or mental retardation. In all of these situations, chromosome analysis of the fetus, the child, and/or the parents should be performed to reveal the precise rearrangement.

Future Risks

The carrier of a structural chromosomal rearrangement has an increased risk of bearing a child with birth defects and/or mental retardation. That risk depends upon which chromosome and which site has broken. For example, a mother with a translocation involving chromosomes 14 and 21 may have a child with an unbalanced set of chromosomes in which there are three segments of chromosome 21, resulting in Down syndrome. Her risk of having a child with Down syndrome is between 11 and 15 percent. If a male partner has a balanced 14–21 translocation, for reasons unknown, his risk of fathering a child with Down syndrome is much lower (2–4 percent). In most other translocations, the risks of having a child with mental retardation and/or malformations are less than 10 percent—usually, around 1 percent.

With a number of less common inversions, an individual's risk of having a child with birth defects or mental retardation ranges between 1 and 10 percent. All of these structural rearrangements, when visible under the microscope, can be detectable prenatally (see Chapter 22).

"FISHing" for Chromosomes

The detection of chromosomal deletions has been facilitated by recent advances in technology. The fine structure of DNA constituting the thousands of genes that make up our chromosomes can be probed by means of tiny fragments of DNA tagged with a chemical signal that fluoresces in the dark. When using this technique, which is called FISH (fluorescent in situ hybridization), one first must unfurl the DNA helix (see Chapter 6) and then position the prepared piece of DNA close to its complementary mate—much as one would fit a jigsaw puzzle piece in place. The location of this tiny DNA probe after it has joined with its mate can be observed under the fluorescent microscope. Its absence indicates a missing gene or gene segment, thereby facilitating a diagnosis of a so-called microdeletion syndrome (see Chapter 2 and pedigree for one such example). By means of a cocktail of such probes—using DNA fragments that are highly specific for a particular chromosome or a region within it—it is possible to literally "paint" individual chromosomes. Different colors can be generated, allowing remarkable photographic evidence of exquisitely tiny chromosomal regions. This technique now also permits precise recognition of the origin of tiny, additional chromosomal fragments. About half of the time, these unusual fragments are familial and of no significance. The other half of the time, they may cause mental retardation or birth defects. Curiously, these additional fragments are most commonly derived from chromosome 15.

The FISH techniques rely on DNA that homes in directly to its complementary, chromosome-specific target. Unlike routine chromosome analysis, which requires that a cell be caught under the microscope in the middle of cell division, FISH analysis can be achieved even in nondividing cells. For this reason it is often used for rapid prenatal diagnosis, in which cells obtained from the amniotic fluid following amniocentesis (see Chapter 20) are viewed to determine the chromosome number for a handful of the most common disorders (e.g., Down syndrome and chromosome 21)—those signaled by the presence of three chromosomes in each cell instead of two.

Chromosome Breaks and Cancer Risks

Most inherited translocations raise no additional risks of cancer. However, chromosomal translocations that develop spontaneously or through stimulation by a cancer-causing chemical in the environment have been linked with cancer. Acquired translocations are especially prone to arise in the bone marrow. As a consequence of two broken chromosome segments exchanging places and reattaching, an ominous development may occur: For example, the juxtaposed new fragment on a different chromosome may bring pieces of two genes into contact that then fuse. This gene fusion produces a cancer-stimulating protein, which in turn activates specific cells, mostly in the bone marrow, causing them to grow out of control and thus to become cancerous. Another mechanism arising from a translocation may operate by landing control genes close to other promoting genes, thereby orchestrating uncontrolled, malignant cell growth.

The vast majority of such translocations in the bone marrow induce leukemias and lymphomas of various types. The most common translocation is between chromosomes 9 and 22. It ultimately leads to chronic myeloid leukemia and acute lymphoblastic leukemia. Various other translocations also are known to cause specific types of leukemia and lymphoma.

Certain well-recognized translocations may be detected prior to the development of any malignancy. Discovery in these circumstances is frequently based on a family history of a specific tumor. The chromosome break usually involves a gene that when defective (mutated) leads to the development of a hereditary tumor. Three important examples include the 11–22 chromosome translocation, which may increase the risk for breast cancer; the 3–8 translocation, predisposing to kidney cancer; and the 3–6 translocation, predisposing to certain bone marrow cancers. It is important that you take note of specific cancers in your family history and seek genetic counseling and, if the counselor recommends it, chromosome analysis. In this way, appropriate surveillance can be initiated, baseline studies done, and very early diagnosis made, facilitating preemptive treatment and possibly saving your life.

Other structural rearrangements that interrupt the normal structure and alignment of genes along chromosomes may also result in the development of a cancer. Once again, involvement of a key gene in a specific location could result

L.M., who was blind, came to her obstetrician for prenatal care. He never inquired about the cause of her blindness. He provided routine care and treatment, her pregnancy progressed without incident, and the labor and delivery were uneventful. The child when born appeared healthy and was discharged from the hospital with the mother.

When the child was about seven months old, her father observed a peculiar whitish appearance behind the pupil of one eye. Upon consulting an eye specialist, the parents discovered that the child had a malignant eye tumor called retinoblastoma, a cancer that shortly afterward was also found in the baby's other eye. Despite treatment efforts with anticancer drugs and radiation, the child's eyes had to be removed to save her life.

in a predictable tumor. An important example that is not infrequently inherited involves a gene on the long arm of chromosome 13. The above poignant case illustrates some important lessons in this regard.

In this case, had the obstetrician inquired about the cause of the mother's blindness, he would have learned that her eyes had been surgically removed when she was an infant. Although the woman did not know the name of the disorder that had made the procedure necessary or that it was a genetic condition (she had been adopted in early childhood), she did know that her own eyes had been removed because of tumors. The obstetrician then would have referred the patient to a clinical geneticist; and subsequent testing would have revealed the involvement of chromosome 13 and dominant inheritance, with a 50 percent risk of bearing a child with this disorder.

Equally egregious was the action of the pediatrician, who failed to have an eye specialist perform an eye examination under anesthesia shortly after birth and every three months thereafter for at least the first two years. This surveillance, instituted because of the mother's history, would have enabled a very early diagnosis of the retinoblastoma tumor, facilitating immediate treatment, with a 90 percent likelihood not only of saving the child's eyes but also of insuring reasonably good vision.

Know your family history, and if there are any serious illnesses in the family that could be genetic, telephone a clinical geneticist to determine whether a consultation is necessary. The name of a Board-certified clinical geneticist can be obtained by contacting the American Board of Medical Genetics or the American Medical Association (see Appendix C).

THE BLUEPRINTS FOR LIFE ················

chapter **6**

Genes

· · · · · · · · · · ·

Your journey through life began when your mother's egg was fertilized by your father's sperm. Both egg and sperm carried coded messages from your parents' families and ancestral relatives, conveyed via genes through the generations. It was through the interaction of genes that all of your cells and organs were structured and now function. The miraculous complexity of our bodies may at first seem difficult to fathom, especially since one cannot see a gene. However, our inability to visualize a gene need not prevent us from understanding critically important information about genes. The quintessential information about genes in this chapter will enable you to understand not only the normal structure and functioning of your body but also how aberrations occur and what consequences follow.

Our genes regulate, modify, control, promote, enhance, and otherwise influence every aspect of our body's structure and functioning. Whether you are tall or short, fat or thin, fair- or dark-skinned, have certain facial characteristics, are allergic, or are affected by a genetic disorder is purely a function of your genes. Your genes largely determine your body's characteristics, your various susceptibilities or resistance to all diseases, how you react to medications, and whether you will transmit certain genetic disorders or fall victim to one or more of them. However, even though all aspects of health depend upon your genes, their interaction with the environment is especially important. In fact, responses to environmental dangers have in an evolutionary way resulted in genetic adaptations for survival.

No advanced knowledge of biology is necessary to form a basic understanding about your genes and what may happen when things go awry.

What Is a Gene?

The 46 chromosomes that you inherited from your parents (23 from each) contain many genes, currently estimated to be between 30,000 and 40,000. Although a single gene is too small to be seen with the naked eye, genes can be chemically extracted from cells, producing a white, stringy, sticky substance. This stringy mass of genes, or DNA, is constituted by very long, narrow threads, which are usually tightly compacted within the chromosome structure. The very long DNA fiber consists of two strands coiled around each other in a spiral (a double helix) and wound clockwise like an ordinary corkscrew. Each strand of DNA is composed of long chains of molecules containing nitrogen, sugar, and phosphate, and the surrounding scaffold is bolstered by calcium, magnesium, and certain proteins, including a type known as *histones*. Over 100 different histones are known that influence gene action.

These long molecular chains, called *nucleotides,* are made up of sugar, phosphate, and nitrogen-containing molecules called *bases.* Together these substances constitute nucleic acids, of which there are two: The one containing the sugar ribose is called *ribonucleic acid,* or RNA. The other, which contains a different sugar (deoxyribose), is referred to as *deoxyribonucleic acid,* or DNA. All of the chromosomes, and hence the DNA, together constitute the nucleus, which is held within a nuclear membrane. This membrane contains many tiny holes, making it porous. The liquid (called cytoplasm) that bathes the nucleus contains the RNA.

A gene is constructed like a spiral ladder or staircase, with the two long side strands being connected at intervals by shorter strands resembling rungs. These rungs are made up of bases, which are always connected to each other in the same way. There are only four possible bases (or nucleotides). Each is referred to by its capitalized first letter: Base A (adenine) always pairs up with base T (thymine), and G (guanine) is always paired with C (cystosine). In RNA there is a slight difference, in that base A pairs up with another base, called uracil (U).

Incredibly, there are about 3 billion base pairs within each nucleus. The largest chromosome (number 1) has about 249 million base pairs of DNA, whereas the smallest (number 21—not number 22, as would be expected) has about 48 million. The X chromosome, of great importance because of its role in many severe X-linked diseases, has about 154 million base pairs of DNA. Individual genes seem to range in size from a few hundred to more than 2 million base pairs. The defective gene causing Duchenne muscular dystrophy is enormous, containing 2,400,000 base pairs!

A gene, then, is a double strand of DNA connected by base pairs and held together within a spiral scaffold of proteins and other elements. Our genes are tightly coiled and packed into the holding scaffold to constitute each of our chromosomes. To give you some idea of how tightly our genes are coiled: If they were unraveled from a single cell, they would measure a few yards. In comparison, the total length of all of the chromosomes in a cell is less than half a millimeter—in other words, less than 0.02 inch.

How Do Genes Function?

Imagine a train with many coaches. Genes are made up of many functioning units (called *exons*) similar to the coaches of a train. The intervening connections between "coaches" are called *introns*. The exons and introns are simply continuous stretches of DNA located within a gene. Each exon is made up of a chain of nucleotides linked together in a pattern that we received from our parents and earlier ancestors. Such a pattern could be TTTAAAAAGGCT. Just three of these nucleotides are needed to construct a single amino acid. Multiple amino acids together make a protein. There are twenty amino acids from which all proteins are made. Each exon, like a factory, makes amino acids, using only three bases at a time to specify the amino acid made. The functions, if any, of the introns have yet to be discovered.

The process in which genetic information is transmitted involves the separation of the two DNA strands and the copying (called *transcription*) of the strand forming a molecule (RNA), which in turn carries the genetic information as a messenger out of the nucleus into the surrounding fluid (cytoplasm). Following transcription, the messenger RNA molecule undergoes further processing in which the introns are removed and the exons are spliced together. In the cytoplasm, the joined exons are transported to tiny factories (called *ribosomes*) where the information they carry is translated into a protein. In fact, even after this translation into protein, further modification or processing can occur, adding to or cutting down the structure of the protein. The completed protein that is made folds itself into a precisely determined shape, depending upon its constituent amino acids. Its physical and chemical structures are key to its proper function. In hemophilia, for example, the presence of the blood clotting Factor VIII protein can be demonstrated, but it is functionally inactive. Remarkably, only about 1 percent of DNA actually works to make proteins!

In sum, the genetic code is the set of bases we have inherited that make up to twenty amino acids. These in turn join to form a protein that executes all of the functions of a gene. It is the proteins that build our bodies and enable them to function. Much of this book addresses the consequences of defects that occur in the pathway just described, or in the chromosomes.

Errors in the Chain

Recall that the four bases are commonly referred to by their initials (A, G, T and C). A gene is a chain or sequence of these four bases, which can be represented by the combination of their letters—for example, TATAATGAAATTCATCCATTAG. The TATA set indicates the start of a gene; the adjacent ATG points to the position of the first building block (amino acid) of a protein; and the TAG signals the end of the working part of the gene, and therefore the end of the protein.

Even the smallest "spelling" error in the DNA chain may cause grave or fatal genetic disease. These errors or defects, called *mutations,* may involve one or more letters (bases) in the chain; and there can be a number of different types of such errors. For example, a single base change (a G instead of a T) could occur; but even with this change, an amino acid can be formed that does not alter the function of a protein. This single base change (called a *point mutation*) may occur without any effect on health. This so-called *silent mutation* is fairly common, accounting for 20 to 25 percent of all possible single base changes. More likely, however, a single base mutation leads to the formation of a different amino acid and an altered protein (in 70 to 75 percent of cases). This so-called *missense mutation* may lead to a severe reduction in the protein or even a complete loss of its function.

In about 2 to 4 percent of single base changes the fault may lead to a shortening of the DNA chain encoding protein. The resulting abrupt termination by this so-called *nonsense mutation* again leads to a protein deficient or devoid of structure and/or function.

More than a single base in the DNA chain or sequence can be involved. A small piece of DNA may be inserted, deleted, or duplicated within the gene or DNA sequence. Such mutations are called *insertions, deletions,* or *duplications*—all of which result in potentially serious deficits in the manufactured protein. If a mutation occurs that involves the insertion or deletion of bases that are not multiples of three, a serious disruption of the DNA sequence or reading frame occurs (called a *frame-shift mutation*), with resultant abnormalities in or absence of the protein product. In cystic fibrosis, the deletion of only 1 amino acid out of 1,483 is the most common mutation, and it results in the most severe form of this disease.

The most recent class of mutations identified has been described as unstable. These *dynamic mutations* consist of repetitive three-letter sequences (triplets) that in a stuttering fashion appear in excess of the normal number within a gene. This form of mutation, which is especially relevant to diseases of the brain and nervous system (see Chapter 7), is subject to expansion into larger numbers of triplet repeats. The larger the number, the more severe the disease and the earlier its onset.

Gene mutations can be inherited directly from one or both parents or arise spontaneously in the egg or sperm. Regardless of their origins, however, mutant genes are transmitted through subsequent generations. The origins of many dominant genetic disorders are new mutations, which frequently account for about half of all occurrences of such disorders. The vast majority of some disorders arise through new mutations, as happens with achondroplasia (characterized by dwarfism, a relatively large head, and bony abnormalities), of which about 7 out of 8 cases diagnosed have their origin in new mutations. Scientists cannot yet fully explain why gene mutations occur, but X-rays, toxic chemicals, and viruses are among the known causal factors.

The Gene Team

The gene is unquestionably the leader; but it is always surrounded and aided by an entourage. Certain letters (bases) in the DNA chain that abut the sequence of letters that constitute the gene, make up the *promoter*. The role of the promoter is to fix the site at which the copying (transcription) of the gene is initiated and to control the quantity of the messenger (RNA). The promoter may extend for several thousand bases, the most important of which are the one or two hundred closest to the gene. Situated within about two dozen base pairs of the start of a gene is another sequence of letters (quaintly called the "TATA box") that is involved in the precise localization of the start of transcription of the gene. A little farther away is another sequence of letters referred to as the CCAAT box, which if present is also involved in securing efficient transcription.

Next in the entourage are the *enhancers*. These DNA sequences are not close but nevertheless increase transcription from a neighboring gene. Some enhancers exercise control only in specific tissues. *Silencers* are DNA sequences that reduce the transcription of a gene. Promoters, enhancers, and silencers interact through signals, even though they are separated by thousands of base pairs, in transcribing DNA.

From lessons learned in the study of the common bacterium *E. coli*, we now know about genes that control the activity of structural genes in determining the amount of a gene product. These are the so-called control genes. These *controllers* are in turn regulated by more distantly located genes (called *regulator* genes), which in turn produce a protein that turns a gene on and off (called a *repressor*).

Structural genes can function only when the regulator is switched off by the repressor gene. Many of these control and regulatory genes, as well as the designer genes (described in the following section) are switched on briefly to direct the development of the embryo. These genes work together in a coordinated sequential cascade to initiate and govern key developmental events, such as the separation of tissues into head and tail, arm and leg, etc.; the initiation of the formation of specific organs; the movement and migration of cells to form the various tissues and organs; and the specialization of cells to form a particular tissue, such as brain, kidney, heart, and so on. Thereafter, these genes are switched off—almost never to function again. In rare instances—for example, in the presence of certain cancers—one or more of these genes may be reactivated in adult life, producing embryonic proteins that are detectable in the blood.

Gene Families

Genes of like structure and function sometimes appear in clusters. For example, a gene cluster on chromosome 6 heavily influences the body's reaction to foreign proteins—such as those introduced via an incompatible blood transfusion or by a heart, kidney, or other organ transplant. These so-called *HLA* genes, or *major histocompatibility* genes, also make us susceptible to certain diseases and infections (see Chapter 10). In addition to an unknown number of single genes, several gene families that have architectural or designer codes are known to play key roles in the development of the embryo, as described above. Most of these genes produce specific proteins that initiate transcription. These so-called *transcription factors* switch genes on and off, activating or repressing their expression.

Designer Genes

Genes that arrange the body's architecture fall into this category. There are at least three such gene families, which together dictate the development of body structures such as a leg or an eye. Genes that dictate the formation of an eye can be placed in any location and turned on, and will then function to produce the structure of an eye. Thus, an eye might even be made to develop on a leg, as was shown by a laboratory experiment on fruit flies. In a similar vein, a gene has been discovered in mice that can be induced to grow active hair follicles, raising for the first time the possibility of a cure for baldness. These designer genes dictate the location and fate of particular types of cells, and along which axis cells move. They are fundamental and have been found to be highly conserved through many species—from the fruit fly, to the mouse, to man. Given their critical role in the development of the early embryo, it is not surprising that a mutation in any of these genes may have profound consequences. For example, one such defective gene discovered by our research group at the Center for Human Genetics is known to cause deafness (as well as widely spaced eyes, eyes of different color, and a white frontal patch of hair—the typical features of Waardenburg syndrome). This particular genetic defect had been identified earlier in certain laboratory mice as well as in humans.

A sex-determining gene that encodes a protein with a specific motif has been discovered in the tiny worm *C. elegans* (see the discussion in Chapter 27), in the fruit fly, and in humans. Remarkably, this gene's conserved human counterpart was not located on a sex chromosome but on chromosome 9—at the very site in which a genetic mutation is found in the human with 46 chromosomes, including an X and a Y, in which sex reversal occurs! Other defects in these designer genes may be so serious as to be lethal within weeks of conception, or they may cause defects of the face, skull, kidneys, limbs, digits, and even absence of the iris (colored portion) of the eye.

Gene Programs

Programmed Cell Suicide

The truism that death is part of life finds no better witness than exists in gene function. The sculpting of internal cavities (e.g., in the heart and the bowels) and channels (through the arteries and veins) requires programmed cell death. In the embryo, for example, the hand develops as a pad, with the fingers firmly knitted together. Around the tenth week of pregnancy, the cells binding the fingers (or toes) together begin to die, allowing the digits to separate. This programmed cell death (called *apoptosis*) is controlled by specific genes. (The word *apoptosis* comes from a Greek root meaning the falling off of leaves from trees or petals from flowers.) A mutation in a gene that governs programmed cell suicide could result in a child's being born with one or more fingers or toes stuck together (syndactyly). Curiously, the second and third toes are the digits most frequently partially stuck together, or webbed.

Defects in genes controlling programmed cell death also play a role in the evolution of cancer (see Chapter 16). Apoptosis is especially important in modulating the number of cells in organs where normal cell turnover rates are high (e.g., bone marrow). Failure of apoptosis would allow an increase in the number of cells—for example, of B-lymphocytes, producing chronic lymphocytic leukemia. Many other disorders, including those of the nervous system, heart, bones, joints, liver, and intestines, are associated with a failure of apoptosis. Fortunately, there are also genes that function to keep apoptosis in check.

Clock Genes

In 1729, a French astronomer, d'Ortous de Mairan, noted that a sun-loving plant could open and close, following the sun's cycle. More recently, internal rhythms have been recognized in fungi, bacteria, frogs, fruit flies, and other organisms. Elegant studies in toads have revealed "clocks" that function in each cell and that are even present during early embryo development. These developmental clocks signal the timing of key mechanisms that control growth and specialization of cells into tissues and organs. In many organisms the molecular clocks are located within the nervous system. The pineal glands of birds, reptiles, and fish contain light-sensitive cells with pacemaker properties, which direct the rhythmic production of hormones. These cells rhythmically send out signals that control sleep, mating, hibernation, and other cyclical behaviors. Humans have similar clock genes, probably in the front portion of the brain called the *hypothalamus*.

Clock genes represent a beautiful example of the need for genetic interaction with the environment: Light-sensitive cells are necessary in order to send signals to help regulate the body's master clock, and clock genes provide the mechanism by which these signals are sent. Interestingly, horseshoe crabs have clock sensors on their tails, whereas swallows have them just within their skulls. Fruit flies have clock genes in their legs, wings, and hair bristles. When biologists at Cornell University Medical College in New York shone a bright light on the backs of the knees of their human experimental subjects, they succeeded in resetting those subjects' master biological clocks. The mechanism by which light signals from behind the knee are detected and transmitted to the master clock cells in the human brain remains to be discovered.

Biological rhythms that we commonly recognize include the daily flux of body temperature and levels of the hormone melatonin. The body's temperature rises throughout the day, peaks in the early evening, and begins to decline around 7 P.M. or 8 P.M., falling to its lowest point at about 5 A.M. Melatonin begins to increase around 10 P.M., inducing a state of sleepiness, and during the day the level gradually drops off.

Cyclical hormone production is also under the control of a master body clock mechanism, since the ages when menses start and menopause begins are strongly influenced by genes. Similar influences exist for uterine fibroids (benign tumors) and for the common condition known as endometriosis (the growth outside the uterus of cells from the uterine lining, causing problems in the pelvic area).

A human clock gene has been identified that regulates daily sleep and activity rhythms. Mutations of this gene and others like it may be responsible for jet lag and various sleep disorders, including the inherited *advanced sleep phase syndrome*, which is characterized by a short sleep cycle. The partners of "day people" or "night people" may now finally have an explanation—once again, it's all in the genes! Mice that lack a certain similar gene suffer from insomnia. Such mice need twice as long to fall asleep as do mice that have this gene; and once they do fall asleep, they sleep 30 percent less than their normal cousins!

Mutations found in two different genes associated with narcolepsy (a debilitating sleep disorder) provide further clues to the genetic basis of sleep. Certain gene mutations in Doberman Pinschers cause them to have narcolepsy; and a mutation in a related gene in mice causes a condition that is remarkably similar to human narcolepsy. Understanding the genetic basis of sleep will undoubtedly lead to the discovery of new treatments for narcolepsy and other sleep disorders. (Who would have thought that when we go to sleep is still influenced by our parents through their DNA?!)

Genes That Distinguish Left from Right

Yes, here we go again. The establishment of left and right, or laterality, is also a function of genes. Your heart and spleen are on your left side, your liver on your right. In the male, the left testis is lower than the right; and in the female, the left breast is often larger than the right. The origins of left-right axis are indeed within the control of genes dedicated to the body's symmetric or asymmetric development. For example, a cascade of noncardiac genes dictate the looping of the embryonic tube that becomes the heart even before the twenty-sixth day after conception. Mutations within such a controlling gene may result in a failure of positioning—for example, in a right-sided heart, a left-sided liver, or a right-sided spleen. This disorder (called *situs inversus*) has been recognized as recessive and follows the inheritance of the same genetic mutation from both parents. A dominant gene also has been identified that causes the same failure to establish normal left-right asymmetry during embryonic development. Frequently in these situations, defective formation also may occur—for example, a child may be born with a structural heart defect.

Brain asymmetry favoring right-handedness and control of speech in the left side of the brain is probably also dictated by genes yet to be discovered.

The Power of Silence

Gene silencers, as mentioned earlier, are DNA sequences that reduce or prevent the functioning of a gene. The most impressive is located on the long arm of the X chromosome, at a spot called the X-inactivation center. Within days after the conception of a female, signals from a single, randomly chosen X-inactivation center silence most of the genes on that chromosome. Due to this random process, about half of the cells in a female have an active, maternally derived X, and the other half have the paternal, functioning X. The cells of both males and females need only one X chromosome to survive.

Imprinted Genes

We inherit our genes in pairs on our 22 paired, non–sex chromosomes. Each pair of genes occupies the same "geographic" location on the paired chromosomes, like rungs on a ladder. Generally, the two genes in each pair are thought to function equally. We now know, however, that in a number of genetic disorders, one of the gene pair has been silenced (inactivated). If the silenced gene is of maternal origin, then the paternal gene is functionally active, and vice versa. This process is called *imprinting*. Imprinting occurs before fertilization and is passed into all of our cells and carried on through our eggs or sperm. But the results of imprinting might later be reversed.

Two disorders in particular serve to illustrate this phenomenon of imprinting. Both occur due to a mutation or abnormality at the exact same spot (although in different genes) on the long arm of chromosome 15. Prader-Willi syndrome (short stature, obesity, learning disorder) and Angelman syndrome (severe retardation, absent speech, seizures, and other features) both arise from a tiny deletion or structural abnormality on chromosome 15. In Prader-Willi syndrome, the deletion is invariably on the father's chromosome; in Angelman syndrome, it is on the mother's.

Various mutations or structural rearrangements of the genes may occur in this imprinted region of chromosome 15, causing either Prader-Willi or Angelman syndrome, depending upon the parent of origin. Deletions account for 70 percent of these two syndromes. Another mechanism with the same parental sex influence accounts for 20 percent of these cases.

Either at the time the egg (or sperm) is formed or one step later (see Chapter 3), a pair of number 15 chromosomes might fail to separate. This would leave the egg or sperm with two number 15 chromosomes from one parent, and one number 15 from the other. An offspring with three such chromosomes could have a syndrome called trisomy 15, characterized by multiple birth defects, or might be miscarried. In other cases, the single number 15 chromosome from the one parent might be deleted, leaving the two number 15 chromosomes from the other parent—a phenomenon called *uniparental disomy*. Either the Prader-Willi syndrome or Angelman syndrome would be the consequence, depending upon the parental origin of the two number 15 chromosomes.

What happens if the same event occurs in a gene on another chromosome in the embryo? If the same process takes place in chromosome 7 and the mutation is in the cystic fibrosis gene, this recessive disorder will occur even if the other parent is not a carrier. However, such occurrences are rare and account for less than 1 percent of cystic fibrosis cases.

Imprinting regions also are found on chromosomes other than numbers 7 and 15. If similar gene duplications occur in these regions, certain rare birth defects and cancers may result. A number of these are explored in greater detail in Chapter 17.

This brief discourse on genes will enable you to make sense of the critically important information in the next chapter, which is aimed at helping you avoid or manage genetic disorders secreted in your blueprints for life.

chapter **7**

You and Your Genes

....

Each of us was created by a matching set of genes contributed by our parents. Often a particular feature such as a chin dimple points to the parent who contributed that particular gene. If you have fair skin and sunburn quickly, this may be a characteristic of only one side of your family. Most such variations are the results of single genes. The interaction of multiple inherited genes with environmental factors more commonly results in recognizable family traits such as short or tall stature, or remarkable intelligence. The Darwin family, which produced outstanding scientists for five generations, is a good example. But genius also springs from completely average families of no particular intellectual distinction, as happened with Newton, Keats, and Einstein.

Certain genes or gene combinations may convey advantages in fertility or survival; but defects within single genes are mostly disadvantageous. Understanding precisely how defects in single genes are transmitted is critically important because it enhances our chances of avoiding or preventing serious genetic diseases.

Harmful Genes from One Parent

We inherit matching genes in pairs, one from each parent. If one of these paired genes has a stronger effect than its mate, it is called a *dominant* gene. If a defect exists in this dominant gene, then the contributing parent is likely to have a genetic disorder as a consequence, and to have a 50 percent likelihood of transmitting the defective gene to each offspring (see Figures 7.1 and 7.2). A dominantly inherited genetic disorder may be due to the manufacture of a defective protein by the aberrant gene, or to the dysfunction or absence of a protein.

Single dominant genes may produce harmless effects—such as large and unusually shaped ears—or result in serious, even fatal genetic disorders (such as Huntington's disease). Thus far, more than 8,500 genetic disorders or traits have been catalogued as due to a single gene. More than half of these are dominantly

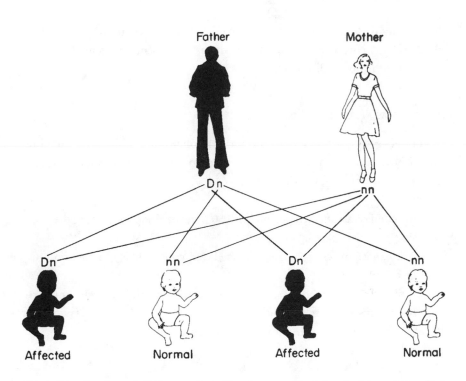

Figure 7.1   Dominant inheritance. One affected parent (the father, in this case) has a defective gene (D) that dominates its normal counterpart (n). Each child has a 50-50 chance of inheriting the defective gene from the affected parent and thus of having the disease.

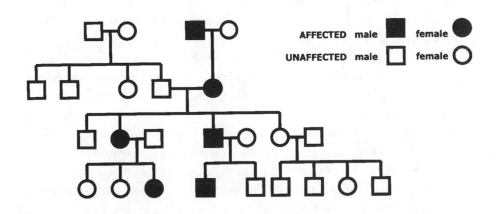

Figure 7.2   A typical family tree involving dominant inheritance—as seen, for example, in Huntington's disease, heart disease due to hypercholesterolemia, and hundreds of other disorders.

inherited, about 36 percent are recessively inherited, and less than 10 percent are linked to the X chromosome. About four-fifths of transmitted gene defects are associated with genetic disorders.

Characteristics of Dominant Inheritance

Certain rules are known to govern this form of inheritance. The gene in question is invariably located on a regular, autosomal chromosome rather than on one of the two sex chromosomes. Furthermore:

1. Any child of an affected person has a 50 percent risk of inheriting the gene from that parent.
2. Males and females are equally affected and equally likely to pass on the harmful gene.
3. Family members who have not inherited the harmful gene do not have the disorder and do not transmit it.
4. The disorder (or a mutation thereof) appears in every generation, with no skipping.

Apart from these rules, considerable variation may be seen in the expression of a dominant genetic disorder within a family. Some affected members may have severe involvement, whereas others may appear not to have any obvious disease— but later give birth to a seriously affected offspring. Explanations of this phenomenon in terms of a "skipped generation" have often been provided by families. However, a better explanation for this apparent lack of *penetrance* is that subtle signs of the defect were missed.

A single defective gene transmitted through the family may have several effects. For example, in Marfan syndrome (a dominantly inherited disorder in which the elastic fibers of the body's connective tissues are defective and weak), blood vessel, bone, and eye defects may all occur. However, one affected individual may simply have very tall stature, long fingers, and one or two other bony defects, whereas his or her offspring with the identical gene defect may suffer from a dislocation of the lens of the eye or from the rupture of a major blood vessel such as the aorta. Reasons for this variability are unknown, but it may in large part reflect the type of mutation, the function of the other inherited normal matching gene, or modifying effects of other genes.

Given the complexity of the human body, it is not surprising that the same dominantly inherited disorders may arise from defects in different genes. For example, profound deafness from birth may be due to a number of different genes. Indeed, dominant, recessive, and sex-linked genes may all result in congenital deafness. About 1 in 10 persons carry one of these deafness genes. A second example is the adult form of a kidney disorder characterized by progressive and slow development of cysts within the kidneys, causing high blood pressure and kidney failure. Thus far, at least three different dominant genes have been discovered that may cause the same adult form of this ailment, which is known as *polycystic kidney* disease.

If a dominant genetic disorder is common, two affected individuals may easily meet and subsequently have children. A child who receives the defective gene from both parents is likely to have a much more serious or a fatal form of the disorder. The resulting disorder could be so severe that death could occur in the womb or in the first weeks or months of life. A common disorder (1 in 500) called *familial hypercholesterolemia* results in early heart disease and is due to a dominant gene. A "double dose" of this defective gene inherited from both parents— if the disorder went untreated—could result in heart attacks even before puberty! However, the potential effects of a double inheritance are not so dire with all dominant genetic disorders: Even a "double dose" of the defective gene does not change the time of onset or the severity of Huntington's disease in the affected offspring.

Gene Mutations

Inherited gene defects are transmitted via the egg and/or the sperm that unites with it. Virtually all of the resultant embryo's cells will contain the same genetic defect (referred to as a *germline mutation*) that exists in the reproductive cells passing from generation to generation. However, mutations may also arise spontaneously in a sperm or an egg and thereafter be transmitted through the generations. For the most part, we do not know what causes mutations. We do know that irradiation from X-rays, atomic energy, or the atmosphere might cause them, as might various chemicals. We also know that the mutations responsible for a handful of genetic disorders (e.g., Marfan syndrome and achondroplastic dwarfism) most frequently occur in the sperm of older fathers.

Current estimates indicate that a new mutation in a given gene occurs in about one germ cell in a million, as that gene is passed from parent to child. This means that 8 to 10 percent of all newborns could carry a new mutation. Fortunately, most of these mutations are silent—probably due to the normal status of the matching gene from the other parent.

The various types of mutations are described in Chapter 6. It is evident that a change even in a single base within a gene can lead to lifelong disease or death (e.g., sickle cell disease). Deletion mutations, not visible under the microscope, may cause recognizable disorders (called *microdeletion syndromes*). Each of these disorders may arise from a new mutation or be transmitted (as a dominant) by one parent to half of the offspring. Features may vary in one person. In addition to the Prader-Willi and Angelman syndromes mentioned in Chapter 6, other examples of microdeletion syndromes and their consequences include:

1. Williams syndrome—Characteristic facial features, aortic valve stenosis (narrowing), variable mental retardation, high level of blood calcium in the newborn period, talkative, uninhibited, distractible, musical.
2. velocardiofacial (DiGeorge) syndrome (see case discussion in Chapter 2)—Heart defects, typical facial features, cleft palate, susceptibility to infection, low blood calcium level, mental illness, and many other variable signs.
3. Miller-Dieker syndrome—Distinctive facial features, profound physical and mental retardation, seizures, spasticity.
4. Smith-Magenis syndrome—Characteristic facial features, mental retardation, short stature, cleft palate, behavior problems.
5. Alagille syndrome—Heart, bone, and eye defects; typical facial features; jaundice due to paucity of bile ducts.

There is considerable variation in each case of these particular syndromes, probably reflecting the additional deletion of one or more contiguous genes.

Another type of mutation—a mutation due to instability in specific genes, or "stuttering"—is discussed later in this chapter, in connection with disease onset and severity.

Harmful Genes from Both Parents

Each of us carries a number of harmful "mutated" genes without showing signs of disease. Defective genes that have no obvious (or very little) noticeable effect on the body are described as *recessive*. Since our genes are inherited in pairs (one from each parent), even if one is defective, the other often has sufficient protein product to keep the body stable and without evidence of any disorder. This is the case with all of the chromosomes except the sex chromosomes, which are discussed below. Measurement of the product of the normally functioning gene in most such cases would show about half the normal function or, if an enzyme, half the activity. For example, a person who carries a defective gene for the degenerative brain disorder called Tay-Sachs disease is likely to have half the normal blood level of the enzyme produced by that gene (hexosaminidase A), and therefore can be diagnosed by measuring this enzyme activity. A carrier of a recessive disorder would have a 50 percent likelihood of transmitting that defective gene to each offspring. A child who receives only this single defective gene would also be a carrier and also show no evidence of any disorder. Only if two individuals who are both carriers of the same defective, recessive gene both transmit their flawed genes will there be a 25 percent risk of having an affected child. As shown in Figures 7.3 and 7.4, there is also a 25 percent chance that two carriers would have a child who is neither affected nor a carrier. The likelihood that two individuals carrying the same defective gene will have children together is increased dramatically if they belong to the same ethnic group (see Chapter 8) or are related (for example, first cousins, or uncle and niece).

Characteristics of Recessive Inheritance

Specific criteria govern recessive inheritance in which an individual receives the same harmful gene located on a non–sex chromosome from both parents, who carry the gene but are unaffected.

1. The parents of an affected child may be related.
2. Males and females are equally likely to be affected.
3. On average, if one child has the disorder in question, the risk of recurrence is 25 percent.
4. Typically, but not invariably, the disorder is not present in the parents or other relatives but only in brothers and sisters.

In a number of cases the mutation in the gene in question may vary from one parent to the other. In such circumstances the disorder will still be manifested, but somewhat differently than expected, as shown by the example on page 76.

Harmful Genes Passed from Mothers to Sons

A female who has a defective gene on one of her X chromosomes will pass it along to half of her male children and half of her female children (see Figure 7.5). The

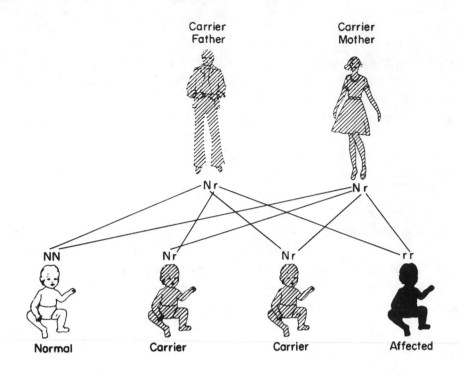

Figure 7.3   Recessive inheritance. Both parents are usually healthy, but each may carry a defective recessive gene (r), which generally causes no problems when paired with its normal counterpart (N). Disease results when a person receives two of these recessive genes. When both parents are carriers, each child has a 25 percent chance of inheriting a "double dose" of the defective gene, a 50 percent chance of being an unaffected carrier, and a 25 percent chance of being neither a carrier nor affected.

daughter who receives the defective X-linked gene also receives a matching gene (usually normal) from the father. This daughter is a carrier, retaining about half of the normal function of the two genes in combination.

A key example among the hundreds of X-linked disorders is the bleeding disorder known as hemophilia (see Figure 7.6). This condition results from an inherited dysfunction of a single factor that enables blood to clot normally. If a female passes along the hemophilia gene to a male child, he will be affected by the disease. Since a woman carries this defective gene on only one of her two X chromosomes, there is a 50 percent chance that her male child will receive the normal chromosome and not even be a carrier (see Figure 7.5). But if the male child does inherit the defective X chromosome and then marries a woman who is a noncar-

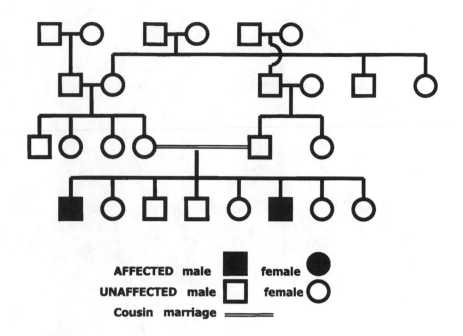

Figure 7.4   A typical family tree involving recessive inheritance—as seen, for example, in cystic fibrosis. Note that there were no affected individuals in the early generations. It is not unusual to find marriages between close relatives (such as cousins) in the family history of individuals with recessively inherited disorders.

rier and has daughters, all of them will receive the X chromosome with the defective gene and become obligatory carriers of this disorder (see Figure 7.7). Their brothers, because they receive their father's normal male (Y) chromosome, will neither be carriers nor have the disease. If an affected male impregnates a female carrier, the probabilities are markedly different (see Figure 7.8). Incidentally, the most common condition transmitted on the X chromosome is red-green color blindness: About 8 percent of white males (less in other races) are affected.

Hemophilia is one of the oldest recognized genetic disorders. Mentions of the illness are found in the Talmud, dating back to before the sixth century C.E.: Rabbis at that time exempted from circumcision a male child whose brother had bled heavily following this ritual. They also exempted the sons of any sisters of a woman who had had a male child who bled excessively after circumcision. In their wisdom, the rabbis did not allow this exemption for the same father's sons born to other women, who—as they must have known—would not have been affected.

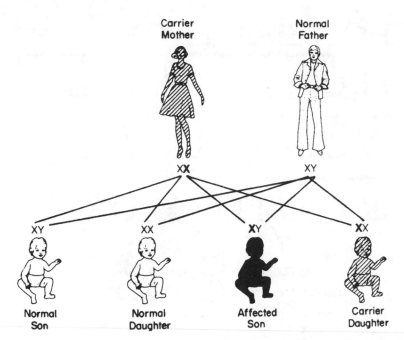

Figure 7.5 Sex-linked (X-linked) inheritance. Here, the defective gene (**X**) is carried on one X chromosome of the mother, who is in most cases healthy. Disease occurs when the X chromosome containing the defective gene is transmitted to a male. The odds of being affected are 50-50 for each male child, and 50 percent of the daughters will be carriers.

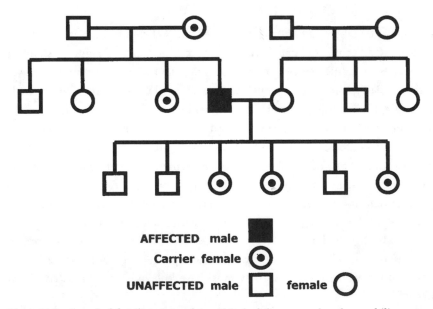

Figure 7.6 A typical family tree involving X-linked disease, such as hemophilia or muscular dystrophy.

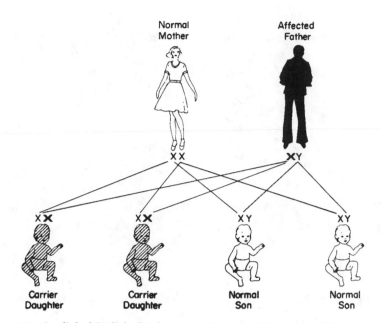

Figure 7.7 Sex-linked (X-linked) inheritance. Here, the defective gene (**X**) appears only on the father's X chromosome. He has the disease (such as hemophilia). He passes his X chromosome to all his daughters, who become carriers, but none of his sons are affected.

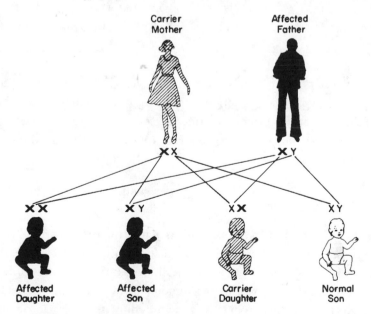

Figure 7.8 Sex-linked (X-linked) inheritance. Here, the defective gene (**X**) is present on the X chromosome of the father as well as on one of the mother's X chromosomes. The father has the disease (such as hemophilia), and the mother is a carrier. There is a 25 percent chance that a daughter will be affected, and a 50 percent chance that a son will have the disease.

J.W. came to me for genetic counseling when she and her husband were planning their second pregnancy. Their pediatrician had referred them for consultation about genetic risks and available tests. Their first child, then three years of age, had cystic fibrosis. His lung function had remained good, but his malabsorption of fats was severe. Even the pancreatic enzyme supplements he took with each meal did not completely keep his intestinal malabsorption under control.

J.W. and her husband were aware of their 25 percent risk for having another child with cystic fibrosis. Their inquiry was primarily focused on the DNA tests necessary to determine the precise defect in the cystic fibrosis genes that they each carried.

Careful review of their family history revealed that J.W. had two brothers who with their wives had adopted children. According to J.W., both brothers "had no sperm" but were otherwise completely healthy. There were no other elements of the family history that had a bearing on future genetic risks in subsequent pregnancies.

I indicated that it was highly likely that each of the brothers had a primarily genital form of cystic fibrosis manifesting only as failure of development of the vas deferens (the thin tubes through which sperm are transported from the testes to the penis). Blood samples for DNA mutation analysis of the cystic fibrosis gene were obtained from both J.W. and her husband and from her two brothers. We found that J.W.'s husband carried the common (F508) cystic fibrosis gene mutation,

The medical history of the British royal family provides a typical "pedigree" for this sex-linked disease. Queen Victoria was a carrier of hemophilia, as was her daughter Princess Beatrice. Two of the princess's sons had hemophilia, and one daughter (Queen Ena) was a carrier. Queen Ena in turn had two sons who were also affected. And so on.

Symptoms and Signs in Carriers

A woman who is a carrier of an X-linked disease rarely carries the harmful gene on both X chromosomes. But even with a single dose of the mutated gene, she might show some signs of being affected. If the disorder is the Duchenne form of muscular dystrophy, she may well have symptoms of weakness when walking, which may become increasingly obvious as she tires during the day. Or she may have weakness when climbing stairs or running, even though her calf muscles look well developed. As many as 84 percent of mothers who carry the gene for whereas she herself carried an uncommon mutation. Each of her brothers carried the same uncommon mutation as well as another mutation that is also uncommon. From this information we were able to deduce that J.W.'s father had transmitted his cystic fibrosis gene mutation to her and each of her brothers, whereas her mother had transmitted the other mutation to J.W.'s brothers but not to J.W.

The poignant question asked by J.W. was whether this family history alone should have prompted gene studies prior to conception. The answer, unfortunately, was in the strong affirmative. Men who are otherwise healthy but who lack sperm should routinely be offered chromosome analysis (for Klinefelter syndrome; see Chapter 4) and cystic fibrosis gene analysis, depending upon the details obtained from family and personal medical history and examination. The discovery of mutations in the cystic fibrosis gene would then lead to tracking of the defect through the family via the testing of partners, siblings, aunts, uncles, cousins, and perhaps other family members. By using all of these options (discussed in greater detail in Chapter 22), serious disorders can be avoided.

In this family, the occurrence of classic cystic fibrosis was due to the parents of the child sharing the same cystic fibrosis gene defect. The mother's brothers had received this same gene defect from one parent and an entirely different defect in the same gene from the other parent, resulting in the failure of development of the vas deferens. They have, therefore, a primarily genital form of cystic fibrosis.

Duchenne muscular dystrophy have some form of heart involvement. The signs include any changes in an electrocardiogram; a tendency toward rhythm disturbances of the heart with potential for serious consequences; cardiac enlargement; and heart failure. If you have a family history of this disease and notice even seemingly minor signs of the kind described here, you should immediately seek a consultation with a cardiologist to arrange for annual surveillance and appropriate preemptive prevention and treatment. If you are a carrier, this knowledge and action could save your life.

Female carriers of hemophilia may not have entirely normal blood-clotting ability. Mothers who are carriers of a white blood cell defect that leads to chronic granulomatous disease, a condition in which the offspring are especially subject to serious infections, may evidence a decreased ability of their white blood cells to kill bacteria. A similar example is the mother of an albino child who shows some pigmentary changes in her own eyes.

Rarely is a female fully affected by an X-linked recessive disease. In order for this to happen, she would have to be the daughter of an affected father and a mother who is a carrier, and she would have to inherit an abnormal gene from both.

Characteristics of X-Linked Recessive Inheritance

Five main criteria help determine whether a disorder is transmitted as an X-linked condition.

1. The disorder is passed from an affected man through all of his daughters to half of their sons.
2. The disorder is never transmitted from father to son.
3. The disorder may be passed through carrier females eventually to affected males who are all related to one another through these females.
4. Female carriers may show signs of the disorder.
5. The incidence of the disorder is much higher in males than in females.

Mosaic Ovaries or Testes

Eggs with normal and abnormal chromosome complements were discussed in Chapters 3 and 4. In a similar vein, some eggs (or sperm) may carry a specific gene mutation whereas neighboring eggs (or sperm) do not. This phenomenon is known as *gonadal* or *germline mosaicism*.

Germline mosaicism occurs in X-linked disorders and is especially important in Duchenne muscular dystrophy (7 to 14 percent of cases arise through this mechanism). Women who have borne a single son with this disorder and who have had DNA studies that revealed they were "not carriers" of the muscular dystrophy gene are highly likely nevertheless to have eggs with and without this gene mutation. It is therefore a safe recommendation that every woman who has borne a son with Duchenne muscular dystrophy be considered a carrier and be offered prenatal diagnostic studies in subsequent pregnancies (see Chapter 22). Gonadal mosaicism is always a concern when a child is born with a dominant or X-linked disorder that has not previously occurred in the family. About 50 percent of cases of certain genetic disorders, such as neurofibromatosis and tuberous sclerosis, are due to new mutations, which may affect subsequent offspring due to hidden gonadal mosaicism.

Harmful Genes from Mothers to Their Children

Situated within the fluid (cytoplasm) bathing the nucleus in every cell are several hundred small bodies called mitochondria, also made up of DNA. Curiously, this DNA is circular and contains the codes of protein subunits called enzymes. During fertilization, the sperm nucleus enters the egg but leaves behind its surrounding fluid, which contains the mitochondria. Hence, the genes contained within the zygote's mitochondria are maternal in origin; this means that they will be transmitted by the mother to all of her children, regardless of their sex. Given this form of mitochondrial inheritance, an affected male would not transmit any defective (or normal) mitochondrial gene to any offspring.

A recent observation that some mitochondrial genes from the father's sperm may indeed enter the mother's egg, if confirmed by future studies, may confound the current interpretation of mitochondrial inheritance. However, the available evidence suggests that adverse consequences from the interchange (recombination) of maternal and paternal mitochondrial DNA are extremely rare. Men with mitochondrial mutations can still be reassured that they will not transmit them to their children.

Further confounding our understanding of maternal inheritance is the fact that a mutation in a mitochondrial gene might not be present in all mitochondria within the same cell. Expected clinical features may therefore be masked or appear less obvious, leading to difficulties in interpretation of the pedigree and possibly to a missed diagnosis.

Since mitochondria play an important role in cell chemistry, it is no surprise that key organs such as the nervous system, heart, and muscle are most frequently involved when a mitochondrial mutation is present. A brand-new defect (a sporadic mutation) may also occur in a mitochondrial gene. One condition that is due to a sporadic mutation in a mitochondrial gene and that is still rarely recognized is signaled by a feeling of fatigue and exhaustion even during normal play or daily activity, worsening with age. Some complain as early as five years of age of a feeling of exhaustion and nausea, and a feeling that their jaw muscles are too tired to chew. These signs sometimes are accompanied by weakness and/or by muscle cramps during exercise.

The characteristic pattern of mitochondrial inheritance has been used to identify human remains—even long-buried bones. Examples of particularly notorious forensic cases include the identifications of a number of "disappeared" children in Argentina who had been abducted under the military dictatorship, and of members of the last Russian royal family, who were executed during the Bolshevik revolution.

Characteristics of Maternal Inheritance

Mitochondrial inheritance (also called maternal, extranuclear, or cytoplasmic inheritance) involves the transmission of a mother's genes to each of her children (Figure 7.9). The primary criteria are:

1. Women with a mutation will transmit it to all of their offspring, of both sexes.
2. Despite being the recipient of a mitochondrial mutation, not all offspring will manifest the disorders in question.
3. Affected and carrier males do not transmit the mutation to any of their children.
4. In families in which there is a recognizable mitochondrial disorder with demonstrable maternal inheritance, the risk to offspring of a carrier female is 50 percent. Where there is no such pattern of inheritance, the risk of recurrence is about 6 percent for offspring and 3 percent for siblings.

Genes from Fathers to Sons

Genes on the Y chromosome are transmitted only to sons, and fortunately there seem to be none that cause death or serious disability. Y-linked (or Holandric) inheritance passes along a set of genes (at least ten) located on the long arm of the Y, all of which participate in the production of sperm. Deletions of one or more of these genes result in no sperm (azoospermia) or few sperm (oligospermia). Such deletions have been reported in 5 to 18 percent of sterile or infertile men. Even in these cases, a single sperm retrieved via a needle from the testis will frequently lead to successful pregnancy. Fertilization is accomplished by the insertion of the sperm directly into a harvested egg (a process called *intracytoplasmic sperm injection*), and implantation of the fertilized egg into the hormonally prepared uterus.

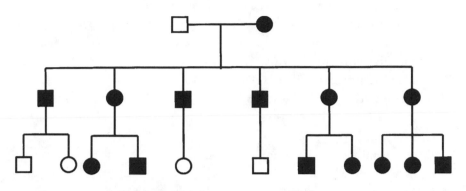

Figure 7.9   Typical pedigree of mitochondrial inheritance, showing transmission of a gene mutation to all children of an affected mother, but not passed along through her sons.

Age of Onset and Severity of a Genetic Disorder

One would think that the presence of a defective gene that causes a serious genetic disorder would be obvious at birth. In fact, the time of onset of serious or fatal genetic disorders varies from the earliest days of embryonic development through to the late seventies or eighties. Many factors influence the expression, penetrance, time of onset, and severity of a genetic disorder. Where a gene is vital to the structure and function of a cell or organ, a mutation may prove lethal within days or weeks of conception. A somewhat less severe mutation might lead to fetal death later in the pregnancy, or to serious defects that become apparent only at birth. In the case of children born with brittle bone disease (osteogenesis imperfecta), many of their bones may be fractured during labor. Other mutations that produce the same disorder may lead, in contrast, to few outward signs until middle age.

Of the more than 900 mutations in the cystic fibrosis gene recognized thus far, the most common is the F508 mutation. When both parents contribute this mutation to their offspring, the more severe form of cystic fibrosis occurs. Some rarer mutations are associated with such mild forms of cystic fibrosis that the disease remains undiagnosed until middle age or even old age! Some individuals affected by one of these mutations are first prompted to seek medical treatment because of chronic bronchitis, problems with the pancreas (pancreatitis), or gallstones. Others are spared the more grave, chronic lung infections.

The health consequences of other mutations—especially those associated with disorders of the nervous system and the muscles—may be conditioned on the sex of the parent transmitting the culprit gene. For example, in the neurodegenerative disorder known as Huntington's disease, a son inheriting the genetic mutation from his father will experience an earlier onset of illness than that experienced by the father. Huntington's disease has manifested itself as early as five years of age. The average age at onset of detectable symptoms, however, is 37 years.

In dominantly inherited myotonic muscular dystrophy (the most common form of adult muscular dystrophy), in which an affected mother transmits the defective gene to an affected offspring (the risk is 50 percent), severe manifestations frequently are present at birth. Hence, pregnant mothers who know they have this disorder should know about the implications of this gene. They should ensure that they deliver in a tertiary care medical center with sophisticated facilities for newborn intensive care, and be certain that a pediatric newborn specialist (neonatologist) is present at the birth. Moreover, because of the increased risk of obstetrical as well as anesthetic complications, high-risk specialists in obstetrics (perinatologists) should be sought out. Prior knowledge of the potential hazards of this defective gene can secure the health and save the life of both mother and child.

The phenomenon of sex-influenced effects on the offspring of affected men or women is due to a process called imprinting. See Chapter 6 for a detailed explanation of this process.

The mutations that are typical in at least fifteen inherited neuromuscular disorders are characterized by a stuttering type of defect in the bases that constitute the gene (see Chapter 7). In normal genes, runs of three base sequences (called triplet repeats)—for example, the CAG or CGG sequence—might be reiterated as many as 10 to 30 times. However, in Huntington's disease, there are an increased number of triplet repeats, mostly in the range of 40 to 121 iterations. Similarly, an increased number of triplet sequences (the number differs according to the particular disease) may be seen, for example, in myotonic muscular dystrophy, Friedreich's ataxia, various types of spinocerebellar ataxia, spinal-bulbar palsy, and the fragile X syndrome discussed in Chapter 9.

Generally speaking, the greater the number of triplet reiterations, the more severe the disease and the earlier its onset. Indeed, with all of these disorders, the phenomenon of *anticipation* comes into play: In other words, the age at onset is progressively younger from generation to generation.

DNA Fingerprints

There are more than 3 billion bits of DNA information in which individual patterns, mutations, or markers can be used for identification. Moreover, in forensic use, the DNA found within tiny specks of blood, for example, can be used to make whatever amount of DNA sequence may be needed as an aid in identifying an individual. DNA fingerprints (not actually fingerprints but telltale patterns in DNA bits) are among the most useful and most accurate tools of personal identification currently available; they can achieve billions-to-one odds—a standard no court requires. DNA fingerprinting can provide a positive identification that is virtually certain and can even lead police to a previously unsuspected perpetrator, whereas other methods serve only to rule out persons under suspicion.

Incest

Sexual intercourse between brothers and sisters, fathers and daughters, or mothers and sons is not only illegal but is taboo in most societies. The published results from studies of incest all note a devastating increase in the occurrence of serious birth defects and mental retardation—by about fivefold—when offspring result from such a union. A 1982 Canadian report on twenty-nine children born of incestuous unions showed that most had severe abnormalities, low birth weight, mental retardation, or medical problems. Four (13.8 percent) had specific, recessively inherited defects.

DNA analysis now provides remarkable accuracy in the determination of paternity and even maternity. Many legal cases of incest between a father and daughter have been successfully prosecuted based on definitive DNA evidence. In these cases it was demonstrated conclusively that unique patterns of genes in the offspring could have occurred only if the fathers had committed incest. The degree of certainty in most of these cases exceeded 99 percent.

Cases of questioned maternity have arisen mostly through immigration services, in response to claims lodged by applicants for immigration. Once again, proof of maternity in such cases was established with 99 to 100 percent certainty.

Rape and Murder

The uniqueness of an individual's set of genes provides a telltale set of DNA "fingerprints." Cells containing DNA from any site can yield identifiable DNA.

Substances successfully mined for DNA in criminal cases have included blood specks, semen stains, hair roots, nasal mucus, saliva, and skin found under the nails of a victim, at a bite site, under a licked stamp or on an envelope, on an instrument of murder, or at other sites. A twenty-five-year-old vaginal swab taken for semen was successfully used in one criminal prosecution. Many convictions of murderers and other felons have been achieved through DNA identification. Also noteworthy and desperately sad is the discovery of the innocence and false incarceration of individuals subsequently released on the basis of DNA evidence.

chapter 8

Ethnic Genes and the Disorders They Cause

Genes not only travel from generation to generation but also drift from continent to continent: Geneticists using sophisticated DNA analyses have tracked population movements from Africa to Asia about 100,000 years ago. Similar population shifts from Asia to Europe are estimated to have occurred about 40,000 years ago, and from Asia to America between 15,000 and 40,000 years ago.

It's no surprise that close-knit populations, migrating together from the same villages and regions, share their genes. For this reason, the likelihood of sharing a defective gene with a mate from the same ethnic group is much higher than with someone from outside the group. The likelihood of a child's receiving the same defective gene from both parents is also higher; this is the reason for the frequency of recessively inherited genetic disorders (see Chapter 7) within particular ethnic groups.

Regardless of our ethnic origins, we all carry harmful genes. The fact that many studies have been conducted and published on genetic disorders among Jews, Finns, and Mormons does not indicate that these groups are particularly or unusually vulnerable to genetic mutations. It does, however, reflect the fact that more information is available about these populations due to their longtime understanding of the health implications of family medical history, their traditional concern for their extended families, and their practice of keeping complete genealogical records.

Coping with the Risks of Being a Carrier

As explained in Chapter 7, those of us who carry a defective recessive gene rarely have any harmful consequences. After a child is born with a recessive disorder, tests may reveal a mutation in the same gene in each parent. In such cases, preconception testing of both parents is highly recommended. When both parents are carriers of a recessive genetic disorder, the chances are 1 in 4 that they will have an affected child; 2 in 3 that their child will be a carrier; and 1 in 4 that the child will be neither a carrier nor affected (see Chapter 7).

Who Needs Testing?

Each of us belongs to a recognizable ethnic or racial group and may be at risk for certain genetic disorders (see Table 8.1). If carrier tests are available for a disorder to which your ethnic group is particularly susceptible, you should consider obtaining them prior to conception. Some may even want to consider such testing prior to marriage. It is self-evident that parents should also be tested after the birth of an affected child.

The Gene Screen

The logic of screening blood samples of ethnically related couples planning to have children gradually has become apparent to public health authorities and private health professionals. The most successful example of such testing has been the routine carrier testing for the Tay-Sachs disease gene, a mutation found most often in the Ashkenazic Jewish population. Potential carriers are tested either for a deficient enzyme called *hexosaminidase A* or for gene mutations detectable through DNA analysis. These two testing procedures together have made the birth of a child with Tay-Sachs disease a rarity in the United States. The few such births that have occurred over the past decade were due to the parents not having been tested; to laboratory error either in the carrier test or the prenatal diagnostic test; or to parents' knowingly having declined to abort an affected fetus.

Dr. Michael Kaback, a pioneer in population screening, reported that between 1970 and 1999 over 1.4 million Ashkenazic Jews had Tay-Sachs disease screening tests. Almost 51,000 individual carriers and 1,379 carrier couples were identified. Prenatal diagnoses of 628 affected fetuses were made. As a consequence, the incidence of Tay-Sachs disease has dropped by more than 90 percent in the United States. Today more babies with this fatal degenerative brain disease are born to non-Jewish couples or to mixed-marriage couples who are unaware of the risks.

One in 27 to 1 in 30 Ashkenazic Jews carries a Tay-Sachs disease gene mutation (see Tables 8.2 and 8.3). Without testing, about 1 birth in 3,600 would be that of an affected infant—100 times the number of such births in the non-Ashkenazic population. Carrier detection rates by either method (enzyme or DNA analysis) exceed 98 percent.

TABLE 8.1 Examples of Genetic Disorders Associated with Particular Ethnic Groups

Ethnic Group	Genetic Disorder	Typical Features
Africans (blacks)	Sickle cell disease and other disorders of hemoglobin	Anemia
	Alpha- and beta-thalassemia	Anemia
	Glucose–6-phosphate dehydrogenase deficiency	Anemia
	African-type adult lactase deficiency	Milk intolerance
	Benign familial leukopenia	Infection
Afrikaners (white South Africans)	Porphyria variegata	Neurological problems and sun sensitivity
	Fanconi anemia	Anemia; leukemia
American Indians (of British Columbia)	Cleft lip or palate (or both)	Facial defect
Amish/Mennonites	Ellis–Van Creveld syndrome	Dwarfism and extra digits
	Pyruvate kinase deficiency	Anemia
	Hemophilia B	Bleeding
Armenians	Familial Mediterranean fever	Inflammation and fever
Ashkenazi Jews	A-beta-lipoproteinemia	Nervous system degeneration and fat malabsorption
	Bloom syndrome	Lymphomas, skin, and growth problems
	Canavan disease	Brain degeneration
	Congenital adrenal hyperplasia	Genital abnormalities; adrenal tumors
	Dystonia musculorum deformans	Disabling muscular and movement disorder
	Familial dysautonomia	Development problems and medical crises
	Factor XI (PTA) deficiency	Blood clotting
	Gaucher disease (adult form)	Blood and bone problems
	Niemann-Pick disease	Brain degeneration
	Tay-Sachs disease	Brain degeneration

(continues)

TABLE 8.1 (continued)

Ethnic Group	Genetic Disorder	Typical Features
Chinese	Thalassemia (alpha)	Anemia
	Glucose–6-phosphate dehydrogenase deficiency	Anemia
	Adult lactase deficiency	Milk intolerance
Eskimos	Congenital adrenal hyperplasia	Genital abnormalities; adrenal tumors
Finns	Congenital nephrosis	Kidney failure
	Aspartylglucosaminuria	Neurodegeneration and mental retardation
	Progressive myoclonus epilepsy	Epilepsy
	Diastrophic dysplasia	Dwarfism
	Dystrophic retinae dysacusis syndrome	Blindness
French-Canadian	Familial hypercholesterolemia	Coronary heart disease
	Spastic ataxia	Muscle and balance problems
	Tay-Sachs disease	Brain degeneration
	Neural-tube defects	Brain or spinal defects (e.g., spina bifida)
Irish	Neural-tube defects	Brain or spinal defects (e.g., spina bifida)
	Phenylketonuria	Mental retardation
	Schizophrenia	Psychosis
Italians	Fucosidosis	Mental retardation
Japanese	Fukuyama congenital muscular dystrophy	Muscle disease
	Congenital plasminogen deficiency	Blood clotting
	Fundus albipunctatus	Blindness
Japanese and Korean	Acatalasia	Severe oral ulcers
	Oguchi disease	Night blindness
	Dyschromatosis universalis hereditaria	Skin disorder

(continues)

Maori (Polynesians)	Clubfoot	Foot deformity
Mediterraneans (Italians, Greeks, Sephardic	Thalassemia (mainly beta)	Anemia
Jews, Armenians, Turks, Spaniards, Cypriots)	Glucose–6–phosphate dehydrogenase deficiency	Anemia
	(Mediterranean type)	
	Familial Mediterranean fever	Inflammation and fever
	Glycogen-storage disease (type III)	Liver disease
Norwegians	Cholestasis-lymphedema	Liver problems
	Phenylketonuria	Mental retardation
Yugoslavs (of the Istrian peninsula)	Schizophrenia	Psychosis

NOTE: This table provides only an indication of the extent of hereditary ethnic disorders. Any specific concern is best discussed with your doctor.

Voluntary population carrier screening programs for Jewish high school and college students are well established in Montreal and elsewhere. A remarkable program was established years ago by an ultra-Orthodox Jewish community in New York. The Dor Yeshorim Program serves a group that have highly restrictive laws about abortion. Marriages in this community are arranged, and the individuals involved are screened for Tay-Sachs and some other disorders before they are introduced to each other. Test results are provided anonymously to a central agency, and no two carriers are ever matched. This program has made prenatal diagnosis and selective abortion moot; it is anonymous and private, avoids any stigma, and, of course, achieves virtually total avoidance of a fatal genetic disease. Similar programs have been established in Israel and elsewhere.

On occasion, a disorder characteristic of one ethnic group might appear in a child born of a couple in an unrelated ethnic group or to a couple of mixed ethnic origin. For example, a non-Jewish French-Canadian couple might have a child with Tay-Sachs disease. Such occurrences were recognized many years ago and led to the hypothesis that a Jewish salesman sailing down the Hudson long ago had "dropped his genes." This hypothesis lasted many years until analysis of the Tay-Sachs gene showed that the characteristic mutations among French-Canadians were different from the common mutations found among Ashkenazic Jews. In fact, about 1 in 30 French-Canadians harbors a gene mutation for Tay-Sachs disease and should have carrier tests if planning pregnancy.

In mixed marriages, testing for a specific ethnicity-related mutation in the individual at risk is recommended first. Only if a mutation is recognized should the partner then be tested.

Common Mutations

Screening, or at least preconception testing, is recommended when the risk of being a carrier is at least 1 percent for the particular ethnic or racial group and when a reliable test is widely available. The likelihood of being a carrier of certain ethnic disorders or of having an affected child is shown in Tables 8.2 and 8.3. The testing of the product of the suspect gene (a protein or an enzyme) would be most efficacious, but for technical reasons it is not practical for ethnicity-linked diseases other than Tay-Sachs. Since we are discussing recessively inherited ethnic disorders, it would not be surprising if within certain ethnic and racial groups some mutations were much more common than others. This is indeed the case for a number of such disorders.

TABLE 8.2   The Chances of Carrying a Particular Genetic Disorder*

If You Are	The Chance Is About	That You Carry a Gene Mutation for
Afrikaner (white South African)	1 in 400	Porphyria†
Armenian, Jewish (Sephardic) Turk, Arab	1 in 7 or 1 in 8	Familial Mediterranean fever
Black	1 in 12	Sickle cell anemia†
	7 in 10	Milk intolerance as an adult
Black and male Black and female	1 in 10 ⎫ 1 in 50 ⎭	Predisposition to develop hemolytic anemia after taking sulfa or other drugs
	1 in 4	High blood pressure
Finns	1 in 36	Aspartylglucosaminuria†
Italian-American or Greek-American	1 in 10	Beta-thalassemia†
Jewish (Ashkenazic)	1 in 27 to 1 in 30	Tay-Sachs disease†
	1 in 15	Gaucher's disease†
	1 in 40	Canavan disease†
	1 in 90	Niemann-Pick disease (type A)†
	1 in 90	Fanconi anemia†
	1 in 38 (women)	Breast/ovary cancer gene mutation†
	1 in 30	Dysautonomia
	1 in 5** to 1 in 7	Familial Mediterranean fever†
	1 in 26	Nonsyndromic deafness†
Oriental	100 percent	Milk intolerance as an adult
White	1 in 10	Hemochromatosis†
	1 in 31	Nonsyndromic deafness†
	1 in 25	Cystic fibrosis†
	1 in 60	Spinal muscular atrophy†

*These disorders occur in other ethnic groups, but rarely.

†DNA carrier test and prenatal diagnostic test are available.

**Because of different mutations, individuals may have milder disease or be asymptomatic.

TABLE 8.3　The Chances of Being Affected by a Particular Genetic Disorder, for Children of Various Ethnic Groups

If You Are	The Chance That You Could Have Been Born With a Defect Leading to	Is About
Black	Sickle cell anemia	1 in 650
	Beta-thalassemia	8 in 1,000
White	Hemochromatosis	1 in 200 to 1 in 250
	Cystic fibrosis	1 in 2,500
	Phenylketonuria	1 in 14,000
Jewish (Ashkenazic)	Tay-Sachs disease	1 in 3,600
	Familial dysautonomia	1 in 3,600
	Canavan disease	1 in 6,400
	Niemann-Pick type A disease	1 in 32,000 to 40,000
Italian American or Greek American	Beta-thalassemia	1 in 400
Armenian, Jewish (Sephardic), Turk, Arab	Familial Mediterranean fever	1 in 250

Cystic Fibrosis.　In *cystic fibrosis* (characterized by viscous secretions and chronic infections of the lungs, and by intestinal malabsorption of fats) the gene is large. More than 900 mutations of cystic fibrosis have been documented. This disorder is most frequently encountered among whites, 1 in 25 of whom are carriers. Laboratory analysis targeted on the 30 most common mutations currently detects between 80 and 90 percent of carriers. New technologies have enabled us at the Center for Human Genetics to routinely test up to about 100 mutations at reasonable cost, increasing the detection rate by a few percent. The frequency of various cystic fibrosis gene mutations varies among ethnic groups within the white population: For example, it is slightly different in those whose ancestors came from England, compared with those from Italy, Spain, South America, and other areas. Testing in Ashkenazic Jews for the 5 most common cystic fibrosis mutations yields a carrier detection rate of 98 percent. A similar detection rate is achievable in those of Celtic origin.

The most common mutation (called F508) accounts for close to 70 percent of all mutations. When both parents carry this mutation, they have a 25 percent risk of having an affected child, and face the most severe, potentially lethal form of cystic fibrosis. In contrast, couples who have less common mutations—possibly even different mutations—might bear a son who in adulthood is found to be lacking the vas deferens (the tubes that convey sperm from the testes to the penis). These men (of whom we have studied more than 300) have no detectable sperm in the ejaculate—a fact that is usually discovered during the evaluation of a couple for infertility. However, recovery of sperm via a needle introduced into the testis allows successful in vitro fertilization. An important caution is of course to test the female partner first for her carrier status, since the male effectively has a genital form of cystic fibrosis and will be transmitting one cystic fibrosis mutation to every offspring he sires.

Sickle Cell Disease.   About 1 in 600 blacks in the United States are affected by *sickle cell disease.* This recessive disease, which is linked to a single point mutation (see Chapter 6) in the hemoglobin gene, results in red blood cells becoming sickle-shaped and breaking up easily, causing anemia and blood vessel blockages. Due to such blockages, individuals affected by sickle cell disease experience recurrent episodes of extremely painful crises in the chest and abdomen and may also suffer strokes. Also because of blood vessel blockage, some 10 to 40 percent of affected men may suffer painful, constant erections, which are difficult to treat.

About 1 in 12 black Americans are carriers of the sickle cell gene. About 5 percent of these carriers may occasionally have some blood in the urine, and most cannot concentrate their urine. These are clinically insignificant abnormalities that require no treatment or occupational restrictions. However, sickle cell carriers participating in military basic training reportedly had an incidence of sudden, unexplained death some 20 times greater than that among recruits who were not carriers. Since this drastic result may be due to exertional heat illness, carriers should undergo only gradual physical training and conditioning, avoid overtaxing their bodies, and be sure to drink enough throughout the day to replace lost fluid and electrolytes.

About 1,000 babies with sickle cell disease are born in the United States each year. The situation is much worse in Africa, where up to one-third of blacks are carriers. About 120,000 affected babies are born annually on that continent, less than 2 percent of whom survive to age 5.

Familial Mediterranean Fever.   Carrier tests for familial Mediterranean fever (FMF) are especially important for individuals of Armenian, Turkish, Arabic, and Sephardic Jewish ancestry, given the high frequency with which this genetic disease occurs among these groups (see Table 8.2). FMF is characterized by recurrent episodes of inflammation affecting the abdomen, chest, and joints, accompanied by fever and eventually by kidney failure if not treated. Many mutations linked to this disease have been discovered, and blood analysis for the four most common provides a carrier detection rate of about 85 percent. Since the advent of mutation analysis for FMF, Ashkenazic Jews also have been found to be carriers, at a similar rate to that among Sephardic Jews (see Table 8.2). However, since clinicians had seen relatively few affected Ashkenazic Jews, a detailed mutation analysis was conducted by Israeli scientists. This study revealed that Ashkenazic Jews frequently have a mutation that causes few, if any, symptoms or signs.

Carrier detection is important in FMF for genetic counseling purposes as well as prenatal diagnosis. In addition, mutation analysis permits early and accurate diagnosis of FMF in those affected, enabling them to receive timely treatment with colchicine to prevent kidney failure. (Kidney failure in the FMF-affected results from deposits of a complex protein called *amyloid*.)

Canavan Disease and Niemann-Pick Disease.   Routine testing of Ashkenazic Jews for Canavan disease (a degenerative fatal disease of the brain, evident from birth or infancy) likewise focuses on the three most common mutations, which appear in 98 percent of all carriers. Similar common mutation analyses are performed among Ashkenazic Jews for Niemann-Pick disease (a neurodegenerative disease evident from infancy) and Fanconi anemia. The recent identification in this same ethnic group of three common gene mutations linked with cancers of the breast and ovary and of a separate gene mutation for colon cancer might also provide grounds for consultation and testing with a clinical geneticist. In this connection, it is important to note that as many as one-third of infants affected by Canavan disease are born to non-Jewish parents.

In families with nonsyndromic deafness that remains unexplained, the most likely culprit gene (Connexin-26) can be analyzed. A mutation known as 35delG is particularly common among whites generally (1 in 31); and another mutation (167delT) is especially common among Ashkenazic Jews (about 1 in 26).

Testing Prior to Conception

Given the serious nature of all of these disorders, strong recommendations have been made for preconception carrier testing (or if missed, very early pregnancy testing). The U.S. National Institutes of Health and the American College of Medical Genetics have strongly endorsed cystic fibrosis carrier tests for couples planning pregnancy or in the earliest weeks of a pregnancy. The American College of Medical Genetics and the American College of Obstetricians and Gynecologists have also issued directives that preconception or early prenatal testing be done to detect carriers of Canavan disease in Ashkenazic Jewish couples, similar to the procedures long in place for detecting Tay-Sachs disease in Jews and sickle cell disease in blacks. Unfortunately, the question of preconception testing is moot in 50 percent of pregnancies, which are unplanned.

Ancestral Founders

From the beginning of time, population groups have migrated for religious, political, climatic, and economic reasons, or simply for the purpose of discovery. The founding members of such groups carried with them both good genes and mutant genes for certain recessive disorders. The genetic consequences of inbreeding in such groups are known as *founder effects.*

A founder effect accounts for the high frequency of a disorder called *porphyria* among the Dutch-derived Afrikaners of South Africa. This dominantly inherited disorder is characterized by attacks of acute abdominal pain, weakness, sun-induced skin rashes, mental illness, and other features. The defective gene causing this condition was introduced by Dutch settlers who disembarked in the area of Cape Town in the late seventeenth century. Today, about 1 Afrikaner in 400 has this disorder.

Another founder effect is responsible for the autosomal recessive disorder called *Ellis–Van Creveld syndrome,* which is characterized by short-limbed dwarfism, extra fingers and toes, and heart defects evident at birth. Although extremely rare in the general population, the condition is relatively frequent among the old-order Amish of Pennsylvania.

Signs in Carriers

Most carriers of defective genes for recessive disorders show no outward signs. An exception to this rule occurs with a rare recessive disorder known as *Bardet-Biedl syndrome,* the main features of which include obesity, extra digits, mental retardation, eventual blindness, and small or poorly functioning gonads. Carriers of this disorder may show a tendency toward obesity and may have high blood pressure, diabetes, and various kidney defects.

Marrying in the Family

In some countries, such as India, Saudi Arabia, and Japan, marriage within the extended family is common. In one Saudi Arabian industrial city, a 1998 study noted that 52 percent of marriages were consanguinous. First-cousin marriages were the most common (39.3 percent of all marital unions). In another study, conducted in the United Arab Emirates, a significantly higher frequency of cancers, birth defects, mental retardation, and physical handicaps was found in the offspring of consanguinous unions.

The consequences and importance of consanguinity—a blood relationship between mates—depends upon how closely the couple is related. First cousins who have two grandparents in common also share one-eighth of their genes. Their risk of having a child with a birth defect, genetic disorder, or mental retardation is 6 to 8 percent in each pregnancy—double the 3 to 4 percent incidence of such defects in the population at large. In uncle-niece unions, one-fourth of the genes are shared. Second cousins have only one-thirty-second of their genes in common and they probably have only a slightly (if any) increased risk of having a genetically defective child. The birth of a child with a rare genetic disorder, in the absence of a family history of such a defect, always raises the question of consanguinity somewhere in the ancestral background.

Most countries prohibit marriages between close relatives, including first cousins. In the United States, fewer than 1 in 1,000 marriages are between first cousins. In some isolated communities around the world, however, unions between close relatives may be as high as 25 percent. Such marriages tend to reduce the number of carriers of defective genes but increase the number of offspring with genetic disease. Genetic counseling is advised for all related couples, not only for risk assessment and carrier detection tests but because of increasing possibilities for prenatal genetic diagnosis.

Avoiding Tragedies: What Should You Do?

There are a number of easy steps you can take to secure a diagnosis of a genetic disorder related to your ethnic origins or to determine your carrier status for any disorders that characteristically occur in your ethnic or racial group (see Table 8.4). If you do not fully know your ethnic origins, search out older relatives to obtain more detailed information. If you were adopted, make an inquiry through the relevant department in your state government. Next, if you are concerned about a particular genetic disorder and have not yet obtained a diagnosis through a physician, consult first with a clinical geneticist. Request a carrier test for any disorder that characteristically occurs in your ethnic or racial group. These tests are best done before conception. If you turn out to be a carrier, then your partner also should be tested. If both of you are carriers, consider prenatal genetic

TABLE 8.4   Checklist for Prevention

---

* Find out your ethnic origins
* Consult a clinical geneticist for a diagnosis
* Request a carrier test(s)
* Have your partner tested as well
* Consider prenatal genetic diagnosis if you are both carriers
* Inform your family
* Urge them to have carrier tests

---

diagnosis in every pregnancy, given a 25 percent risk of having an affected child. Be sure to inform your siblings, parents, grandparents, uncles, aunts, and first cousins if you are a carrier. Strongly encourage them to be tested: Even if they do not plan to have more children, they might already have passed on the harmful gene to their offspring. These easy steps can help you secure your life and health and avoid the tragedy of having a child with a serious or fatal genetic disease.

chapter 9

When Threads or Blueprints Go Awry

···

Mental Retardation and Birth Defects

The perplexed look on the young woman's face rapidly changed to one of anguish. She had been referred to me for a consultation concerning a previous malformed, stillborn child and two prior miscarriages. Her initial response to a remark about the weather brought a broad smile to her face. I instantly recognized that she had a single central upper incisor tooth (we normally have two upper central front teeth). Consternation followed my observation, which apparently had not been made before by her dentist or anyone else in her lifetime! She had been unaware of this feature, given an otherwise perfectly good set of teeth. I explained that this disorder was due to a single defective gene controlling development of the midline, which when transmitted could cause a profound midline defect of the face and the front of the brain. This defect had caused her son to be malformed and stillborn and most likely also had precipitated one or both miscarriages. The risk of recurrence was 50 percent. Fortunately, my observation regarding what might have seemed an incidental physical trait prompted this young woman to seek high-resolution ultrasound surveillance during her next pregnancy, which provided reassurance that the fetus was free of any such brain and facial defect. A happy outcome ensued.

The phenomenon of a single gene having several effects, from mild to severe, was noted in Chapter 7. Because many carriers of a mutant gene show mild effects or no effects at all, for the most part we have little or no advance warning of potential major birth defects in our children. On average, all couples have a 3 to 4 percent risk of bearing a child with a major birth defect, mental retardation, or genetic disorder. If one includes cases diagnosed later in a child's life—by 7 to 8 years of age—the risk is 7 to 8 percent, the difference being due to genetic disorders not detected or diagnosable in infancy that become more evident in early childhood.

Extent of the Problem

In the United States alone, over 160,000 pregnancies a year could end with serious birth defects or genetic disorders, with or without subsequent mental retardation. About 3 percent of the population here and in other Western countries have mental retardation of varying degrees. There are over 8 million individuals in the United States who are mentally retarded (i.e., have an IQ of 70 or less). The public health implications are self-evident.

Major and Minor Birth Defects

Birth defects (also called *congenital malformations*) account for significant morbidity and mortality in the early weeks of life. In the Western world, such defects account for almost one-third of all admissions to children's hospitals. Most major birth defects, including mental retardation, lead necessarily to medical attention due to various impairments in body structure and/or function. Minor birth defects occur even more frequently but by definition do not threaten either health or appearance. However, the presence of three or more minor defects (see Table 9.1) cannot be ignored, because the chances are strong (about 20 percent) that a related, major birth defect is present even if not obvious.

When Normal Development Goes Awry

The normal development of the human embryo is an incredibly complex and finely orchestrated process directed by genes and their products and influenced by environmental factors. It is not the purpose of this book to delve into the fine details of the embryo's development. Rather, emphasis is given to a broad understanding of key factors that adversely affect embryonic and fetal development, particularly with an eye to avoidance and prevention of birth defects and mental retardation.

When fetal development goes awry, the most critical period is in the first twelve weeks following conception, especially weeks 3 through 8. Throughout this period, three major processes occur simultaneously: formation of many cells, specialization of cells into specific organs, and migration of cells to all developing organs. Part of the process includes programmed cell death (see Chapter 6), fusion between tissues, and step-wise development in a specific sequence for each tissue or structure. Inherited or environmental (or both) influences may cause defects in one or more of these processes. Interference with programmed cell

TABLE 9.1   Examples of Minor Anomalies

Cranium and scalp	Redundant neck skin
Triple hair whorl	Droopy eyelids
Persistent fontanel	
Malpositioned fontanel	*Skin*
Back of head flat/prominent	Shoulder dimple
Prominent forehead	Sacrum dimple
	Prominent sole crease
Ears	Single palm crease
Very small	Skin tags (front of ears)
Lack of folds	Birthmarks
Ear lobe crease/notch	Pigmented spots
Lop ear	Hypopigmented spots
Cup-shape ear	
Low-set ears	*Trunk*
	Extra nipples
Face and Neck	Two-vessel umbilical cord
Flat bridge of nose	Umbilical hernia
Prominent bridge of nose	
Eyes too close	*Limbs*
Eyes far apart	Tapered fingers
Eyes slant upwards	Overlapping fingers/toes
Eyes slant downwards	Broad thumb, great toe
Cleft gums	Incurved 5th fingers/toes
Small mouths	Nails underdeveloped
Big mouth	Increased space between 1st, 2nd toes
Prominent tongue	Webbed toes
Small chin	Overlapping digits
Webbed neck	Heel prominent

death in the hand could result in fingers remaining fused together (syndactyly); interference with migration of cells could result in failure to close the canal housing the spinal cord, causing spina bifida (extrusion of the spinal cord and nerves through a hole in the back); and interference with the migration of cells could eventually lead to mental retardation, epilepsy, and other brain dysfunction.

Failure of fusion between adjacent tissues could lead to defects such as cleft lip and palate. Disturbances in the sequence of development could interrupt the delicate but orderly cascade of developmental processes, even though the visible signs seem unrelated. A good example is an abnormally shaped ear on the same side of the body as a missing or abnormal kidney. Since cells signal to each other, abnormalities in one area may affect another part of the developing embryo, resulting in other abnormalities. Noxious agents such as the German measles (rubella) virus, the cytomegalovirus, toxoplasmosis, medications, fever, X-ray treatment, and other poisons may all adversely affect embryonic or fetal development and cause multiple defects. An inherited or new gene defect may have similar consequences at this critical phase of development.

Hence, malformations may be limited to one organ or may appear in many different body systems, depending upon the timing of an acquired "insult" or the presence of a defective gene that acts only at a given time and possibly only on a specific developing organ(s).

Syndromes

Congenital malformations (i.e., malformations existing at birth) may occur singly or in combination. A pattern of abnormalities is described as a *syndrome*. Over 2,000 syndromes have been recognized and catalogued. Diagnosis of a syndrome depends on the ability of the clinical geneticist to detect characteristic features (sometimes subtle) and to recognize an abnormal developmental pattern. Interruption in the sequence of development may lead to defects in other tissues downstream in the developmental cascade.

Deformation

Babies are sometimes born with a noticeable physical deformation such as a clubfoot. Such anomalies occur after an organ or tissue is normally formed, and are due to mechanical forces distorting the normal structure. Deformations mostly occur in late pregnancy, largely arise from malposition, and mostly involve cartilage, bone, and joints. The majority of deformations correct spontaneously and slowly after birth, with treatment necessary in some. Common examples include clubfoot, asymmetric skull, small lower jaw, and bowing of the leg below the knee. An abnormally shaped uterus, insufficient amniotic fluid cushioning the fetus, multiple fetuses producing crowding, and abnormalities in the mother's pelvis are examples of the causes of intrauterine constraint promoting deformation.

Dysplasia

When a structural defect is produced by an abnormality in the organization of the cells within a specific tissue or in their function, this is called *dysplasia*.

Dysplastic tissues do not self-correct over time and can cause serious problems, including tumors, later in life.

Disruption

A tissue developing normally on its own may also experience interference from an internal source. For example, tears in the amniotic sac inside which the fetus grows may result in one or more strands of tissue traversing the cavity within the womb, much like elastic bands. These amniotic bands may slice across tissues (for example, the face) and cause dreadful defects in facial structure, or ensnare a limb or digit, cutting off its blood supply. The latter process may result in that digit or distal part of the limb not developing or actually being amputated. There have been cases in which the birth of the infant has been followed by the delivery of a missing digit or part of a limb—thankfully, rare occurrences. Disruptions may also occur if the blood supply to a tissue is interrupted by clots or showers of clots or by a hemorrhage into the tissue. Death or abnormality of the tissue or part may be a consequence.

Causes of Birth Defects and Mental Retardation

The list of known causal factors that result in birth defects and/or mental retardation is long and categorically includes defective genes, chromosome abnormality, and environmental factors. A lack of oxygen to the fetus during labor and delivery may account for between 8 and 14 percent of cases of cerebral palsy. The causes of this nonprogressive neurological disorder are mostly not known.

The most common causes of mental retardation are various chromosome abnormalities (e.g., Down syndrome), discussed earlier (Chapters 3, 4, and 5). Second-most common are subtelomeric gene deletions, covered in Chapter 5. In third place is the fragile X syndrome, which, because of the many opportunities for avoidance and prevention, is discussed here in detail.

The Fragile X Syndrome

For about a century we have known that mental retardation occurs more commonly in males than in females. A genetic explanation for this observation was suggested by studies performed in Australia of an unexpectedly common disorder with striking features, which we now recognize as the fragile X syndrome. Once thought to occur only in males, this disorder has since been recognized in females; the signs in females, however, are considerably less severe. One male in 5,500 to 6,000 has the fragile X syndrome; and about 1 in 187 boys with learning disorders have it. Since the occurrence of mental retardation within a family pedigree is not at all unusual, it is important to recognize the other telltale signs of this syndrome. This is especially crucial because the effects of this disorder become progressively more severe with each subsequent generation. Moreover, most individuals who are affected by fragile X have never been diagnosed.

The Main Features.   The most common and important features of the fragile X syndrome in males are shown in Table 9.2. Similar but less striking features are seen in female carriers, if they show any signs at all. No single feature is diagnostic. Rather, it is the combination of a majority of the main features that suggest the diagnosis, which can then be precisely determined by DNA analysis. The wide range of features, however, also may contribute to many with this syndrome going undiagnosed. Should you recognize that a family member possesses both the intellectual deficits and characteristic physical features, you could (with tact and sensitivity) recommend a consultation with a clinical geneticist.

Particularly striking physical features include large, prominent ears; a long midface; a large head; hyperextensible finger joints and, frequently, double-jointed thumbs and flat feet. The other physical features mentioned in Table 9.2 are usually observed only during physical examination by a physician. Not all features are present in an affected individual.

Boys with this disorder have enlarged testicles, which become obvious around puberty. Examination of testicular tissue under the microscope reveals normal cells, the extra size of the testicles being largely due to an increase in fluid within the tissues. Sexually, these males function normally and may produce sperm and therefore father offspring.

A highly variable set of intellectual and behavioral characteristics are found in this syndrome. The most important of these characteristics, mental retardation is mostly mild to moderate but may on occasion be severe, especially in males. Attention deficit and hyperactivity disorders are very common; autistic behavior is uncommon. Both males and females may exhibit repetitive movements (e.g., hand flapping).

Delayed speech is common, and the quality and nature of the speech problem becomes highly characteristic. Affected males have difficulties in articulation and leave out words, resulting in grammatically poor sentences that tend to sound like the recitation of a litany. This speech pattern is so characteristic that it has alerted many physicians to consider a fragile X diagnosis. There is also a tendency for

Feature	% Estimated Frequency	Comments
Nervous system and skull		
Mental retardation	80	Mostly moderate but varies from mild to profound (see text for normal transmitting males)
Learning problems	VC2	Mild to severe; IQ tends to decline with age
Speech and language problems	VC	Unclear, repetitive, echoes others; poor language content; usually, delayed speech
Attention deficit disorder and hyperactivity	80	May be the first sign of brain dysfunction in affected boys
Shyness early on	C3	Later often cheerful, friendly, and cooperative
Autistic behavior	20	Repetitive mannerisms including hand biting, hand flapping, rocking, poor interpersonal interaction, and gaze aversion
Seizures	20	Less frequent in adults
Large head	C	With prominent forehead remaining relatively large into adulthood
Tics and twitches	19	
Psychotic behavior	UC4	Occasionally; may resemble paranoid schizophrenia
Facial Appearance		
Prominent ears very common (85 percent)	70	Often large, soft, and protruding; middle ear infection
Long midface	70	
Prominent lower jaw	VC	
Overcrowded teeth with overbite	VC	
High-arched palate	52	Usually a mouth-breather
Crossed eyes	36	
Genitals		
Large testicles	70	After puberty; in most adults; occasionally in children

Muscles, Ligaments, and Bones

Floppy	VC0	As infants
Loose-jointed	67	
Stiff walk	UC	Only in some
Flat feet	71	
Long, thin arms	C	
Curved spine	C	Occasionally
Prominent shoulder blades	C	
Caving-in of breastbone	UC	Occasionally; mild to moderate
Downward slope of the shoulder	C	
Protuberant abdomen	C	
Hernias of groin and diaphragm	15	

Skin

Velvety soft texture	VC	Often
Hand calluses	29	From self-abuse
Abnormal palm creases and fingerprints	25	

Heart

Prolapsed mitral valve	35	
Slightly dilated aorta	UC	
Heart murmur or click	18	Occasionally

1. Only where good data exist
2. VC = Very common
3. C = Common
4. UC = Uncommon

affected males to repeat words, phrases, and sentences continually and to echo someone else's words automatically when spoken in their presence. Frequently, they may simply talk continuously and compulsively. When young, the boys are shy and anxious, though later they may be very friendly, cheerful, and cooperative. However, distressing behavior also may be observed, ranging from hyperactivity and inattention to outright psychotic episodes. Seizures occur but are not common.

The Pattern of Inheritance.   Transmission of the fragile X syndrome was originally thought to conform to patterns seen with other X-linked disorders such as Duchenne muscular dystrophy (see Chapter 7). Unexpectedly, a unique pattern of inheritance was uncovered that does not conform precisely to what is usually seen in X-linked inheritance. While the fragile X mental retardation gene is indeed transmitted on one of the X chromosomes, the nature of the defect within the gene makes for some unusual patterns of transmission and personal risk, which can be summarized as follows:

1. An intellectually normal female who inherits the gene for fragile X syndrome from her carrier mother has a 50 percent risk of having an affected son, whose risk of being retarded is 40 percent. Half of her daughters will carry the gene, but only 16 percent of them will be retarded.

2. If such a daughter is retarded, her risk of having an affected *and* retarded son is 50 percent. If she has a daughter, the risk is 28 percent that the child will also be mentally impaired.

3. Men who are seemingly entirely normal and do not even show the fragile X chromosome when tested may nevertheless transmit the gene to all of their daughters. These females are usually intellectually normal. However, when they reproduce, 50 percent of their sons will be affected, and 40 percent will be retarded. Half of their daughters will be carriers, among whom 16 percent will be retarded.

4. Normal-but-transmitting males may account for 20 percent of all cases of the fragile X syndrome. Unfortunately, such males will remain undiagnosed until new techniques reveal their cryptic endowment or until one of their children or grandchildren is diagnosed with this disorder.

5. Interestingly, half of the women carriers who bear a son who is a normal-but-transmitting male are themselves intellectually impaired.

These carrier females also have a 50 percent risk of having carrier daughters, but these girls have only a 5 percent risk of being intellectually deficient.

Several other observations are important. For reasons that remain speculative, women who carry the fragile X mutation more frequently experience premature ovarian failure (very early menopause). About 3 percent of women who have very early menopause (for example, in their early thirties) are carriers of the fragile X defect. There also seems to be an increased rate of numerical chromosome disorders (such as Turner and Klinefelter syndromes; see Chapter 4) among the offspring of female carriers. An unexplained increased rate of twinning by carrier females also has been noted.

A remarkable example of X chromosome inactivation (see Chapter 6 for an explanation of this process) occurred in identical Dutch twin sisters, both of whom have the fragile X mutation. One sister is mentally retarded, and the other, of normal intellect. DNA studies at Erasmus University in Rotterdam revealed that, by chance, the X chromosome in blood cells containing the fragile X mutation in the normal sister had largely been silenced. The opposite had occurred in the intellectually impaired sister. Random X-silencing had taken place shortly after the separation of the embryos.

The Nature of the Gene Defect.   The stutter-like reiteration of base triplets that is also seen in the gene responsible for the fragile X syndrome is described in Chapter 7. The phenomenon of anticipation is characteristic of disorders produced by this type of dynamic mutation, with the severity of the disorder increasing through the generations. Less than 4 percent of affected individuals have a different type of mutation, which is not routinely tested for.

The concurrence of a male relative with mental retardation on the mother's side of the family together with a sister, aunt, or grandaunt with very early menopause should prompt a DNA analysis for fragile X syndrome in related daughters planning to have children.

Drugs That Cause Defects

The medications that indisputably cause birth defects are shown in Table 9.3. You may have thought that no pregnant woman need be reminded of the dangers inherent in taking drugs. However, it is remarkable that a majority of women take

TABLE 9.3 Medications Conclusively Shown to Cause Common Birth Defects

Medications	Uses	Defects and Consequences
Antibiotics		
Tetracycline	treatment of infections	pigmentation of teeth and underdevelopment of enamel
Streptomycin	treatment of infections (e.g., tuberculosis)	deafness
Anticancer drugs (e.g., busulfan, chlorambucil, cyclophosphamide, 6-mercaptopurine, aminopterin [methotrexate])	treatment of cancer	fetal death, or multiple and variable serious birth defects
Anticoagulants (e.g., dicumarol, warfarin)	prevention of blood clots	underdevelopment of cartilage and bone (especially in nose); blindness; mental retardation; possible fetal or newborn bleeding
Anticonvulsants (e.g., Dilantin [phenytoin], Tegretol, carbamazepine, trimethadione, phenobarbital, valproic acid [Depakene])	Treatment of epilepsy	multiple and variable serious birth defects; mental retardation; for valproic acid, spina bifida
Antithyroid preparations (e.g., propylthiouracil, methimazole, radioactive iodine)	treatment of hyperthyroidism	enlargement or destruction of fetal thyroid gland
Hypoglycemic drugs	treatment of high blood sugar	low blood sugar in newborns
Psychiatric drugs		
Lithium	Treatment for depression	right-heart defects
Psychoactive drugs (barbiturates, opioids; benzodiazepines)	Treatment for anxiety states	right-heart defects

Drug	Use	Effects
Sex hormones (e.g., progesterone, estrogen, Progestoral, diethylstilbestrol [DES], methyltestosterone, Norlutin, Danazol)	oral contraceptives; formerly, prevention of miscarriage	masculinization of a female fetus; for DES, vaginal cancers in adolescence, miscarriage, tubal pregnancy, and structural abnormalities of the genital tract
Miscellaneous		
Antimalarial medicines (e.g., quinine, chloroquine)	treatment of malaria and certain heart rhythm disturbances	miscarriage; marked reduction in the platelet count in the newborn; possible deafness or blindness
Misoprostol	treatment of duodenal ulcers	could cause miscarriage; major risk of causing Möbius syndrome (congenital facial paralysis with/without limb defects)
Accutane (Isotretinoin or other derivatives of vitamin A), Etretinate	treatment of serious acne or psoriasis	hydrocephalus or microcephaly; deformed or absent ears; heart disease; abnormally small eyes; cleft palate; mental retardation; other defects; miscarriages
Vitamin A in excess	harmful diet fads	miscarriage; defects of head, face, brain, spine, and urinary tract
Penicillamine	to rid body of excess copper	hyperelastic skin
Angiotensin-converting enzyme inhibitors	treat hypertension	kidney damage and failure in newborns; skull calcification defects
Thalidomide	formerly used overseas for treatment of nausea and vomiting during pregnancy	absent to profoundly deformed limbs; multiple and variable serious birth defects

NOTE: Check with your physician about any medication with a different name that may be similar in function and consequences to those listed here.

between 4 and 14 different prescription or nonprescription drugs during pregnancy. Every woman planning a pregnancy who must take a specific medication should first consult a physician about the potential risks of birth defects, hazards of breast-feeding, and possible complications in the newborn. Changes in medication or temporary periods with no medication may be viable options, but only in consultation with your physician.

The drug thalidomide, taken for nausea in pregnancy, resulted in thousands of children in West Germany and England being born without arms and legs or with catastrophic deformities of the limbs as well as other serious birth defects. In determining whether a drug has caused a birth defect, a primary consideration is the specific stage of pregnancy at which the drug was administered. Mothers who took thalidomide on the thirtieth day after conception had children with the most deformed arms and legs, whereas those who took the drug on the thirty-fifth day had babies who only had lower limb defects. Many women who took this drug also had perfectly healthy babies. The timing turned out to be more important than the dosage in producing these defects. Thalidomide has since been reintroduced in some countries, for the treatment of leprosy and AIDS. One can only hope that pregnant women are not again subject to the devastating experiences women have had with this drug in the past.

Genetic as well as other factors determine the way a drug is degraded in the body. In some individuals, certain drugs are broken down more quickly than in others. The slower the breakdown, the higher the concentration in the body and the likelier that the drug will become toxic. Drugs are routinely tested in animals before use in humans, but the interactions in the different species may vary. For example, no abnormalities appeared in the offspring of pregnant rats and mice that were given doses of thalidomide. However, the drug did cause serious defects in the offspring of monkeys and rabbits. Drugs also might interact with one another in damaging ways or might compete with the body for essential nutrients such as folic acid, disadvantaging the developing embryo and resulting in defects such as spina bifida.

Vitamin Supplements

A high proportion of women take multivitamin supplements during pregnancy. Remarkably, too many fail to take supplements until two menstrual periods have been missed. Such omissions suggest that many women (as well as the physicians, midwives, and nurses caring for them) are oblivious to the fact that vitamin supplements can aid normal embryo development. By the time most women schedule their first appointment (8 to 10 weeks after conception), the die is cast. For example, the neural tube (which later becomes the vertebral column containing the spinal cord) closes between 26 and 29 days after conception. Failure to close completely results in spina bifida (or the skull defect known as anencephaly). At least 70 percent of such cases could be prevented if all pregnant women took a folic acid (vitamin B) supplement. Yet a 1998 survey conducted in the United States found that only about one-third of women of childbearing age took vitamin supplements containing the needed daily dose (0.4 mg) of folic acid. Most were unaware of its protective effects against spina bifida and anencephaly. A major study published by our research group in 1989 and well reported in the mass media concluded that 70 percent of cases of spina bifida were preventable by taking folic acid supplements during the three months prior to conception and during the first three months of pregnancy. Subsequent clinical trials confirmed these observations; and it is now the standard of expected care among medical practitioners to recommend that women planning pregnancy take folic acid supplements. This is a rare, major step forward in the prevention of a common birth defect. As yet, the causes of about 60 percent of all birth defects remain undetermined.

Unfortunately, the U.S. Food and Drug Administration (FDA) delayed acting on this information for four years, and then opted to institute an inadequate standard of folic acid fortification of cereal grains, which would provide only about 25 percent of the necessary protective dose. The FDA's hesitation was motivated by concern that folic acid supplements might mask or delay the diagnosis of pernicious anemia due to vitamin B_{12} deficiency. But the latter condition is much rarer, and the FDA's decision therefore seemed illogical, ignoring the balance between an enormous benefit and a relatively minimal risk of potential harm. An effective national program of folic acid supplementation would have protected some 20,000 pregnancies in which neural tube defects subsequently developed!

On the opposite end of the spectrum, some pregnant women take too much of a particular vitamin and unintentionally jeopardize the health of the fetus. The idea that if some vitamins are good, more are better, is seriously flawed. Indeed, excess doses of vitamin A taken in the earliest weeks of pregnancy are associated with a significantly increased risk of birth defects, as our research group discovered.

Although little is known about the similar effects of other vitamins and nutritional supplements, clearly folic acid supplements in moderate amounts can be beneficial when planning pregnancy and during the early stages of pregnancy. In addition to preventing neural tube defects, there is evidence that folic acid reduces the frequency of other defects, such as cleft lip and palate. Equally clearly, large doses of vitamin A should be avoided. The need for all other vitamins remains to be proven.

Maternal Illness and Birth Defects

The immediate environment of the fetus is the mother. Consequently, any illnesses, drugs, toxins, dietary aberrations, or other environmental stresses (such as excessive heat) may potentially harm the developing fetus. For example, my research group found that use of the hot tub or sauna or the occurrence of a high fever during pregnancy were all associated with an increased risk of a neural tube defect such as spina bifida, if any of these heat exposures occurred during the first six weeks following conception. Hot tubs (or hot baths in some countries) are to be avoided, since excessive heat increases the risk of conditions like spina bifida occurring by two- to threefold. The problem is that in the first 2 to 6 weeks, women may not realize they are pregnant. Hence, if pregnancy is possible, avoid hot tub use.

Women who allow their insulin-dependent diabetes to get seriously out of control and who become pregnant during this time have 2 to 3 times the usual risk of having a child with a major birth defect or mental retardation (increasing that risk to about 7 to 10 percent). The tighter the control, the fewer the birth defects, and the less severe they will be. Typical defects are spina bifida, lower spine abnormalities, heart malformation, and kidney and intestinal defects. Well-controlled diabetes is not associated with an increased risk of birth defects.

Epileptic women, primarily (but not only) because of the anticonvulsant drugs they take, also face a 7 to 10 percent risk of having a child with major birth defects or mental retardation. Even without anticonvulsant medications, women with epilepsy face a slightly increased risk. Special medical attention is necessary to select a drug that presents the least risk of harm to the fetus at the same time as it prevents endangerment to the mother's life from uncontrolled seizures. Any changes in medication should be made only under the care of a physician and at least three months prior to a planned conception. Some of the more potentially harmful (teratogenic) drugs can be restarted 3 to 4 months after conception, with the precise fetal age having been established by ultrasound study. These mothers also need careful prenatal surveillance, including high-resolution ultrasound studies.

Women with lupus erythematosus (a chronic disorder of the body's immune system, which affects virtually all tissues and organs) have an increased risk of having a child with congenital heart block. This condition, which is detectable on the electronic fetal heart monitor strip in late pregnancy, results from the blocked transmission of nerve impulses to the heart muscle. Affected children are otherwise normal but may suffer fainting spells or dizziness and may be subject to sudden death. Treatment after birth may involve implantation of a permanent pacemaker.

Many genetic disorders may be concurrent with pregnancy, and all require specialist care and attention. Discussion of these many disorders is beyond the scope of this book. Mothers with a known genetic condition are exhorted to see a clinical geneticist prior to pregnancy to secure their own health and that of a future child.

Birth Defects and Multifactorial Influences

If you grew up in a desert climate and later moved to a locale with much grass and foliage, you might be affected by hay fever, sinus problems, or asthma. You inherited a set of genes that predisposed you to interact with various pollens, thereby precipitating allergic reactions. Even though you inherited the predisposition, it was only after exposure to a different environment that your inherited liability became apparent.

There are in fact a huge number of medical disorders and certain birth defects that are due to this type of interaction. Environmental factors may be dietary, infectious, occupational, toxic, drug-related, seasonal (possibly viral), and so on. One remarkable example is the interaction between dietary folic acid (part of the vitamin B group) and in many cases (but not all) an inherited predisposition to have a child with spina bifida (see discussion above).

Many other common birth defects, such as cleft lip and palate, heart defects in which there is a hole between the lower two chambers (called ventricular septal defect), and congenital dislocation of the hips, fall into this category. Thus far, most of the relevant environmental factors have yet to be identified.

One important, treatable cause of mental retardation is malfunction of the thyroid gland. About 1 child in 4,000 are born with hypothyroidism, which, if untreated, results in irreversible mental retardation. Most infants in first-world countries are screened soon after birth and treated (for life) with thyroid hormone if found deficient. Treatment prevents mental retardation, but it must be started before the third week of life.

A relatively uncommon though potentially serious cause of irreversible mental retardation (discussed further in Chapter 27) is the biochemical genetic disorder phenylketonuria. An untreated pregnant mother with this disorder, unless she follows a special diet, has a 90 to 100 percent risk of damaging the brain of the developing fetus.

Defects of the developing brain, which are occasionally genetic in origin (e.g., hydrocephalus) but are more often of uncertain origin (e.g., microcephaly and cerebral dysgenesis), account for up to 15 percent of cases of severe mental retardation. The risks of recurrence of particular brain defects can be as high as 25 percent (occasionally, 50 percent) but most often are between 2 and 5 percent. Genetic evaluation is mandatory for any child with mental retardation, since at least half of such cases are of genetic origin.

Infectious diseases and other environmental causal factors are not discussed here. Despite the advances in medicine, the cause of mental retardation remains undetermined in 40 to 60 percent of cases.

Risks of Having a Second Child with a Birth Defect or Mental Retardation

A firm diagnosis must first be established and the mode of inheritance clarified before any estimate of subsequent risk can be provided. It is easier to state this principle than it is to put it into practice. Diagnoses of many birth defects may be straightforward but are often confounded by wide variation, association with other defects, and mimicry of other defects (called *phenocopies*). Determination of the mode of inheritance can also be confounded by the fact that similar and even identical disorders can arise via different genetic routes. Earlier, single gene inheritance through dominant, recessive, X-linked, and mitochondrial genes was discussed (Chapter 7). The most common form of inheritance, however, is the form in which multiple genes interact with one or more environmental factors, resulting in a defect or disorder.

In general, for these gene-environment (called multifactorial or polygenic) disorders, the risk of a second child being affected ranges between 2 and 5 percent. The risk is slightly higher for a third affected child. Mothers who have had a child with unexplained mental retardation (even after the latest subtelomeric deletion analysis, described in Chapter 5) must still rely on previously published data. Some of these risk figures are given here for guidance and shown in Tables 9.4 and 9.5. They are based on two studies: a Minnesota study of 80,000 individuals with IQs of less than 70, and an Australian study of 2,000 consecutive families* who

---

Consecutive families in this case are families in which mental retardation appears in each successive generation without skipping.

TABLE 9.4   Minnesota Study: Risks of Mental Retardation

Parent Union	Risk That Next Child Will Be Retarded
For Persons Who Have Not Had a Mentally Retarded Child	
Normal person with one retarded sibling × normal person with all normal siblings	1.8%
Normal person with two or more retarded siblings × normal person with all normal siblings	3.6
Normal person with one retarded sibling × retarded person	23.8
Normal person with all normal siblings × normal person with all normal siblings	.5
For Persons Who Have Had One or More Mentally Retarded Children	
Normal person, one of whose parents had one or more retarded siblings × normal person with all normal aunts and uncles	12.9
Normal person with all normal aunts and uncles × normal person with all normal aunts and uncles	5.7
Normal (or unknown) person × retarded person	42.1

attended a clinic for the mentally retarded (individuals with Down syndrome, and families in which information was incomplete, were excluded).

These older studies lacked the benefit of the advanced genetic tests now available. Even very recent data from multiple reports are far from accurate, but remain good estimates for recurrence rates. After one son with severe mental retardation, the risk of having a second affected child is between 3.5 to 14 percent. Following the birth of a daughter with severe mental retardation, the risk of recurrence (for either sex) is between 2 to 14 percent.

Common Superstitions

Despite the dawning of the twenty-first century, many superstitions still surround birth defects. Throughout the ages, men and women have reacted to the birth of malformed offspring with awe, fear, or admiration—depending on whether the defect was considered an omen of evil and imminent disaster or a sign of good fortune. Artifacts depicting "monsters" or other species of conjoined twins occasionally served as prototypes of gods or demigods with magic powers.

TABLE 9.5   Australian Study: Risks of Mental Retardation

Severity of Retardation in One Child	Future Risk of Having a Retarded Son	Future Risk of Having a Retarded Daughter
Profoundly retarded (IQ below 20)	5.6%	4.4%
Severely retarded (IQ 20–35)	2.1	3.6
Moderately retarded (IQ 36–51)	9.1	4.2
Mildly retarded (IQ 52–67)	7.7	3.9
Borderline retarded (IQ 68–85)	4.4	1.5

NOTE: All risks listed are for couples who have had one child with mental retardation of unknown cause.

It seems that in very ancient times, grossly deformed infants were regarded as divine and probably worshiped. Our fascination with major birth defects has been expressed in art for thousands of years. A sculpture of a double-headed twin goddess was discovered in southern Turkey that dates back to the Stone Age.

From time immemorial, some have believed that a major birth defect was God's punishment for sins committed or a sign of punishment to come. It is not unusual, even today, to encounter parents of a child with a birth defect who confide their innermost belief that God is punishing them for their sins or misdeeds.

One common belief is that a shock or stress or excessive worry has produced the birth defects in a child. During the Second World War, the bombing of London and other major English cities did not lead to an increase in the frequency of birth defects, despite overwhelming stress. However, a Danish study published in 2000 focused on 3,560 pregnancies in which mothers experienced severe emotional stress (defined as death or first hospital admission for cancer or heart attack in partners or in children). These researchers compared this group of pregnant mothers with 20,299 pregnancies without such stress. They observed a slightly higher risk of spina bifida and associated defects in the offspring of stressed mothers, especially following the death of an older child in the first three months of pregnancy. One wonders how much poor appetite (and therefore reduced folic acid intake) had to do with these results.

Another age-old idea, that the thoughts of the pregnant mother may affect the development of the fetus, also has not been substantiated. The power of "mental impressions," as they were called, was believed to act from the moment of con-

TABLE 9.6   How to Avoid Birth Defects, Mental Retardation, and Genetic Disorders

1. **Know** your family history.

2. **Determine** your ethnic or racial origin and ask your doctor which carrier test(s) you and your partner should have.

3. **Consult** a clinical geneticist for genetic counseling before pregnancy.

4. **Seek** obstetric consultation before conceiving.

5. **Plan** your pregnancy.

6. **Change** anticonvulsant drugs for epilepsy to the least harmful to the fetus in consultation with your doctor.

7. **Control** diabetes fully before becoming pregnant.

8. **Check** with your doctor about the safety of your medications before pregnancy.

9. **Take** folic acid (0.4 mg daily) for 3 months prior to and after conception.

10. **Avoid** excess doses of vitamins (especially vitamin A).

11. **Avoid** working with toxic chemicals such as organic solvents when planning and during pregnancy.

12. **Limit** caffeine intake.

13. **Reduce** alcohol use.

14. **Stop** smoking.

15. **Do not** use illicit drugs.

16. **Avoid** use of the hot tub and sauna and reduce fever promptly when planning pregnancy and in the first 3 months.

17. **Schedule** a maternal blood screening test at 16 weeks for defects such as spina bifida and Down syndrome.

18. **Consider** prenatal genetic studies for all of the indications described in Chapters 3–5, 7–9, and 22.

19. **Check** the results of the newborn screening test.

ception. Children conceived out of wedlock, allegedly in great passion, were thought to be artistically gifted. Many centuries ago, in Greece, expectant mothers were encouraged to look at beautiful statues and pictures in order to make their children strong and beautiful; indeed, the Spartans made laws to that effect. Many mothers have blamed their negative thoughts, moods, and emotions during pregnancy for children born with cleft lip and/or palate.

Avoidance and Prevention

The steps all prospective parents need to take in order to avoid having a child with a birth defect, mental retardation, or a genetic disorder are summarized in Table 9.6. Most of these already have been discussed. The exhortation to plan pregnancy is commonly missed, given that some 50 percent of pregnancies are unplanned. Smoking during pregnancy is linked to many adverse outcomes, including poor fetal growth, premature delivery, and an increased likelihood of death soon after birth. Smoking also decreases fertility and increases the risks of sudden infant death syndrome as well as child illnesses. Perhaps even more ominous is my research group's discovery in 1999 of a tobacco-specific carcinogen (that is, a cancer-producing agent) in the amniotic fluid surrounding the fetus of mothers smoking in early pregnancy. The future implications of this early exposure to a cancer-causing substance are the subject of our ongoing study.

The link between individual men's exposure to hazardous chemicals and subsequent birth defects in their offspring has remained a concern, but the evidence is still inconclusive. Men who are scheduled to undergo treatment for cancer are advised to store sperm samples in a sperm bank for later use. Those who decide to father children after chemotherapy are usually advised to wait at least six months. However, their risks of fathering a child with a birth defect appear to be no higher than random risks.

Excess caffeine use by pregnant women (some of whom have reported drinking eleven cups of coffee per day) is linked to an increased likelihood of miscarriage. Fewer than four cups per day might seem reasonable; but remember that caffeine is also present in tea, soft drinks, and chocolate. Occasional use of alcoholic beverages such as wine does not seem to be problematic. However, although no absolute limit for alcohol use during pregnancy has been established, this is not the time for binge drinking, becoming drunk, or even having alcohol daily. Some studies have reported birth defects, learning problems, and/or mental retardation occurring in the children of 40 to 50 percent of women imbibing serious amounts of alcohol daily in pregnancy.

There are few opportunities in life to control what might otherwise have been called your destiny. Clear opportunities do exist to prevent tragedies in childbearing, and the steps outlined in Table 9.6 will help you recognize and take advantage of these opportunities. Remember, not knowing does limit your choices but it doesn't alter your chances!

chapter 10

Genes That Make You Susceptible to Various Disorders or Complications

Imagine growing up in the desert. One fine day you take a trip to visit a relative in an area with a high pollen count. Within days after your arrival, you begin wheezing, and the diagnosis made at the local hospital is asthma. Despite your perfect health before your trip, you had in fact inherited a set of genes whose presence was signaled only after environmental exposure (to pollen). The same thing might occur with first exposures to cats, dogs, medications, latex gloves, and an endless list of other items. On occasion, reaction to an outside stimulus (penicillin or a bee sting) could result in immediate collapse, shock, and death (anaphylaxis). Sometimes, even after long exposure—for example to a cosmetic—an allergy unexpectedly makes its first appearance.

Even more worrisome than environmentally induced reactions are the large number of common illnesses that arise as a consequence of the interaction between a set of genes we have inherited and some environmental factor(s). The extensive list of disorders in this group is collectively termed *multifactorial* or *polygenic*. In contrast, defects within single genes and easily traced through their pattern of transmission through a family constitute the group of *monogenic* disorders discussed in Chapter 7. Although defects in single genes do directly cause a genetic disorder, in some cases environmental influences may also play an important or even critical role (e.g., the biochemical genetic disorder of phenylketonuria, in which ingestion of protein by the affected infant can cause irrevocable brain damage).

Discovery of multiple genes and interacting environmental factors has been slow, given the enormous complexity of the task, as well as the fact that such disorders may also be caused by defects in single genes. Common diseases like coronary heart disease, hypertension, diabetes, mental illness, senile dementia, and certain cancers are important examples. These and many other multifactorial disorders aggregate within families. The risk for a first-degree relative such as a child or sibling generally ranges between 3 and 15 percent for different disorders. Risks vary not only from one disease to another but even for the same disease, from one family to another. Within a family, the specific risk is influenced by the severity of the disorder, the number of affected family members, and the nature and contribution of the environmental factor(s). The risks to relatives of an affected person increase as the frequency of the disorder or defect decreases in the general population. For disorders in which there is a significantly higher frequency in one sex compared to the other, offspring of affected individuals of the less frequently affected sex are at higher risk.

The complex interaction of genes with environmental factors in a disorder such as coronary heart disease is more easily understood when one recognizes that genes control blood fats, blood clotting, cells in the bloodstream, structure of arteries, and so on. A staggering total of more than 200 genes are involved in controlling cholesterol's absorption in the gut, transfer into and through the bloodstream to the liver, and processing in the liver and other cells of the body, as well as its excretion from the body.

An extremely important realization is that in multifactorial disorders, despite the inherited predisposition, health can be secured and lives saved by altering environmental factors—for example, eating a healthy diet, exercising, and not smoking. Even in some diseases caused by a single gene—such as alpha-1-antitrypsin deficiency, which predisposes to pulmonary emphysema—serious lung problems may be averted by not smoking and by avoiding certain occupational inhalants.

Among the estimated 30,000 to 40,000 genes we possess in every cell of our bodies are an infinite number of combinations of varying groups of genes that endow susceptibility to a large number of diseases, in which many environmental agents may interact. These conditions and the many environmental factors that together cause disorders would fill an encyclopedia. I have chosen only the more common disorders and exposures for discussion here, focusing especially on conditions in which knowledge could help secure your health and even save your life.

Blood Groups and Disease

Through blood-group studies it has been possible to come to some conclusions about the movement of the Celts, a group that includes the Cornish, Welsh, Irish, and Scottish, as well as the Norwegian Vikings (Norsemen). The Vikings invaded Ireland and Britain and temporarily settled there until their final eviction in the late tenth and early eleventh centuries C.E. Some Norsemen, after leaving Ireland

and Britain, settled in Iceland, taking with them Celtic wives and slaves. Genetic evidence shows that most of the early settlers in Iceland were Celts rather than Norsemen. The genetic evidence is based on the frequency of particular blood group systems in peoples of those areas today. From similar studies it has been deduced that the inhabitants of the west coasts of Norway, Scotland, and Ireland share a common ancestry. The most likely explanation appears to be the importation of Celtic women to Norway as wives and slaves of Norsemen in the Viking period.

Our uniqueness is reflected by the remarkable diversity of our genes and their products. One person can be distinguished from another by analyzing the genes or their protein products. A gene on chromosome 9 is responsible for our ABO blood group system. The specific protein (antigen) that adheres to the surface of our red blood cells allows us to determine whether we belong to the A, B, AB, or O blood group. This blood group system was discovered by Dr. Karl Landsteiner in 1900, since which time more than 250 different proteins on and in red blood cells have been recognized.

You have inherited a pair of number 9 chromosomes with the ABO genes, one from each parent. At the exact spot (locus) on chromosome 9 where this gene is located, alternate forms of the gene (called alleles) may occur. These alleles may encode different antigens. For example, an individual with blood group B could have two B genes or one B and one O gene. Each of these blood groups can be recognized through opposing proteins (antibodies) in laboratory tests. That is why a person with blood group B who marries someone with the same blood group may have a child with blood group O. (See Table 10.1 for a listing of the various combinations that are possible.)

The A and B antigens are inherited as single genes. Hence, if you have a B gene and an O gene on each of your number 9 chromosomes, you will transmit only one to a child. In North American whites the ABO blood group frequencies are O (45 percent), A (40 percent), B (11 percent), and AB (4 percent).

A and B antigens are located on the outer surface of the red blood cells. To give you some idea of the system's complexity: Group A cells have about a million antigen sites. Moreover, these cells have yet another protein deriving from a completely different gene on chromosome 19 that makes a substance that is the immediate precursor of the A and B antigens. Group O red cells have about 1.7 million sites for this substance on each cell.

The ABO blood group system affords great accuracy for correctly matching blood transfusions. In addition to the ABO system, there are some 23 different blood group systems with more than 257 antigens identified. Transfusions of

TABLE 10.1   Possible ABO Blood Groups in Parents and Their Children

Parents' Blood Group			Children's Blood Group	
Partner 1	×	Partner 2	Possible	Impossible
O	×	O	O	A, B, AB
O	×	A	O, A	B, AB
O	×	B	O, B	A, AB
O	×	AB	A, B	O, AB
A	×	A	O, A	B, AB
A	×	B	O, A, B, AB	None
A	×	AB	A, B, AB	O
B	×	B	O, B	A, AB
B	×	AB	A, B, AB	O
AB	×	AB	A, B, AB	O

incompatible blood (including maternal-fetal transfers) due to mismatching in the ABO system can result in serious clinical problems including severe anemia of the newborn (due to red blood cells being destroyed); autoimmune hemolytic anemias in which red blood cells are destroyed by the body's own protein interactions; and serious, antigen-caused reactions that may result in death.

Not unexpectedly, ethnic differences occur in the frequency of blood groups. The B gene, for example, is three times more frequent in Orientals than in whites, whereas it is virtually absent in Native Americans.

ABO Blood Groups

Scholars have long pondered the origin of the European gypsies. The majority of these people once lived in Hungary, in isolated groups within which there was much inbreeding. A remarkable difference has been noted between the frequencies of the ABO blood group system in gypsies as compared with Hungarian non-gypsies. The blood group systems in gypsies, however, are remarkably similar to those in certain Asiatic Indians. It is therefore likely that gypsies have Indian origins.

The search for associations between ABO blood groups and disease has been relatively unrewarding. Perhaps the most firmly established association is between the blood group O and duodenal ulcer. Even then, Type O occurs only about 1.4 times more often in people with duodenal ulcer. Pernicious anemia and stomach cancer are slightly more common in those with Type A blood, whereas stomach (as opposed to duodenal) ulcers are a little more common in Type O individuals. Type A is associated with a rare birth defect called the nail-patella syndrome, which is characterized by totally absent or abnormal nails, kidney problems, and absent or defective kneecaps. Young women who take oral contraceptives have a higher incidence of blood-clotting complications if they have Type A, B, or AB blood rather than Type O. A dominantly inherited muscle disease known as *myotonic muscular dystrophy* is associated with the Secretor blood group, in a different classification system.

Rh Blood Groups

The Rh blood groups were discovered in 1939, and are thus named because the blood of rhesus monkeys was used in the experiments. There are two Rh genes, both located on chromosome 1. Rh-positive individuals have both genes, whereas Rh-negative individuals have only one. Persons who have the specific Rh protein, or antigen, on their red blood cells are designated "Rh positive"; they constitute about 85 percent of all whites. The other 15 percent do not have this protein and are called "Rh negative." Worldwide, Orientals are rarely Rh negative, and only about 1 percent of blacks are Rh negative.

If an Rh-negative woman becomes pregnant by an Rh-positive man, the main hazard is that the fetus may be Rh positive. In such a case, fetal red blood cells with the Rh antigen cross over into the mother and are recognized by the mother's immune system as "foreign invaders." The Rh-negative body mounts an attack by producing a protein *antibody* to counter the entering foreign protein. Forever after, the mother's body makes the anti-Rh antibody, which will cross over into the fetus in subsequent pregnancies. Every time a fetus is Rh positive (after the first pregnancy, which in most cases will be fine), the Rh-negative mother's antibody will attack and destroy the fetal red blood cells. Fetal death, stillbirth, or a severely ill newborn with hemolytic anemia and jaundice may be the result. Blood transfusion of the fetus in the womb or exchange transfusion after birth is usually lifesaving, and will also prevent later mental retardation, deafness, or cerebral palsy.

An Rh-negative mother also might be sensitized during a transfusion of incompatible blood, which would affect the very first and subsequent pregnancy. Similarly, sensitization of the mother may occur from fetal red blood cells crossing over during an unrecognized miscarriage very early in a first pregnancy; in the second pregnancy, which the mother thinks is her first, the fetus may be affected. Sensitization may also occur immediately after an amniocentesis, if not prevented.

If you plan to have a child, it is not only enormously important that you know your genes but also that you find out your gene-encoded Rh group. The reason is that it is possible to prevent an Rh-negative woman from becoming sensitized simply by giving her an injection of a specific protein, immunoglobulin, upon the birth of her first baby. Because of this immunoglobulin treatment, hemolytic jaundice in the newborn is steadily being wiped out. Remember, however, to find out your Rh group (and that of your mate) before pregnancy; and ensure that the first visit to the doctor occurs early in pregnancy, rather than two to four months after conception. Rh immunoglobulin should be given immediately (within seventy-two hours) in situations in which fetal red blood cells may enter the Rh-negative mother's circulation—for example, following delivery, miscarriage, and amniocentesis.

Incompatibility between mother and baby in the ABO blood groups only occasionally causes problems (anemia, jaundice, or both) after birth. Curiously, ABO incompatibility between mother and child is strongly protective against Rh sensitization: For example, if the fetal cells are A or B and the mother's cells are not the same, the mother's antibodies destroy them so rapidly that there is not sufficient time for the slower process of Rh sensitization to occur.

Among these 23 blood group systems, most commonly recognized are the ABO and Rh systems. Well known among the rest are the MNSs, Lutheran, Kell, Lewis, and Duffy systems.

The HLA System

Like red blood cells, white blood cells are coated with proteins, which are called *human leukocyte antigens* (HLA). These antigens are encoded by a cluster of genes on the short arm of chromosome 6 at a location called the major histocompatibility complex (MHC). The leukocyte antigens are not confined to white blood cells but are also distributed in many tissues. This turns out to be critically important, since these antibodies fulfill important functions in regulating the body's immune response to foreign invaders (infection) or foreign tissue, as occurs in transplantation and in mismatched transfusion.

The MHC has three classes of closely linked genes whose protein products coat the surface of white blood cells and other tissues. There is a complicated nomenclature for the three classes of genes that code for different antigens. For example, class I genes code for HLA-A, -B, -C, -E, -F, and -G. Class II genes code for HLA-DP, -DQ, -DR, -DN, and -DO. Certain of these HLA types are associated with particular diseases (see Table 10.2).

TABLE 10.2   HLA-Associated Diseases and Their Relative Risks

Disorders with at Least 10-fold Increased Risks		
	HLA Type	Relative Risk
Ankylosing spondylitis	B27	96
Uveitis and arthritis (Reiter disease)	B27	40–69
Narcolepsy	DRB1*1501	29
Pemphigus	DR4	25
Rheumatoid arthritis (Caucasian)	DR4	6–14
Behçet's disease	B51	16
Diabetes (insulin dependent)	DR3/4	15
Psoriasis	Cw6	13
Goodpasture syndrome	DR2	13
Sjögren disease	DQB1*0201	12
Primary sclerosing cholangitis	B8	4–12
Celiac disease	DRB1*0301	11
Disorders with Less Than 10-fold Increased Risks		
Hemochromatosis	A3	8
Cancer of cervix	DQw3	7
Multiple sclerosis	DRB1*1501	6
Myasthenia gravis	DR3	7
Alopecia areata	DQw7	6
Hashimoto disease (thyroid)	DQw7	2–5
Primary biliary cirrhosis	DRB1*08	5
Graves' disease (thyroid)	DR3	3–4
Optic neuritis	DRB1*1501	4
Crohn disease	DRB3*0301	6
Ulcerative colitis	DRB1*0103	5

The cluster of genes in the MHC (containing over 220 genes) are inherited and transmitted as a unit (called a *haplotype*) on one chromosome. Each parent has two haplotypes or clusters of HLA genes in the MHC site. The children inherit one haplotype from each parent, just as in the autosomal recessive disorders described in Chapter 7. There is therefore a 1 in 4 chance that any two siblings are HLA identical and a 1 in 2 chance that they at least have the same haplotype. There is also a 1 in 4 chance that they do not share an HLA haplotype. The high probability of two siblings sharing an HLA cluster of genes is the reason why brothers or sisters are looked at first when one sibling needs a transplant. Many readers will remember the recent, much-publicized case of the couple who decided to have another baby when their daughter (then in her late teens) developed leukemia and needed a matching bone marrow transplant. They did have a successful pregnancy (albeit heavily criticized), and their child did indeed have a matching HLA cluster of genes. A bone marrow transplant was later performed and has apparently been successful. There was no prior guarantee of a match, however, since genes have been known to "jump" from one location to another: Genes might have crossed over from one MHC cluster to the other number 6 chromosome (a rate estimated at 0.8 percent for two of the clusters), possibly generating a new haplotype—and thus confounding the hopes of a match.

The distribution of HLA genes among different races varies widely. For example, HLA antigens that are common in the white population (HLA-A1, B8) are rarely seen in Orientals. Other antigens (e.g., HLA-A34) occur virtually only in blacks, and HLA-B46 is found almost exclusively in Orientals. Disease associations aside, you can easily recognize the use of such information in homicide investigations. Disputed paternity used to be solved by HLA-typing as well, until this method was superseded by the much more informative analysis of DNA. (See Chapter 7 for a discussion of DNA testing in cases of paternity, homicide, and incest.)

Associations between HLA genes and disease have been known since 1967. The list of disorders associated with HLA antigens is believed to exceed more than 500, but many linkages are unproven or weak. Examples of the most common and important of these associations are shown in Table 10.2. Note that the mere presence of a particular HLA antigen in the blood does not necessarily translate into having the associated disease. It does, however, indicate a greater likelihood (or *relative risk*) of the disorder's eventually developing. If you have the HLA antigen B27, it is 96 times more likely that you will develop ankylosing spondylitis (arthritis of the spine).

Complex hypotheses have been offered to explain the association of HLA with disease. Some point to the production of disease-causing proteins by HLA genes or to interactions of these genes with other proteins or cells, leading to abnormal tissue reactions. For some disorders (hemochromatosis, congenital adrenal hyperplasia, narcolepsy, and spinocerebellar ataxia), the actual gene causing each disorder is linked directly to HLA genes on the same chromosome. The fact that someone with a particular HLA type does not develop the disorder in question points to the role that other genes probably play in the face of obvious genetic susceptibility.

HLA testing is frequently used to screen organ donors, particularly in the case of bone marrow, kidney, pancreas, heart, lung, and liver transplants. In general, a close if not identical HLA match or ABO blood group compatibility assures the best prognosis for most transplants.

Allergy, Asthma, and Emphysema

The body's immune system protects it against invasion by foreign proteins. Upon entry into the body (via breathing, swallowing, touching, or intravenous injection), foreign proteins are met by the body's own protective proteins (called antibodies), and a reaction ensues. The reaction may become manifest as allergies (e.g., skin eczema or skin eruptions), respiratory problems (e.g., asthma), or occasionally a state of shock (anaphylaxis), the last of which could be fatal. A large number of genes are involved in the recognition of foreign proteins and in signaling factory cells to manufacture antibodies as well as other cells to transport, capture, and destroy the foreign proteins. The allergic reaction or the disease results from the immune reaction between the outside protein invaders and the body's protective mechanism.

Typical allergens include the house dust mite, grass pollens, and animal dander (particles of shed skin and fur). Allergens themselves are harmless, causing symptoms of disease only because of the intensity of the immune reaction that they provoke.

Asthma and eczema are the most serious of the allergic diseases. Asthma has become a worldwide epidemic affecting an estimated 155 million people. Eczema is also common and affects between 10 and 20 percent of children in the Western world. Most childhood asthmatics also have eczema or allergic skin reactions of varying degrees of severity. The house dust mite is regarded as a primary allergen for asthmatics.

There has been a dramatic increase in the frequency of asthma and other allergic disorders in the past century. Factors in the environment are thought to be important. Curious but significant observations include reports that allergic disease is more common in cleaner, Westernized environments and that asthma and allergic skin disorders are less common in the children of animal farmers, less common in younger siblings, and less common in households with dogs as pets! Multiple studies have concluded that the increase in the frequency of allergic disease may be due to factors associated with a Western lifestyle.

Asthma and allergy are strongly familial and clearly are the consequences of a genetic predisposition. Identical twins have a much higher frequency of both having asthma than do nonidentical twins. The same can be said for eczema in twins. Eczema in one parent (compared to asthma) confers a higher risk of eczema in a child. Multiple interacting genes are regarded as responsible for the interactions that ultimately result in asthma or allergy. Some asthma syndromes, including aspirin-induced asthma and industrial asthma, may be related to other genes.

Asthma may also develop from materials inhaled at work. Indeed, asthma has emerged as the principal occupational lung disease in the United States. More than 250 substances used in the workplace may cause occupational asthma. Among these many substances are the isocyanates used in polyurethane foams and paints, which are a particularly frequent cause of asthma. Once again, predisposing genes have been noted by studies in France, in which a particular HLA group made asthmatic illness more likely and a different HLA group appeared to some extent protective. Some fortunate souls who smoke throughout their lifetime never develop lung cancer (they do, however, develop other diseases of the heart and lungs); the protective genes involved here have yet to be identified. In contrast, people exposed to asbestos are known to have even greater risks of lung cancer if they have inherited the same defective gene (GSTM1) from both parents or have another set of genes (NAT2) that delay the processing of environmental chemicals or drugs in the body.

People who have a history of allergy or who smoke are both at greater risk than others for asthma developing in response to occupational exposures. Certainly a number of genes appear to be associated with susceptibility to emphysema (loss of elasticity of the lung, overexpansion, and difficulty in breathing out, with resultant shortness of breath). Given the inevitable development of emphysema after long years of smoking, recognition of susceptibility genes for smoking is important. Recent studies show that people who carry a particular variation in one specific gene (the dopamine transporter SLC6A3) are less likely to be smokers. This gene variation was found among those who were more likely to quit smoking as well as among people who were not novelty seekers, a personality trait characterized by the need for instant gratification. A different gene, which codes for a liver enzyme (CYP2A6), is partly responsible for breaking down nicotine. A defect in this gene results in nicotine remaining undegraded longer. Smokers with this defective enzyme were found to smoke 20 percent fewer cigarettes, probably due to their elevated level of residual nicotine, which diminished their need to "refuel."

Emphysema can be directly inherited via a single gene defect. The genetic disorder, known as alpha-1-antitrypsin deficiency, results from a defective gene transmitted by each parent equally to the affected offspring. This gene codes for the enzyme antitrypsin, which, when deficient, results in the loss of normal lung elasticity and in progressive overinflation and destruction of lung tissue. The common gene defect occurs in as many as 1 in 20 Caucasians, being decidedly rare among Orientals, Asians, American Indians, and blacks. Emphysema develops insidiously, mostly between the ages of 30 and 40 years, and progresses to severe physical limitations and incapacitation within five years of onset. About 10 to 15 percent of those who have progressive lung disease also have skin allergy and asthma. Those bearing the defective gene linked to this disease should strictly avoid smoking and passive smoke inhalation as well as occupations that require work in environments with dusty or polluted air. Lung transplantation, which is the only method of significantly prolonging life in the presence of the disease, is not often achievable because of the shortage of suitable donor organs.

Antitrypsin deficiency is also the most common genetic cause of childhood liver disease (cirrhosis) and the most common reason for liver transplantation in children. A family history of early onset emphysema or childhood liver disease points toward this diagnosis, which can be confirmed by DNA analysis. DNA testing can be used to detect carriers of alpha-1-antitrypsin deficiency as well as to facilitate prenatal diagnosis for a couple found to be carriers, who face a 25 percent risk of having an affected child.

A Predisposition to Clot

By now it will be self-evident that any condition affecting your health necessarily involves your genes. But even in cases where a single gene defect predisposes you to a particular disorder, an environmental factor may tip the scale. Defects in two common, important genes are illustrative. A mutation in the Factor V Leiden gene predisposes a person to clots forming in veins, especially in the legs (a condition known as *deep vein thrombosis*). The danger is that a clot may break away from the vessel wall and be carried through the bloodstream into the lung, where the resultant blockage to circulation (called *pulmonary embolism*) can cause sudden death or eventual right heart failure. Thromboembolism ranks as the most important cause of maternal death in the United States. Stroke at a young age (younger than 30 years) is another hazard. Between 1 in 20 and 1 in 25 whites have a mutation in this gene!

This defect is likely to have been inherited from one parent as a dominant disorder, but most who carry the defective gene are unaware of its presence. However, certain "external" events may precipitate thrombosis. These include the use of oral contraceptives; any surgical operation, including cesarean section; or a long period of immobilization (such as a long trip by air or by car; an unrelated, debilitating illness; or a normal pregnancy during which one is relatively inactive). A high level of the amino acid homocysteine in the blood (a risk factor for heart disease; see Chapter 16) is a further precipitant. A single, common mutation in the Factor V gene accounts for more than 90 percent of cases.

A similar story exists for the prothrombin gene, a defect in which also predisposes to thrombosis. Although this defect is somewhat less common (between 1 in 50 and 1 in 100), deep vein thrombosis and pulmonary embolism are potential serious consequences. One particular mutation in this gene accounts for about 90 percent of all such defects. In about 40 percent of cases of thrombosis, defects in both Factor V and prothrombin genes occur together, albeit originating from different chromosomes. Where thrombosis is suspected, tests should always be performed for mutations in both genes. The frequency of defects in both the Factor V and prothrombin genes is considerably lower among blacks and even more so among Orientals.

To secure your health against these disorders, take the following steps:

1. Know your family history. If any family member has had deep vein thrombosis or pulmonary embolism, or an episode of thrombosis or stroke at an early age (younger than 50), request a DNA test for both mutations.

2. If you have had a deep vein thrombosis or pulmonary embolus or early stroke, request DNA testing for both the Factor V and the prothrombin genes.

3. If you take oral contraceptives and have a family history of thrombosis, be tested immediately for both genes. A recent study conducted in Spain found that several dozen women taking oral contraceptives who had strokes before the age of 30 all had a defect in one of these genes. Women who have inherited one of these gene defects have a risk of thrombosis that is 30 times that of other women. Those who inherited a gene defect (in either gene) from *both* parents have a risk that is more than 100 times greater. *Do not take oral contraceptives if you have a family history of early thrombosis until you have been tested for these two genes.* A number of other genes also predispose their carriers to thrombosis, but at present they are not as easily identified in tests.

4. If you are to have any surgery and have a family history of thrombosis, see to it that your doctor first tests for these two gene defects.

5. Prophylactic anticoagulant medication is indicated when there is deep vein thrombosis or pulmonary embolism. These medications

may also be indicated for long-term or even lifetime use in situations where a gene defect has been found and thrombosis has already occurred.

6. If you have suffered unexplained repeated miscarriages, DNA analysis of these two genes should be considered, given the association of these mutations with pregnancy loss.

7. If you have one of these gene mutations, consult your physician about the need for anticoagulants, particularly if you are taking long trips or are confined to bed for weeks to months.

Both the Factor V and the prothrombin genes are dominantly inherited. Hence, an affected person has a 50 percent risk of transmitting this gene to each of his or her offspring.

Genes and Drugs

The body systematically breaks down every medication taken: Various genes produce enzymes that degrade drugs, eventually rendering them inactive. Some people's genes metabolize drugs quickly, whereas others' work more slowly. In some cases a much larger dose of an anticoagulant (blood thinner) is needed in order to achieve the same effect; in others, even a small dose of the same drug causes bleeding due to the body's extreme sensitivity to that substance. Certain gene patterns also have been recognized in schizophrenics who failed to respond to antipsychotic medications, different from the patterns present in those who did respond to the same drugs.

Babies who inherit a particular mitochondrial gene defect (A1555G)—especially those born prematurely—are at risk of deafness due to a class of antibiotics (aminoglycosides) that are frequently used during the first few days and weeks of life. Any maternal family history of deafness should serve as an early warning to avoid the use of these antibiotics and to seek immediate DNA testing following delivery.

Blacks and people of Mediterranean origin (Italians, Greeks, Sephardic Jews, Armenians, Turks, Spaniards, and Cypriots) have a significant likelihood of having a defect in a gene that predisposes a carrier to develop a severe hemolytic anemia (called G6PD, or glucose-6-phosphate dehydrogenase deficiency) after taking sulfa, antimalarial, or other drugs or after eating fava beans. Knowledge of such a family history should precipitate testing, and if the test results are positive, lifetime avoidance of such medications is recommended.

Variations (or *polymorphisms*) in the structure of certain genes dedicated to making products that disassemble medications explain why individuals in some ethnic groups metabolize particular drugs much more slowly. For example, Asians (Japanese, Chinese, and Koreans), more frequently than whites, have a slower breakdown of certain anticonvulsant drugs used to treat epilepsy. Consequently, higher levels of the administered medication can collect in the body while "waiting" to be metabolized, increasing the risks of toxicity and side effects in the same manner as do excessive dosages. Differences in the speed with which our bodies metabolize drugs (you may be a fast or slow metabolizer) influence the choice of the medication and its dose. These choices may also be influenced by your ethnicity. The list of drugs the metabolism of which may vary according to structural gene variations include drugs used for pain, angina, heart rhythm disturbances, depression, hypertension, asthma, and psychosis. The pharmaceutical industry is now using the "new genetics" to help identify medication responsiveness and to tailor drugs toward groups of individuals with known genetic susceptibilities. The metabolism of all drugs is influenced, more or less, by the specific gene variations we have inherited.

Genes, Sunlight, and Cancer

Born without the protective skin pigment melanin, the pale white albino can easily fall victim to the sun's rays. Besides serious sunburn, there is a very much higher risk of skin cancer. There is also a high risk of cancer within the eye because of a lack of the protective melanin in the colored part of the eye (iris). Both recessive and X-linked (see Chapter 7) forms of this disorder are known. Other genes may predispose carriers to the common skin cancer melanoma. Again, cognizance of the family history puts a person on notice about his or her personal risk even before testing for melanoma genes is routinely available. Other disorders (e.g., xeroderma pigmentosum) predispose individuals to skin cancer when they are exposed excessively to ultraviolet radiation from the sun.

Genes and Alcohol

Alcoholism also is genetically influenced, as has been shown by family, twin, and adoption studies. Again, particular sets of genes produce enzymes that metabolize alcohol at different rates. Variations in these genes influence the body's response to and "addictive need" of alcohol. These very same polymorphisms dictate how long alcohol will remain unprocessed in the body as well as an individual's sensitivity. Asians, for example, more often become flushed after consuming alcohol than do whites. Perhaps partly for this reason and others related to alcohol metabolism, the proportion of abstainers and infrequent drinkers is higher among Chinese, Japanese, and other Orientals than among other world populations.

Apart from the multiple genes influencing the body's response to alcohol, many other factors might contribute to the development of alcoholism in a particular individual. A family history of alcoholism is clearly important, as are certain heritable forms of behavior that might include antisocial personality, impulsiveness, sensation-seeking behavior, and extreme emotional volatility. Certain environmental factors—including culture, age, diet, health, lifestyle, and social traditions—also may contribute. In genetically influenced or caused conditions such as attention deficit hyperactivity disorder and Tourette's syndrome (uncontrollable utterances and twitches, learning disorder, and sleep problems), affected persons appear to be at greater risk for the development of problems with alcohol and drug abuse or dependence.

Genes and Diet

Foods, too, are digested and their chemical constituents processed by our body's gene machines. It's a common experience, after eating asparagus, to notice a peculiar smell in the urine. Although the smelly chemicals in the urine are passed by everyone, only some can actually detect the odor. This ability to recognize the smell of asparagus is a dominantly inherited trait.

Milk intolerance is especially common among adults and is distinctly different from milk allergy. The milk sugar lactose is normally broken down by the enzyme lactase, which is produced by a specific gene. Insufficient or deficient lactase activity results in incomplete digestion of lactose in the bowel. This defective intestinal digestion leads to excessive gas, cramps, and diarrhea. This type of lactase deficiency is extremely common among Mediterraneans, blacks, Asians, Native Americans, Ashkenazic Jews, and Eskimos. The removal of milk and lactose-containing foods from the diet, or their consumption in lesser quantities, rapidly eliminates the symptoms of this condition. The mode of inheritance of the culprit gene is recessive, and variations in the gene product add to the high variability of symptoms, especially with increasing age.

Folic acid, an essential vitamin B group nutrient, plays a key role in many aspects of the body's chemistry. Deficiency is linked not only to anemia but also to an increased frequency of heart disease and cancer of the cervix. Through mechanisms not yet understood, dietary folic acid supplements can ward off the birth defect known as spina bifida. Those who have already had a child with a neural tube defect such as spina bifida are advised to take 0.4 mg of folic acid daily in all subsequent pregnancies for three months prior to pregnancy and for the three months immediately following conception.

Hereditary fructose intolerance, a very uncommon genetic disorder, becomes apparent after eating fructose-containing foods (cane sugar, honey, fruits, and baby formula). Symptoms and signs include abdominal pain, vomiting, sweating, low blood sugar, and even collapse into coma. If the symptoms are left untreated, damage can result to the liver and kidneys. As adults, those who suffer from this disorder typically show a strong aversion to candy and perhaps as a result have few if any dental cavities. This disorder is inherited as an autosomal recessive trait. Three common mutations in the culprit gene account for some 87 percent of cases of European ancestry.

Various dietary and other factors also may adversely impact diseases caused by single-gene defects. Examples include phenylketonuria (here, protein is the problem), Wilson's disease (in which excessive copper is the problem), and a number of rare disorders listed in Table 10.3.

Genes and Infection

Susceptibility to infections is directly controlled, regulated, or influenced by a large number of genes in many organ systems. A discussion of the many genetic disorders in which susceptibility to infections may occur is beyond the scope of this book. Suffice it to say that any infectious agent that enters the body faces a veritable army of protective mechanisms, all of genetic origin. A fault in any one of these complex systems may result in the successful entry of a foreign organism through the protective net. We have all encountered people who frequently suffer from upper respiratory infections. In contrast, there are now well-known cases of individuals who carry the HIV virus but remain free of AIDS symptoms. The "secret" of their immune system defense appears to be an advantageous gene. This gene also has been discovered in a few individuals who remained uninfected by the HIV-1 virus despite exposure to it through sexual intercourse. Their resistance has been tracked to a gene (the chemokine receptor CCR5) inherited from each parent, which contains a tiny deletion conferring resistance to HIV-1 infection.

Contained within the cluster of HLA genes is the tumor necrosis factor (TNF) gene, inherited from each parent, the role of which is to respond to inflammation.

TABLE 10.3   Examples of Single-Gene Disorders in Which Environment Plays an Important Role

If You Have This Genetic Disorder	The Type of Inheritance Is*	Precipitating Factors Are	Health Consequence	How to Secure Your Health— Avoidance, Prevention, and Treatment
Albinism	AR; X/L	Sun exposure	Skin cancer	Use high-grade sunscreen and protective clothing and sunglasses
Alpha-1-antitrypsin deficiency	AR	Smoking; occupational/ environmental polluted air	Lung emphysema and liver cirrhosis	Do not smoke or work in polluted atmospheres
Anhidrotic ectodermal dysplasia	X/L	Heat	High environmental temperatures or high temperature from infection, both life threatening because of inability to sweat	Avoid overheating or high temperature; rapid cooling
Celiac disease	AR	Gluten-containing foods	Malabsorption of foods and diarrhea	Avoid gluten-containing foods
Cystic fibrosis	AR	Fatty foods; infections	Chronic lung infections and malabsorption of foods	Avoid infections; low-fat diets with pancreatic supplements; fat-soluble vitamins A, D, E, K supplements needed
Deafness due to mitochondrial gene (A1555G) mutation	M	Antibiotics such as streptomycin, Gentamycin	Deafness	Especially important for premature infants to avoid these antibiotics
Diabetes insipidus (kidney type)	X/L	Dehydration	Rapid dehydration leading to confusion and death if not treated	Rehydrate with water; vasopressin hormone treatment

(continues)

TABLE 10.3

If You Have This Genetic Disorder	The Type of Inheritance Is*	Precipitating Factors Are	Health Consequence	How to Secure Your Health—Avoidance, Prevention, and Treatment
Factor V Leiden deficiency	AD	Oral contraceptives, surgery, immobilization	Thrombosis in veins, especially of the legs, and risk of pulmonary embolism and sudden death	Avoid precipitating factors (see text); anticoagulants for short or very long periods
Fructose intolerance	AR	Sugar; honey, fruit, sucrose	Abdominal pain, vomiting, sweating, low blood sugar, and collapse	Avoid cane sugar, honey, fruit, and sucrose
Galactosemia	AR	Milk and its products	Untreated newborn develops liver and kidney failure, cataracts, and mental retardation or death if infection supervenes	Avoid milk and its products
Glucose 6–phosphate dehydrogenase deficiency	X/L	Sulfa drugs; fava beans	Hemolytic anemia	Do not take sulfa drugs or eat fava beans
Lactase deficiency	AD/AR	Milk and its products	Cramps, diarrhea, abdominal distention, and excess flatulence	Avoid milk and its products
Malignant hyperthermia	AD	Certain anesthetic agents	Extremely high temperature during surgery, and death if untreated	Emergency treatment with Dantrolene, dramatic cooling; steroids
Phenylketonuria	AR	Protein	Mental retardation if not detected and treated in the first few weeks of life	Special, very low protein diet

(continues)

TABLE 10.3 *(continued)*

If You Have This Genetic Disorder	The Type of Inheritance Is*	Precipitating Factors Are	Health Consequence	How to Secure Your Health—Avoidance, Prevention, and Treatment
Porphyria	AD	Barbiturate drugs	Acute abdominal pain, hypertension, psychosis, many other features	Avoid barbiturates
Prothrombin gene deficiency	AD	Oral contraceptives, surgery, immobilization	Thrombosis in veins, especially of the legs, and risk of pulmonary embolism and sudden death	Avoid precipitating factors (see text); anticoagulants for short or very long periods
Sickle cell disease	AR	Infections; high altitudes	Hemolytic anemia followed by possible arterial blockage; life-threatening infection especially from pneumococcal bacteria	Pneumococcal vaccine; avoid high altitudes (oxygen lack)
Wilson's disease	AR	Copper-containing foods	Liver cirrhosis and brain damage	Avoid nuts, cherries, and chocolate
Xeroderma pigmentosum	AR	Sun exposure	Skin cancer	Use high-grade sunscreen and protective clothing

*AD = autosomal dominant; AR = autosomal recessive; X/L = sex-linked; M = mitochondrial

A single base mutation in the promoter region of this gene (see Chapter 6 for technical explanation) is found more often among those who collapse due to bacterial infection (that is, who go into septic shock and die) than among the general population. Eventually, routine tests will become available for genes that can cause sudden death in cases of infection, enabling preemptive therapeutic action.

Genes and Behavior

At this point you will need no further convincing that the cells of the brain and their interconnections as well as their architectural layout are consequences of genes and their products. Yes, undoubtedly, some environmental influences also operate, even at the stage of the early embryo. It is no surprise, then, that there is significant heritability in intelligence (cognitive ability), personality, and of course psychopathology. Many studies of twins, including those reared separately, as well as of adopted siblings, have confirmed the importance of genetic influence. The high correlation of intelligence between identical twins is striking, compared to that between nonidentical twins. Evaluations of personality traits have focused on extroversion, agreeableness, conscientiousness, neuroticism, and openness. These studies have indicated genetic influences for all five of these personality dimensions—none, however, as strong as those for intelligence. Twin and adoption studies of males and females alike have pointed to substantial genetic influences also on human sexual orientation, although which specific genes are involved is not yet known.

Additional insight into the genetics of behavior has been gained from studies of mice. A gene in the mouse (called *Mest*) known to be responsible for influencing motherly behavior has been studied. By knocking out this single gene in mice that have been genetically engineered, researchers created negligent mothers. Curiously, the Mest gene passes in its active form only from the father. Other studies with mice focused on personality. When normally socially aloof laboratory mice were given the gene from a very sociable rodent, the prairie vole (a stocky, hamster-size rodent), the mice became much more agreeable creatures and spent more time cuddling one another. Prairie voles mate for life and spend about half their time sitting on a nest together, nuzzling, sniffing, and grooming each other. The key to their social behavior involves the hormone vasopressin. For the hormone to exert its effect, it has to attach to the surface of brain cells. This attachment is accomplished by specific proteins (receptors) that clip onto the vasopressin. When the gene for the vasopressin receptors was introduced into specially bred mice and they were injected with vasopressin, they became much friendlier toward one another.

We are only beginning to develop insight into the genes that dictate, regulate, modify, or otherwise influence all brain functions. But the early results of these laboratory studies in mice suggest that we eventually will find susceptibility genes that influence mood, behavior, personality, and even psychopathology.

Susceptibility Genes and Brain Injury or Hemorrhage

Studies performed in Israel and in Portugal finger a particular gene (apolipoprotein E) as a major factor adversely influencing the outcome from traumatic brain injury as well as the recurrence of cerebral hemorrhage. In other words, possessing this gene and suffering either a brain injury or a cerebral hemorrhage significantly increases the likelihood of a poorer outcome.

"Athletic" Genes

The remarkable endurance of some athletes is amazing. In all of us, genes produce enzymes that regulate the circulation; but a more plentiful blood supply reaches the muscles of some athletes. Developing evidence points to the product of a single gene as the explanation—that is, angiotensin-converting enzyme (ACE), which normally degrades the chemicals that dilate blood vessels, thus cutting down the supply of blood and oxygen to the muscles. A particular advantageous change in the ACE gene present in some individuals diminishes the activity of the circulating gene, facilitating the longer dilation of blood vessels and thereby promoting a better oxygen supply to muscles. Those with this variant in the ACE gene show an improved response to exercise.

Recognition of genetic susceptibility for many common disorders (e.g., heart disease, diabetes, obesity, hypertension, mental illness, severe allergies) and adverse drug reactions will become commonplace in the coming decades. Perhaps one day, early identification of newborns' propensity to develop certain diseases in childhood or in adulthood will facilitate lifestyle changes or medical interventions that may forestall, delay, or ameliorate illness. Knowledge of our genetic susceptibility to respond to specific medications (the science of pharmacogenomics) will make it possible to produce drugs tailored to our genetic makeup and honed to achieve optimal results.

HARMFUL GENES ·····································

chapter 11

Genes and Heart Disease

....

Jim Fixx, the athlete who is credited with initiating the worldwide jogging craze in the early 1970s, collapsed and died of a heart attack on July 20, 1984, while jogging in Vermont. For the fifteen years prior to his death he had been running about 80 miles per week. He knew he had a family history of early death from heart disease. Friends reported that Fixx had complained of exhaustion and tightness in his throat (angina) while running. A physician friend tried to persuade him to have a stress test. He refused. An ex-wife mentioned that he avoided doctors and appeared to deny the risk he ought to have acknowledged from his family history. It is noteworthy that seventeen years earlier he had been a two-pack-a-day smoker and had been overweight (220 pounds). He took up jogging and slowly saw his weight drop to 159 pounds. Autopsy following his sudden death showed two major coronary arteries completely blocked. Apart from that problem, he was in "fine, excellent shape"! Sadly, Jim Fixx had failed to recognize the significance of his family history and to heed multiple signals indicating his need for immediate medical attention, and had refused to see a physician.

It is extremely disturbing to realize that a longtime marathon runner can suddenly drop dead from a heart attack while running. We tend to think that an athlete with such extraordinary fitness and such a lean, wiry build could not possibly have accumulated any fat in his coronary arteries. Genes, however, rule supreme, and your genetic health supersedes all other considerations. Fortunately, by knowing your genes, you can secure your health and save your life, especially when it comes to heart disease. The key, again, is to know your family history, be aware of risk factors, be alert to symptoms and signs, obtain regular medical care, and have the recommended screening tests. There is much you can do to prevent heart disease, whatever your genetic endowment.

Virtually all heart disease results from susceptibility genes, defective single genes, or genes interacting with various environmental factors. This chapter focuses on typical coronary artery disease—the condition that results in heart attacks. Other cardiovascular disorders, which result from single defective genes including disorders of heart muscle, heart rhythm, and arteries, are discussed in the next chapter.

The Extent of the Problem

Diseases of the heart and blood vessels (cardiovascular diseases) are the main cause of death in developed countries. In the United States alone, about a million deaths annually are attributed to cardiovascular disease. According to the American Heart Association, about half of these deaths are due to coronary artery disease (CAD), 27 percent to other heart disorders, 16 percent to strokes, 4 percent to disease of the blood vessels, and a few percent to other cardiovascular problems. About 20,000 women in the United States under age 65 die from CAD each year, and more than one-third of such deaths are those of women younger than 55 years. Yet most women are much more fearful of breast cancer than of cardiovascular disease; perhaps few are aware that about 1 in 28 women die of breast cancer, whereas 1 in 2 women die of cardiovascular disease. The odds of developing cardiovascular disease before age 60 are 1 in 3 for men and 1 in 10 for women.

Death rates from coronary artery disease vary from country to country. The highest death rates currently are those in Russia, Eastern Europe, and Scotland, and the lowest, in Japan, France, and Spain. Among 35 industrialized nations, the United States ranks twelfth in mortality of women and sixteenth in men's. We have a long way to go. It is certainly high time that the giant fast-food corporations began to act in a health-conscious and responsible way by cutting their high-fat offerings. They should know that early signs of fatty deposits in arteries are readily apparent in adolescents and young adults in the United States.

The good news is that there has been a greater than 50 percent decline in death rates from CAD over the past 30 years—almost certainly due to the many advances in emergency care and long-term treatment. However, there is evidence of an actual increase in the prevalence of CAD over the same period. This suggests that we are paying inadequate attention to critically important risk factors.

Causes and Risk Factors

The vast majority of cases of CAD are due to a number of genes interacting with multiple environmental factors (so-called *polygenic* or *multifactorial disease*). The implicit message for prevention is to control the risk factors. Relatively few cases of CAD are caused directly by a single defective gene. These conditions are discussed in the following section. This section surveys the recognizable risk factors for which intervention has clearly been shown to reduce the incidence of CAD, including cigarette smoking, high cholesterol levels, high blood pressure, left heart enlargement, and factors that cause blockage in coronary arteries. Steps you can take to reduce your known risk factors include maintaining control of diabetes, exercising, reducing cholesterol and other blood fats, losing weight, reducing the level in the blood of the amino acid homocysteine, and—for women who have gone through menopause—taking estrogen supplements. (However, note that estrogen use in postmenopausal women with a family history of breast cancer or an increased risk of thrombosis is *not* recommended.)

Smoking

Cigarette smoking is the most important preventable cause of premature death and disability, being responsible for between 17 and 30 percent of all deaths due to cardiovascular disease. Nevertheless, close to 1 in 4 Americans still smoke, despite the horrendous list of consequences (see Table 11.1). Notwithstanding myths to the contrary, cigar smoking also significantly increases mortality. Some cigars contain as much tobacco as an entire pack of cigarettes! Worldwide estimates indicate that about half a billion of the world's current population will eventually die from tobacco-related illness between the ages of 35 and 69 years. Nonsmokers who live with smokers also have a 20 to 30 percent increased risk of death from CAD, and a 30 percent increased risk of lung cancer. In addition, children whose parents smoke will be more likely to contract ear infections, bronchitis, and pneumonia as well as to have more frequent attacks of asthma than children who live in a smoke-free environment. Most smokers nowadays are aware of the increased risks of lung cancer; but many do not realize that the risks also are higher that they will die of other diseases of the lungs (such as emphysema) or of the heart.

The incidence of fatal heart attacks among smokers is about 70 percent greater than among nonsmokers. Although there is correlation between the dose and the outcome (the more cigarettes smoked, the worse the prognosis), smoking as few as 1 to 4 cigarettes per day increases the risk. Women who both smoke and use an oral contraceptive have a thirteenfold increase in risk of fatal CAD compared to those who take the pill but do not smoke. Smokers also have a significantly increased risk of strokes—almost five times higher than that of nonsmokers. Finally, impaired circulation as well as amputation and rupture of the aorta due to aneurysm (see Chapter 12) all occur with much higher frequency among

TABLE 11.1   Consequences of Smoking

1. Premature death
2. Coronary artery disease
3. Lung damage (emphysema)
4. Lung cancer
5. Stroke
6. Bronchitis
7. Pneumonia
8. Aggravation of asthma
9. High blood pressure
10. Increased heart rate
11. Damage to internal lining of coronary arteries
12. Harmful effects on platelets
13. Harmful effects on cell chemistry
14. Peripheral arterial disease
15. Amputation of toes, fingers, feet, or legs
16. Abdominal aortic aneurysm
17. Failure of vascular surgery (e.g., heart bypass)
18. Accelerated atherosclerosis
19. Predisposition to clotting
20. Coronary artery spasm
21. Arrhythmias
22. Cancers of the larynx, mouth, esophagus, and bladder
23. Cancers of the colon, kidney, pancreas, stomach, and cervix
24. Smoking during pregnancy: low birth weight, miscarriage, fetal death, prematurity, and death near the time of birth
25. Earlier menopause
26. Duodenal and gastric ulcers
27. Fractures with osteoporosis
28. Infertility
29. Cataracts
30. Macular degeneration

NOTE: Smoking causes a significant increased risk for all disorders listed.

smokers. Nicotine directly damages the internal lining of the coronary arteries and the aorta, promotes blood clotting, causes spasm of the coronary arteries, and predisposes the smoker to serious rhythm disturbances (arrhythmias) of the heart.

Fortunately, those who stop smoking by 65 years of age reduce their risk of an initial or repeated heart attack by about 50 percent. Within three to four years after quitting smoking, the ex-smoker's risk of CAD begins to resemble that of a person who has never smoked. Even those who stop smoking after age 65 can significantly reduce their risks of CAD.

Some Facts About Fats

Atherosclerosis is the condition in which the arteries harden due to the deposits of cholesterol and other substances in the walls of the arteries—especially of those leading to the heart, brain, and aorta. When arteries are blocked in the heart, CAD is the outcome; blockages in vessels to the brain cause stroke; and poor circulation to the legs causes pain and eventually gangrene, leading to amputation. The common fats (lipids) that deposit in the arteries include cholesterol, saturated fats, unsaturated fats (polyunsaturated and monounsaturated, as you can read on your food labels), and triglycerides.

Cholesterol is an important substance used to build cells as well as to make hormones. Major cholesterol sources are of animal origin and include butter, cheese, whole milk, eggs, and meat. Saturated fats are found especially in red meats, cheese, whole milk, ice cream, coconut oil, and palm oil. Unsaturated fats come mainly from plants; monounsaturated fats originate in olive oil and canola oil. Polyunsaturated fats are those derived from corn oil, sunflower seed oil, and safflower oil. Triglycerides contain three fatty acids that circulate in the bloodstream, and they may be saturated, monounsaturated, or polyunsaturated. Fats travel around in your blood combined with protein in packages called lipoproteins. Low-density lipoproteins (LDL) are the cholesterol carriers in your blood and frequently are called the "bad cholesterol," since high levels result in cholesterol deposits in the arteries. The major protein carrier of LDL is called apolipoprotein B (apoB). High-density lipoprotein (HDL) is considered the "good cholesterol" because it transports cholesterol from the arteries back to the liver for disposal. HDL is therefore regarded as protective against CAD. The main protein carrier of HDL is apoA-1. Very low-density lipoprotein (VLDL) mostly carries triglycerides in your blood, and its two major proteins are apoB and apoE. Subtypes of apoE (discussed in Chapter 19) reveal that men with apoE-43 and apoE-44 are particularly susceptible to CAD, and that women with apoE-33 enjoy protection from CAD.

Those with high levels of good (HDL) cholesterol have a lower risk of CAD, whereas those with low HDL cholesterol levels have a higher risk. High levels of LDL cholesterol are associated with higher risks of CAD, and low LDL levels, with lower risks. Men in their twenties who have cholesterol levels over 200 mg need to take action to reduce their risks. Cholesterol over 240 mg, even at that age, will require a statin drug (see below), since men of this age who already have high LDL cholesterol face a 3.5-fold greater risk of death from heart disease later in life.

Coronary Artery Disease and Inflammation

Fully half of all heart attacks occur in people who have normal cholesterol levels. In figuring out why this is the case, scientists recognized that the fatty deposits in our arteries evoke an inflammatory reaction. This chemical irritation and inflammation somehow signal the liver to react by producing certain proteins in response. A number of these proteins circulating in the blood have been analyzed and include C-reactive protein (CRP), serum amyloid A, interleukin-6, and an intercellular adhesion molecule. Elevated levels of CRP have been found to be the strongest predictor (even better than cholesterol and other lipids) of the risks of heart attack and stroke as well as the need for heart bypass surgery or angioplasty of the coronary arteries.

Heart attacks in those whose cholesterol levels are normal may result from blockage of the coronary arteries when fatty-inflammatory deposits in the arterial wall or lining rupture into the artery. Statin drugs exercise their beneficial effect at the very sites of developing fat deposits (atherosclerotic plaques), in the lining or the wall of the coronary arteries. They modulate the body's immune defense response to infection at these sites, where they also prevent overgrowth of smooth muscle cells and keep blood platelets from forming clots. Similar mechanisms explain the effects of aspirin in protecting against heart attacks. An additional benefit of statin drugs may well be that they decrease levels of CRP (that is, the degree of inflammation), an effect that appears to occur largely independently of the level of LDL cholesterol in the blood. Statins result in much higher protective rates (54 percent versus 25 percent) against heart attacks when administered to those with high levels of CRP (evidence of inflammation) than to those with low or normal levels. Most recently, individuals over age 50 who took statin drugs were noted to have a significantly lower risk of developing dementia.

Meanwhile, the hunt is on to identify the infectious agent(s) at the root of the chronic inflammation causing or contributing to the fatty-inflammatory deposits that may lead to a heart attack. A number of organisms are suspects, but final judgments will need much more work.

Hypertension

Hypertension is a major risk factor for CAD, and it is fully discussed in Chapter 13. The higher the blood pressure, the greater the risk.

Clots in the Coronaries

A huge amount of data shows that clots forming in the coronary arteries (and other arteries) result from aberrations in the clotting cascade. Without getting too technical: Increased platelet reactivity or high concentrations of one or more of the following clotting promoters are associated with CAD—fibrinogen, homocysteine, lipoprotein (A), factor VII, and plasminogen activator inhibitor. All, of course, are products of genes.

Diabetes Mellitus

Diabetes appears to double the risk for CAD in men and may triple the risk for middle-aged women. Among those with insulin-dependent diabetes, there is a risk that one-third may die of CAD and another third of kidney disease. More than half of the deaths of non-insulin-dependent diabetics appear to be attributable to atherosclerosis. Diabetes is discussed fully in Chapter 14.

Obesity and CAD

Obesity clearly increases the risk of CAD, more especially if the distributed weight is mainly abdominal. Obesity is fully discussed in Chapter 15.

Miscellaneous Risk Considerations

Various other causal or contributing factors to CAD are recognized. The more risk factors present, the greater the risk of CAD. The rates of CAD increase in postmenopausal women. Estrogen replacement therapy has been associated with a decrease in overall rates by approximately 50 percent, but more recent data indicate *increased* risks in the first two years of treatment. The eventual favorable effect of the hormone is probably due to its impact on HDL cholesterol and LDL cholesterol as well as on the clotting factor fibrinogen. Concern remains, however, about an increased risk of breast cancer resulting from estrogen replacement.

For a while, stress was thought to be related to CAD. The hard-driving, impatient, so-called type A personality was thought to be at much greater risk than the easygoing type B personality. Although some still advocate this view, the evidence has not supported the classification of stress as a risk factor for CAD.

Alcohol intake has been associated with a reduced risk of CAD. However, more than one drink per day in women or two drinks per day in men have been reported to increase the risk of other diseases, such as cancer and traumatic death. The protective HDL cholesterol level is raised by alcohol.

The use of the antioxidant vitamin E may reduce the risk of heart attacks and the rate of progression of CAD, as does the use of beta carotene (in carrots, orange juice), but the available data conflict.

Fetal-Infant Programming and Adult Consequences

From experiments with dietary manipulations in animals (e.g., pigs), we have learned that by interfering with how the body deals with cholesterol during development we can affect the body chemistry permanently. Similar evidence has accrued for humans. A significant number of men and women who at birth had a reduced liver size (as measured by abdominal circumference) have been found to have raised levels of total and LDL cholesterol. The implication has been that impaired growth of the liver near the end of pregnancy may result in a permanent alteration of LDL cholesterol chemistry. Elevated LDL cholesterol with an associated increased risk of CAD has also been reported among men who were breast-fed longer than one year. Prolonged exposure to maternal hormones in breast milk has been implicated. Elevated levels of the clotting factors fibrinogen and factor VII have also been noted in men who were short at birth, had impaired liver growth, and failed to gain weight in infancy. The postulate is that some programmed impairment of liver structure and function may occur late in pregnancy, resulting in these elevated levels later in life.

Directly Inherited Coronary Artery Disease

Multiple genes interacting with the various environmental factors we have already discussed are responsible for most cases of CAD. However, a number of other diseases accompanied by atherosclerosis (fat deposits in the inner lining of arteries) are inherited via a single defective gene(s); these diseases are discussed in greater detail below. Although they are much less common than multifactorial CAD, the risks of inheriting one of these single-gene disorders is high—

between 25 and 50 percent. In all of these diseases, the elevation of total cholesterol, LDL or triglycerides, is associated with increased risks of cardiovascular disease, CAD, and heart attacks. High total blood cholesterol (or blood fats or lipids) is very significantly associated with CAD and fatal heart attacks. The higher the level of cholesterol, the higher the risk of CAD.

Familial Combined Hyperlipidemia

This is the most common single-gene lipid disorder, found in about 1 person in 200. About 1 in 5 of those with CAD under age 60 have familial combined hyperlipidemia. Typically, moderately elevated levels of triglycerides and cholesterol (and apoB) are found in such individuals, whereas their HDL cholesterol is low. In most cases a first-degree relative is known to be affected, and there is often a family history of early heart attacks. This disorder is dominantly inherited, with a risk of 50 percent that the offspring of an affected person will inherit. The condition is not generally manifested until adulthood.

Familial Hypercholesterolemia (FH)

FH ranks among the most common genetic disorders, affecting about 1 in 500 persons. In this condition, the gene responsible for a protein called the LDL receptor is defective. The remarkable discovery of the LDL receptor was made by Drs. Michael Brown and Joseph Goldstein at the Southwestern University School of Medicine in Dallas, who were awarded the 1985 Nobel Prize for Medicine and Physiology for this achievement. The function of the LDL receptor is to remove LDL from the blood. Its failure to do so results in the accumulation of large amounts of LDL cholesterol.

FH is transmitted as a dominant disorder in which one individual with a single defective gene has a 50 percent likelihood of transmitting the defect to each of his or her offspring. In the event that an affected individual has inherited a defective LDL receptor gene from each parent equally, all of their offspring will be affected. In these fortunately rare cases, astronomical levels of cholesterol are encountered together with the expected catastrophic consequences of very early cardiac disease. The cholesterol levels may be so high as to cause deposits under the skin (called *xanthomas*), which often become evident in the early twenties. These round, hard lumps of fat also may be found around areas of increased friction, such as in the tendons of the elbow, the Achilles tendons in the heels, and even over the knuckles. Cholesterol may also deposit in the cornea of the eye, appearing as a half-moon shape (called *corneal arcus*). Smaller deposits may appear as tiny, orange-colored lumps under the eyes or on the eyelids and are called *xanthelasmas*.

Since the cloning of the LDL-receptor gene, over 600 different mutations have been described. An individual with one defective LDL receptor gene is described as heterozygous for FH and has the disorder. Those who inherit the defective gene from both parents are described as homozygous for FH and are much more severely affected. Typically, heterozygous FH is characterized by high total and LDL cholesterol levels with normal triglyceride levels and usually a family history of high blood cholesterol or premature cardiovascular disease.

Homozygous FH is rare, occurring in about one in a million people worldwide. Special tests for the LDL receptor are needed for diagnosis, requiring the cultivation of cells derived from a tiny skin biopsy. In contrast to those with the heterozygous form, those with homozygous FH show evidence in childhood of cholesterol deposits in the body parts described above. Because of its severity, those with *untreated* homozygous FH rarely survive beyond their twenties. Treatment is discussed below.

Familial Hypertriglyceridemia (FHTG)

This is a relatively common, dominantly inherited disorder occurring in about 1 person in 500. It is characterized typically by moderate elevations of triglycerides without significant elevation of total cholesterol. HDL-cholesterol levels are mostly lower than normal. The precise molecular basis of this disorder is still being elucidated, but it is known that excess fat in the diet or excess carbohydrates, no exercise, obesity, alcohol use, and estrogens all may increase the levels of triglycerides. High triglyceride levels in at least one first-degree relative are usually found. FHTG is generally not associated with a significantly increased risk for CAD.

Familial Defective Apolipoprotein B-100 (FDB)

This disorder very much resembles heterozygous FH, described above, and it affects about 1 person in 600. Typically, levels of LDL cholesterol are elevated and triglycerides are normal. Cholesterol deposits may occur and there is an increased risk of premature atherosclerosis. FDB is caused by mutations in the apolipoprotein B-100 gene. This defect interferes with the normal function of the LDL receptor, resulting in the accumulation of LDL in the blood. Analysis of the FDB gene is possible but usually unnecessary to initiate treatment.

Familial Dysbetalipoproteinemia (FD)

About 1 in 5,000 individuals are affected by this genetic disorder, which can be transmitted as recessive or dominant. Typically, moderate elevations of triglycerides and cholesterol are found in the blood of those who have the disorder. HDL-cholesterol levels are usually normal. However, the levels of both triglycerides and cholesterol may on occasion be very high. The signs of FD—deposits of cholesterol and premature atherosclerosis—first appear in adulthood. These deposits may be quite startling in appearance, like clusters of small grapes on elbows, knees, or buttocks. Even the creases of the palm of the hand appear orange-yellow, and the peripheral arteries of the limbs are commonly affected.

Defects in the gene for apolipoprotein E result in FD. This protein's function is to remove fat constituents from the blood, and its failure results in their harmful accumulation. Precise diagnosis of FD is achieved by apolipoprotein analysis or through gene analysis.

Other Genetic Lipid Disorders

There are a few other genetic lipid disorders (e.g., familial chylomicronemia, familial hepatic lipase deficiency), which because of their rarity require specialist care. These conditions are also characterized by premature atherosclerosis, cholesterol deposits, and involvement of the pancreas (pancreatitis). There are also a number of rare genetic disorders in which extremely low levels of the protective HDL cholesterol are found. Curiously, for most of these disorders, CAD is not a major problem—a fact that points to the complexity of cholesterol chemistry in the body. It would seem, at least for some of these rare conditions, that HDL cholesterol is being processed extremely rapidly, leaving very low levels circulating in the blood. Both dominant and recessive forms of these conditions are recognized.

Are You at Risk of a Heart Attack?

It makes no sense to wait for a heart attack when you can take steps to avoid one. Yet there is a natural tendency, especially among men, to delay medical attention when they appear and feel well. A change in this mind-set is long overdue. Men in particular need to realize that early consultation with medical professionals about the risks of heart attack (and other potentially fatal, latent genetic disorders) could save their lives.

Three unarguable risk items stand out. Do you know your cholesterol level? Have you had your CRP level checked? What was your blood pressure reading the last time it was taken? It is now well established that early recognition and treatment of high cholesterol levels reduces CAD. Studies of some 37,000 apparently healthy men with high cholesterol levels showed a 20 to 30 percent reduction in CAD, after only a 10 percent reduction in serum cholesterol. Men and women with recognized risk factors including a family history of CAD, hypertension, diabetes, obesity, and smoking should have their cholesterol and lipid profiles and homocysteine levels tested forthwith, and thereafter on a regular schedule as recommended by their physicians.

Expert panels are not in full agreement about when to screen for cholesterol levels in individuals without any recognizable risk factors. The National Cholesterol Education Program recommends total cholesterol and HDL testing at age 20 and every five years thereafter. The American College of Physicians and the United States Preventive Services Task Force recommended such tests on all men between the ages of 35 and 64 and women between the ages of 45 and 65. If levels are found to be normal, screening at least once every five years is recommended. Obviously, if cholesterol levels are close to the upper limit of a normal reading, more frequent testing would be appropriate.

Despite the fuss made about the ability of the electrocardiogram and the stress electrocardiogram to predict cardiac problems in healthy men, there is no good evidence that these tests affect the ultimate prognosis. (This statement applies to entirely healthy individuals, and *not* those with recognizable risk factors.)

There is abundant evidence that early detection of hypertension is lifesaving. (See Chapter 13.) Experts agree that the blood pressure should be checked beginning at least at age 21, and, if normal, tested periodically thereafter. The Joint National Committee on Detection, Evaluation and Treatment of High Blood Pressure recommends measurements every two years if the diastolic pressure has been below 85 and the systolic pressure below 130. Annual measurements are recommended if the diastolic pressures has been between 85 and 90 or the systolic pressure has been between 130 and 139. Obviously, those with higher blood pressure should have their pressure checked more frequently.

What Should You Do?

The single most important strategy is to prevent or interdict the development of CAD. The reality of success in such an endeavor is evidenced by the approximately 50 percent decline in the mortality rates from CAD over the past three decades. However, there has not been anywhere near that rate of decline in the incidence of heart attacks. The striking advances demonstrated by these figures heavily reflect remarkable medical and surgical progress in saving lives, but not in preventing or avoiding CAD. What should you do if you are at risk?

Regardless of your genetic endowment, there is much that you can do to avoid CAD, to secure your health and save your life (see Table 11.2).

First, know your family history with special reference to heart attacks, hypertension, and strokes and the age at which they occurred. Pay close heed to your own past medical history and your present symptoms (if any). Do not put off a visit to your doctor if you suffer chest pains, shortness of breath, palpitations, puffy (swollen) feet, or other worrisome signs. Reassurance by your physician that your worries are groundless surely would be welcome. More important, early attention may save your life. Be aware of the three most important risks predisposing individuals to CAD: (1) hypertension, (2) high cholesterol or lipid levels, and (3) smoking. Specific guidelines for dealing with hypertension are provided above and in Chapter 13. Steps to dramatically reduce total cholesterol and other lipid levels may be necessary and can be taken. The advent of statin drugs has made it possible to lower these levels by as much as 40 percent—a decrease that has been associated with a significant reduction in the numbers of first and subsequent heart attacks. Medical therapy needs to be accompanied by a low-fat diet. The more vigorous Step 2 diet recommended by the American Heart Association restricts total fats to less than 30 percent of energy (calories), with saturated fats accounting for less than 7 percent of energy. Dietary intake of cholesterol is restricted to less than 200 mg per day. Those with extremely high triglyceride levels may benefit by adding foods rich in omega-3 fatty acids to their diet (e.g., some kinds of fish and fish oils).

Women whose triglyceride levels are extremely high are advised not to take estrogen or the vitamin A–related drug Isotretinoin (or Etretinate). Such women might also benefit from consulting a dietitian, since restriction of total fat as well as simple sugars may be necessary; and alcohol should be avoided entirely. Drugs used to lower triglycerides include fibric acid derivatives (such as Gemfibrozil), nicotinic acid (niacin), and fish oils.

In those with homozygous familial hypercholesterolemia and with cholesterol levels well in excess of 1,000, various techniques are available to drain the blood of these fats (a process called *apheresis*). For those with familial heterozygous hypercholesterolemia who already have established CAD and cannot tolerate the fat-lowering drugs and who decline apheresis, a surgical approach is available. A surgical operation called partial ileal bypass, by creating a direct connection to

TABLE 11.2   Steps to Avoid or Prevent a Heart Attack

1. Know your family history.
2. Pay attention to your own medical history and any new symptoms or signs.
3. Do not delay seeing a physician.
4. Check your blood pressure (see Chapter 13 for schedule).
5. Check your cholesterol and blood fat profile.
6. Stop smoking.
7. Lose weight if you are obese.
8. Exercise (preferably aerobic) for 30 minutes at least 3 or 4 times per week.
9. If male, take an aspirin on alternate days.
10. If cholesterol is high, take a statin drug.
11. If homocysteine is high, take folic acid daily.
12. Take 400 mg of vitamin E daily (but see text).
13. Postmenopausal women, consider estrogen supplements (but see text).
14. If diabetic, gain tight control.
15. Have a glass or two of wine daily.
16. Establish a leisure schedule for relaxation.

the lower end of the small intestine (the ileum), allows the contents of the upper intestine to bypass much of the small intestine. The internal processing of bile and other intestinal contents by the liver is thus interrupted, stimulating the action of residual LDL-receptors.

The Nurses' Health Study at the Harvard School of Public Health has tracked the medical history of 84,129 women since 1980. One of the many key findings from this study has been that women who do not smoke, who drink an average of less than 1 alcoholic beverage daily, who get at least 30 minutes of physical activity daily, and who eat a diet rich in fruits and vegetables and low in saturated fat and trans-fatty acids have a heart disease risk rate 82 percent lower than that of women who follow some or none of these healthful habits. The message is clear: Know your genes, determine your risks, establish healthful habits, and secure your life. For now, it's up to you. One day, perhaps, a gene therapy might be devised in which a functioning LDL receptor gene is introduced into the body of a person who lacks this gene. Meanwhile, many other steps can also be taken.

Insulin-dependent diabetics should maintain strict control over insulin dose, diet, and exercise. The obese need the personal discipline not only to minimize their fat intake but also to maintain a moderate total caloric intake. Crash diets simply do not work. Rather, a gradual but steady change in lifestyle and eating habits that can be maintained for a lifetime is recommended. Aerobic exercise is needed at least 3 or 4 times per week. Physical training over the long term will put you farther ahead than occasional, intensive, but short-lived bursts of activity.

Various supplements also have been found to reduce the frequency of CAD. A glass or two of wine or equivalent alcohol amount daily appears to be helpful. Estrogen for postmenopausal women was advocated for a time; but recently, 27,000 participants in the Women's Health Initiative Study were warned that they would have slightly higher risks of heart attacks, strokes, or blood clots during the first two years of the hormone's use. More study on this subject can be expected. Folic acid supplements are recommended if homocysteine levels are elevated, even though definitive proof of a significant reduction in atherosclerotic vascular disease is lacking. Some reports have advocated vitamin E supplements in the amount of 400 international units daily, but proof of benefit in this case also remains absent. Eating a diet rich in beta-carotene (which is most abundant in carrots and green leafy vegetables) may also reduce the risk of heart attack. Last, at least for middle-aged men, one aspirin every other day has been proven to significantly protect against heart attack.

chapter 12

Genetic Disorders of Heart Structure and Rhythm, and of the Arteries

··

The sudden deaths of athletes on the basketball court or during another competitive event have been witnessed by millions of television viewers. The Irish Olympian Noel Carroll collapsed and died while running. Others seemingly in their prime have suffered the same fate. The distress and dismay we all feel at witnessing any sudden, unexplained death is heightened by the apparent paradox of a world-class athlete—theoretically, the fittest of the fit—meeting with such an end. The fact is, there are many different cardiovascular disorders, mostly genetic in origin and mostly uncommon, that account for these catastrophes. The vast majority (although not all) of these disorders would have been detected, and some avoided, if detailed and timely evaluations had been sought and obtained.

This chapter focuses mainly on genetic disorders of heart muscle, arterial structure, and heart rhythm, all of which may lead to sudden or premature death or to disability. The emphasis is on avoidance, prevention, early detection, and timely treatment. Attention is also given to genetic syndromes involving cardiac structure for which the reader may wish to seek genetic counseling.

Many different genes encode the proteins whose function it is to construct and regulate the functioning of all constituent elements of the heart—the muscle, the connective tissue "cement" between cells, the nerves conveying electrical impulses, and the inner lining of the heart and its valves. The structure and architectural layout of the coronary arteries and other arteries of the body are also under genetic control. A defect in any of these genes might cause serious disability or death. Remarkably, despite a genetic mutation being evident from the time of conception, the vast majority of these cardiovascular disorders first become manifest in adulthood. These will be considered first.

Genetic Disorders of the Heart Muscle

The heart is composed almost entirely of muscle the function of which is to beat rhythmically, pumping blood through the body, without resting, for a lifetime. Diseases that affect the heart muscle (that is, *cardiomyopathies*) may be either genetic or acquired. Most cardiomyopathies are nongenetic and arise from other (usually unknown) causes. Many causes of acquired cardiomyopathies are recognized, including autoimmune disease (e.g., lupus erythematosus), many types of infectious diseases, vitamin B_1 (thiamine) deficiency, selenium deficiency, carnitine deficiency, excess alcohol consumption, and a large number of toxins. This chapter focuses on genetic cardiomyopathies.

"Large Muscle" Cardiomyopathy

Several defective genes have been discovered that cause "large muscle" or hypertrophic cardiomyopathy. This dominantly inherited condition occurs in about 1 in 5,000 adults, and is characterized by a larger than normal heart muscle mass containing muscle fibers that are in disarray. The result is impairment of the heart's pumping power.

Symptoms of familial hypertrophic cardiomyopathy (FHC) include breathlessness during exertion, pain in the chest (angina), fainting spells, and palpitations. FHC is one of the conditions that can result in sudden death in apparently healthy young athletes. In these cases, it is frequently possible to demonstrate heart abnormalities in first-degree relatives by imaging studies (echocardiograms). The reason for sudden death is thought to be either rhythm disturbances or heart pump failure. The diagnosis is usually clinched by using both the electrocardiogram and the echocardiogram.

A handful of genes leading to FHC, each with multiple mutations, have been discovered thus far. A specific mutation found in one or another of these genes has already yielded prognostic information in that its detection may spell either a benign course or likely early death if the condition is left untreated.

Diagnosis of FHC should be followed, where possible, by gene analysis aimed at identifying the specific mutation. That information will help determine prognosis and the appropriate levels of surveillance and treatment. In addition, given the dominantly inherited nature of FHC (with 50 percent risks for transmission of the defective gene), genetic counseling can be provided. First-degree family relatives also can be tested for the specific gene mutation, and where necessary, surveillance can be instituted and preemptive treatment initiated. Again, full knowledge of the family history can provide a forewarning of a possible sudden death.

"Floppy Muscle" Cardiomyopathy

Weak and floppy heart muscle is the problem in familial dilated cardiomyopathy (FDC). The result is impaired heart pumping action and progressive dilation of the heart chambers. FDC occurs in about 1 in 2,700 adults. Symptoms and signs, including abnormalities of cardiac rhythm (arrhythmias), are similar to those in cases of hypertrophic cardiomyopathy. FDC accounts for about a third of all cases of heart failure in those between 25 and 65 years of age. Familial cases of FDC may occur through various modes of inheritance including X-linked, dominant, and mitochondrial (see Chapter 7). A handful of likely gene culprits for the dominant form of FDC have been identified, but their structures have not yet been determined.

Whereas specific defective genes may account for about 25 percent of FDC occurrences, other cases of FDC occur as part of a generalized neuromuscular disease, such as muscular dystrophy of the Duchenne, Becker, or Emery-Dreifuss types; myotonic muscular dystrophy; and Friedreich's ataxia. Various other rare genetic conditions, the presence of which is signaled by certain characteristic features, also can impair the function of the heart (because these conditions are rare, they are not discussed in detail here). The recommendations listed above for obtaining diagnosis and treatment for FHC apply equally to FDC.

Genetic Disorders of the Blood Vessels

Many genes are involved in determining the structure of arteries and veins—including their inner lining, muscular walls, and connective tissue (collagen) construction—as well as the structure and function of heart valves. Here we focus only on those genetic disorders in which anticipatory knowledge could help secure your health and/or save your life.

Marfan Syndrome

This is primarily a disorder of connective tissue involving mainly the bones, eyes, heart, and aorta. About 1 in 10,000 have this condition. Those affected are typically tall, have long fingers, may have a curved spine (scoliosis) and a dent in or prominence of the breastbone, loose joints, and skin stretch marks. They are usually nearsighted and have a tendency toward detachment of the retinas as well as dislocated lenses. Heart findings include abnormalities of the mitral (between left upper and lower heart chambers) and aortic valves (between left ventricle and outflow tract, i.e., the aorta) and dilation of the aorta near its origin with the heart. This dilation may be progressive and cause the adjacent aortic valves to leak, potentially resulting in heart failure. This aortic dilation may bulge outward like a small balloon (in which case it is described as an *aneurysm*) and may eventually rupture, possibly causing sudden death. Aortic aneurysms tend most often to develop near the heart but may also form in any segment of the aorta. Fortunately, impressive advances in cardiac surgery save many with aortic aneurysms that have not yet ruptured.

Marfan syndrome is due to a defective gene transmitted as a dominant disorder. An affected person has a 50 percent risk of transmitting this condition to each of his or her offspring. The culprit gene makes a faulty protein that ordinarily would stabilize the elastic fibers within connective tissue—but this abnormal protein cannot accomplish its function. As a result, the weakened fibers stretch, bulge, and rupture. Mutations in this same gene also have been found in several individuals with aneurysms of the aorta but who do not have any other features of Marfan syndrome.

Clinical diagnosis is sometimes difficult, given the great variation of symptoms and signs even within families. Gene linkage analysis (see Chapter 6) may be valuable in identifying individuals within a family with Marfan syndrome. The huge size of the gene makes mutation analysis too costly.

Knowledge of this condition in the family is important and should prompt a discussion with your physician about your potential risks. Studies including aortic scans, echocardiograms, and a complete visual exam by an eye doctor may be advisable if you are clearly at risk. Gene linkage studies provide high degrees (about 99 percent) of diagnostic certainty in informative families, if at least two linked members can be studied together with other family members.

Early and accurate diagnosis of this disorder is important to secure the best possible outcome. Steps an affected person could take in consultation with a physician might include prevention or early treatment of curvature of the spine; limiting adult height in girls; seeing an orthopedic surgeon to diagnose and correct flat feet and other abnormalities including joint dislocations; and addressing eye complications. The cardiovascular risks, however, dominate all other concerns, given their potential life-threatening nature. Early and precise diagnosis leads to lifelong surveillance of the heart and the aorta in particular. Death may occur from rupture of the aorta or leakage of the aortic valve in the heart. At least annual echocardiograms (heart studies by ultrasound), supplemented by CT scans and MRI, allow accurate measures of the aorta. Such studies will enable cardiothoracic surgeons to operate preemptively if dilatation of the aorta reaches a certain size, averting the potential catastrophe of aortic rupture and certain death.

Early diagnosis is important, since the use of specific medication (beta-blockers) has been clearly shown to decrease the rate of aortic dilatation (and ultimate rupture) and may even eliminate the need for aortic surgery. Of course, even after surgical repair of a dilated aorta (aneurysm), continued surveillance is necessary because additional aneurysms may form elsewhere in the aorta. In addition to complications in the aorta, aortic and mitral valves may be leaky and may need replacement.

Pregnancy in Marfan syndrome cannot be undertaken lightly. Careful assessment of the aortic diameter is necessary, because at a particular width, pregnancy would be contraindicated. If pregnancy does occur and the fetus is carried to term, the decision about the route of delivery should be predicated on an assessment by specialists of the mother's cardiovascular status.

Additional advice for the individual diagnosed with Marfan syndrome includes a prohibition against contact sports, weight lifting, scuba diving, and competitive athletics. You should know your family history and pay special attention if there are one or more family members who have or had an aortic aneurysm, with or without dislocated lenses of the eyes. If such individuals are also tall and slender, the likelihood increases that they are afflicted with the syndrome. Consultation with a clinical geneticist is recommended for clarification or diagnosis and treatment.

The Ehlers-Danlos Syndromes

The Ehlers-Danlos syndromes (EDS) are a group of inherited disorders of connective tissue characterized mainly by excessively lax joints, very stretchable skin, easily torn tissues, and a tendency to bruise easily. More than eleven separate forms of EDS are recognized, and great variability exists among them even within families. It has been suggested that Niccolo Paganini, the famous virtuoso violinist, had EDS.

For our purposes here, a discussion of the form of EDS known as Type IV is especially important, since knowledge about this form of the disorder can be life-saving. EDS Type IV involves a defect in the gene responsible for a particular type of collagen (Type III). This is a dominantly inherited disorder that the affected person has a 50 percent risk of transmitting to each of his or her offspring. The gravest danger with this disorder is the potentially lethal complication of rupture of the aorta or another large artery. Teenage boys are at high risk of arterial rupture. Sudden perforation of the bowel also may occur. Skin fragility is common, and thin scars that become pigmented are not unusual. Dilatations (aneurysms) may occur in certain key arteries (such as those leading to the brain) and in the aorta. Any family history of aneurysms occurring in a relative who also had extremely hyperextensible joints should prompt immediate attention and possibly a consultation with a clinical geneticist. Familial aneurysms of arteries within the head, or familial abdominal aortic (or other) aneurysms, may truly represent this form of EDS. Because of the fragility of blood vessels, varicose vein surgery in unrecognized cases may result in loss of a limb and even a life.

Steps similar to those recommended for Marfan syndrome would be appropriate for those with EDS Type IV, including close attention to pregnancy and delivery decisions. Tragedies can be avoided by knowing your family history, establishing a diagnosis, and preventing complications.

Genes and Abnormalities of Heart Rhythm

At first blush, you might not think genes would have much to do with abnormal heart rhythms. Genes, however, control the structure of the heart and its conducting system.

There is no known more efficient pump than the heart—a muscle the size of your fist, that pumps for a lifetime despite heaps of abuse and no maintenance. Genes dictate the structure and layout of the cable (nerve) system enabling transmission of the electrical impulses that power the heart. The main power station (called the *sinoatrial* or *SA node*) is located in the right upper heart chamber (atrium). A substation (the *atrioventricular* or *AV node*) sits in the lower portion of the wall between the two upper chambers of the heart. Bundles of nerve fibers connect the SA node to the AV node and thereafter continue down into the left and right lower chambers (ventricles). The SA node discharges spontaneously and is the cardiac pacemaker. It is also connected to the brain and nervous system (via the vagus nerve), as is the AV node.

Measurable electrical activity can be demonstrated in heart muscle, even at rest. When the SA node emits an electrical pulse, the wave spreads through the upper chambers of the heart (like a pebble causing concentric waves after being dropped into water), reaching the AV node and triggering an electrical discharge through the nerve fibers traversing both ventricles. The to-and-fro electrical process at the level of the heart muscle (called *depolarization* and *repolarization*) results from the movement of common ions (sodium, potassium, and calcium) across cell walls. These ionic movements are effected through channels between muscle fibers, the structural and functional integrity of which rests mainly on genes and their products. Therein lies the potential for disruption by genetic defects.

If the pathway for an electrical impulse is blocked, the heart may simply stand still, and sudden death may occur. One rare familial, dominantly inherited condition involves AV-node heart block. If only a partial block occurs, then treatment can be initiated with anti-arrhythmic medications and use of a battery-powered pacemaker.

Normally, heart muscle cells do not spontaneously discharge electrical impulses. However, there are genetic (and acquired) disorders in which precisely this happens. At any one spot, an irritable heart muscle (called an *ectopic focus*) may suddenly begin discharging electrical impulses and completely disrupt the regular heartbeat. This ectopic focus could occur in the upper or the lower chambers of the heart, causing all manner of abnormal rhythms. Included among the inherited arrhythmias are disorders originating in the upper chambers (familial atrial fibrillation; familial total atrial standstill), in the SA node (congenital sinus node dysfunction; AV-node bypass [Wolff-Parkinson-White syndrome]); and other, less common arrhythmias. Most can occur from acquired disorders, but they also can be dominantly inherited. Our focus is on familial arrhythmias that may threaten both health and life and for which early attention can be lifesaving.

The Long QT Syndrome (LQTS)

Fainting during physical or emotional stress is the most common feature that draws attention to this syndrome, often as early as childhood or adolescence. Seizures may also be the presenting feature—and unfortunately, so might sudden death. The LQTS is characterized by a longer-than-normal interval between the Q and the T waves of electrical conduction through the heart as seen on a heart rhythm tracing (an electrocardiogram). This disorder is due to arrhythmias of ventricular origin that arise from a defect in heart ion channel function (discussed below). Up to 24 percent of individuals with this syndrome may experience life-threatening ventricular arrhythmias. Exercise or strong emotions are the factors that most often precipitate arrhythmias in this syndrome. However, a study in the Netherlands that focused on six families in which one of the LQTS mutations was identified found that members of these families were especially sensitive to sudden loud noises, such as alarm clocks or doorbells. These acoustic stimuli precipitated arrhythmias, leading to fainting in some members and sudden death in another, who died after being awakened by an alarm clock.

Thus far, both dominant and recessive forms of the LQTS have been identified. The recessive form is rare and is associated with deafness from birth (the Jervell and Lange-Nielson syndrome). The cardiac malfunction in the Jervell and Lange-Nielson syndrome is transmitted as a dominant disorder despite the recessive nature of the deafness. In fact, the LQTS has been noted in about 1 in 100 children deaf from birth, which gave rise to a recommendation that an electrocardiogram be performed on all children with congenital deafness. An even rarer form of LQTS has been noted in association with webbed fingers. In one series, 3 of 5 affected children died. The dominant form (Romano-Ward syndrome) is not associated with hearing impairment.

Diagnosis of the LQTS is made with the electrocardiogram. However, much expertise is necessary, since a number of factors can interfere with the interpretation, leading to both false positive and false negative results. A definitive diagnosis of the LQTS rests on the symptoms, the family history, exercise stress testing, and, where necessary or possible, analysis of the gene. Potentially compounding the difficulty of diagnosis is the fact that six different genes thus far have been identified as causing the LQTS, and it is clear that unknown others exist. DNA linkage analysis or mutation analysis is possible only if the exact gene in a family is known. Almost 200 different mutations have been identified so far. Other genes that may cause the syndrome are being sought.

Considerable evidence shows that the LQTS also may be caused by extraneous factors such as a range of different medications; chemical imbalance disturbing potassium, magnesium, and calcium ion levels; starvation diets (doing the same thing to these ions); and injury to the nervous system. Available data also point to some underlying predisposition to the LQTS. Family studies suggest that silent gene carriers may not be as rare as once thought and that they only come to light after exposure to one or another medication.

In this context, a report from Italy about a six-week-old infant who nearly died of SIDS (sudden infant death syndrome) takes on special meaning. This healthy infant was found blue, not breathing, and without a pulse. Immediate resuscitation was successful, and later, the LQTS was diagnosed on the electrocardiogram. Subsequently the child was discovered to have a genetic mutation that causes the LQTS. Neither parent had the mutation (paternity also was proved), which meant the child's mutation was new. Early studies of 34,000 newborn electrocardiograms by the same research group at the University of Pavia concluded that 50 percent of infants who later died of SIDS had had a prolonged QT interval on their third or fourth day of life. Since SIDS remains a leading cause of death in the first year of life, further molecular studies can be expected. Meanwhile, parents are reminded to avoid putting babies to sleep on their stomachs.

Accurate diagnosis and preemptive treatment of the LQTS can literally be life-saving. Predictors of high risk include associated deafness, very prolonged QT interval on the electrocardiogram, slow heart rate, associated webbed fingers or toes, and a history of fainting or ventricular arrhythmia. In those with the diagnosis who have already had symptoms or signs, medications (beta-blockers or sodium channel blockers) are usually recommended to reduce the risk of potentially fatal arrhythmias. Those who have already suffered cardiac arrest and have been fortunate enough to be resuscitated should seek surgical implantation of a defibrillator to monitor and immediately correct an arrhythmia. Other treatment options include a permanent pacemaker, or surgery to cut the sympathetic nerve chain transmitting the harmful electrical stimuli.

Other Familial Heart Rhythm Disorders

A host of other inherited arrhythmias have been identified. Family history studies show that virtually all, if not appearing for the first time (a new mutation), have been transmitted as a dominantly inherited disorder. An affected individual therefore has a 50 percent likelihood of each of his or her children inheriting the culprit gene. All rhythm abnormalities have the risk of suddenly decreasing the flow of blood exiting the heart, leading to a cascade of symptoms including light-headedness, palpitations, collapse, fainting, chest pain, and even sudden death. Disorders in this group include familial ventricular tachycardia, bidirectional ventricular tachycardia, familial total atrial standstill, Wolff-Parkinson-White syndrome, familial atrial ventricular heart block, and other familial conduction disorders. Inherited structural defects of the heart (including the cardiomyopathies discussed earlier in this chapter) also predispose individuals to arrhythmias.

You should be well informed about any history of cardiac arrhythmias in your family. Any such history should prompt further inquiry with your physician. If more than one individual in your family is affected by the disorder, you should also visit a cardiologist and a clinical geneticist. Specialist diagnosis and treatment for these disorders is necessary. The likely range of treatment is medications, especially beta-blockers, anti-arrhythmic medications, implantable pacemakers or defibrillators, and the possible use of anticlotting (anticoagulant) drugs.

Mitral Valve Prolapse (MVP)

The valve between the left upper (atrium) and lower (ventricle) heart chambers is called the mitral valve. When its leaflets are floppy, the pressure in the left ventricle just before contraction makes the floppy valve leaflets bulge into the left atrium. A mitral valve leak is often the result. The vast majority of people with mitral valve prolapse have no symptoms. The relatively few that do may experience anxiety and panic attacks, palpitations, easy fatigue, and possibly some dizziness when standing up suddenly from a lying position. About 1 in 10 experience progressive leakage at the mitral valve, ultimately requiring surgical repair or valve replacement. Cardiac rhythm disturbances also are relatively common with a mitral valve prolapse.

The main cause of MVP is a degeneration in the structure of the mitral valve. Given the connective fibrous content of the valve, genetic disorders in which there are defects of connective tissue, such as Marfan syndrome and the Ehlers-Danlos syndromes, among others, may also be evident. Dominant inheritance accounts for the vast majority of cases. Although an affected person has a 50 percent risk of having an affected offspring, there is considerable variability (even in families with MVP) in the clinical signs and symptoms.

If your family has a history of MVP, you need to know about it. If either of your parents or any of your siblings have MVP, you should be checked by your physician. In addition to a physical examination, you should expect to undergo a test by ultrasound imaging (an echocardiogram). Once a diagnosis is made, most people who are not experiencing symptoms of impaired valve function require no treatment. However, if you are having any of the symptoms noted above, you might wish to discuss with your doctor the use of medications such as beta-blockers, which are very effective. Those with severe MVP are at risk for stroke. They're also at risk of an infection on the heart valve causing clots that may shoot into the brain (causing strokes) or progressively destroying a leaflet of the mitral valve. This complication (called *infective endocarditis*) must be avoided at all costs. One of the most important steps would be lifelong vigilance and the use of an antibiotic (in high single dosage) just before dental work, or under specialist care before undergoing genital, urinary tract, or gastrointestinal tract surgery. Antibiotic coverage is not thought to be necessary unless there is marked MVP with or without valve leakage.

Structural Heart Defects

Among the most common serious birth defects are heart malformations, which affect about 1 in 125 newborns, occurring for the most part between the fourth and tenth weeks of fetal development. These heart defects (referred to as congenital heart disease [CHD] because they are present at birth) are caused by environmental factors (some well known, such as the rubella virus, others postulated), by specific single-gene or chromosomal defects, or by gene-environmental interactions.

Chromosomes and Heart Defects

Numerical abnormalities of the chromosomes (see Chapter 3) are very often associated with CHD. The most common is Down syndrome (trisomy 21). Over 40 percent of children born with Down syndrome have heart defects, the most common of which is at the central junction of the upper and lower chambers (called *atrioventricular septal defect*). Interestingly, the mouse chromosome 16 has a considerable number of genes that are similar to those on human chromosome 21. It is no surprise, then, that the mouse with trisomy 16 has heart defects that closely resemble those found in the human with trisomy 21. Various other structural defects may occur in Down syndrome, the next most common being a hole between the ventricles of the heart, and called *ventricular septal defect*.

The vast majority of babies born with trisomy 18 (see Chapter 3) have CHD, with ventricular septal defect and atrioventricular septal defect being the most common. There is a similar high incidence of CHD in trisomy 13 (see Chapter 3), again with the same two heart defects being the most common. In these infants, however, the heart is frequently located on the right side instead of in the normal position, on the left. In Turner syndrome (see Chapter 4), some 50 percent have cardiac or aortic defects, the most common of which is a dramatic narrowing of the aorta. Other defects may occur in the aortic, mitral, or pulmonary valves, causing either narrowing or leakage. Structural defects between the heart chambers also may occur. Much rarer chromosome disorders do occur, and are frequently associated with a range of different cardiac defects.

Structural chromosome defects (see Chapter 5) are also frequently associated with CHD. A missing snippet of a chromosome (called a deletion) or a duplication of a small chromosomal segment may have serious consequences including CHD, mental retardation, and other defects. A number of microdeletion syndromes are well recognized (see Chapter 7), and are distinguished one from the other mostly by characteristic physical signs with associated cognitive defects or mild to moderate mental retardation. I highlight two of these syndromes here (the Williams and the velocardiofacial syndromes) because of their associated heart defects. Notwithstanding their generally remarkable characteristics, many who suffer from these defects have not yet been diagnosed, since the precise DNA diagnostic techniques have only been available for a few years. You may well recognize a family member with one or more of these syndromes. If you do, it is important to warn the individuals involved of the possible complications and to encourage them to seek diagnosis and preemptive treatment. Taking such action will at the very least ease parental frustration and anxiety about not knowing the cause of cognitive defects in their children, and it may well save a life. Those who are planning a pregnancy should obtain genetic counseling and specific diagnostic tests for carriers, or if pregnancy has already occurred, they should seek prenatal diagnosis.

Williams syndrome refers to a combination of distinctive signs that include a broad forehead, flattened midface, a flat nasal bridge and broad nasal tip, thick lips, wide mouth, small chin, wide-spaced eyes, teeth abnormalities, and sagging cheeks. Serious cardinal signs include marked narrowing just above the aortic valve (supravalvular aortic stenosis), a raised blood calcium level in early infancy, and intellectual deficits. Particularly striking are the typical engaging personality, musical interest and ability, and hoarse voice, all in the face of mild-to-moderate mental retardation. As with other syndromes, the signs may be variable.

The defect causing this syndrome is a tiny snippet of missing DNA (a microdeletion) involving at least one gene (the elastin gene) on the long arm of chromosome 7 in at least 96 percent of cases. Abnormalities in at least two other chromosomes (11 and 22) are responsible for a number of other cases. The resulting features of this gene deletion are widespread abnormalities of elastic fibers throughout the body, later visible in the face, which appears prematurely aged.

Williams syndrome occurs in about 1 in 20,000 live births and almost always is the first occurrence in the family. Only a few reports describe familial cases—which means that the vast majority are due to new mutations. The considerable variation in features almost certainly reflects the simultaneous deletion of contiguous genes, only a few of which have been identified.

Velocardiofacial syndrome (also known as the *DiGeorge syndrome*) is accompanied by a wide range of signs that makes diagnosis difficult (see the case discussion in Chapter 2). The classic signs include particular facial characteristics; possibly a cleft palate; speech, learning, and intellectual disabilities; heart defects; and behavioral/psychiatric problems. Typical facial features include a bulbous tip of the nose, underdevelopment of the midface, small eyes that may appear wide-set, and low-set ears. Up to one-third of those affected have major congenital heart defects that almost invariably require reparative cardiac surgery. Low blood calcium levels are frequently noted in infancy. Later in childhood, attention deficit disorder compounds the learning and behavioral difficulties. In adulthood and sometimes even in childhood, obsessive-compulsive disorder appears, only to later merge into an increased frequency of mental illness (of the schizo-affective type).

The cause of this disorder is a gene microdeletion on the long arm of chromosome 22, detectable only by special studies and not by routine chromosome analysis. Since both Williams syndrome and velocardiofacial syndrome are transmitted as dominant disorders, risks are 50 percent for an affected person having an affected child. Hence early detection of both the velocardiofacial and Williams syndromes is important especially if more pregnancies are anticipated.

More than two dozen microdeletion syndromes are already recognized, and many may cause cardiac defects. All children (or adults) known to have congenital heart defects accompanied by any other defects should obtain high-resolution chromosome analysis and, where indicated, special deletion studies by FISH (see Chapter 5). Consultation with a clinical geneticist will assist in determining which of these tests are applicable.

Single gene defects that interfere with the heart's rotation during fetal development, resulting in right-sided hearts with or without major anatomical defects, are discussed in Chapter 6.

Multifactorial Heart Defects

Multiple genes interacting with one or more environmental factors are thought to be the cause of the most commonly occurring structural defects of the heart (e.g., ventricular septal defect; atrial septal defects; patent ductus). The empirical risk for having a second affected child is 2 to 5 percent. Similar risks for having an affected child are likely when one parent has been affected by a congenital heart defect of multifactorial origin.

Syndromes and Heart Defects

A large number of malformation syndromes are known in which cardiac defects also may occur. The range of associated features is vast and includes typical facial features, other birth defects, mental retardation or learning disabilities, and impaired growth. Again, anyone with a cardiac defect and any other major or multiple minor anomaly is advised to see a clinical geneticist.

Environmental Causes of Heart Defects

The word *environmental* in this context encompasses not only factors extraneous to the body (viruses, heat, toxins, radiation) but also the internal environment of the fetus, which is affected by maternal illness as well as by medications taken by the mother. Regardless of the factor, the first four to ten weeks of fetal development are the main period of vulnerability and of greatest risk under exposure. Maternal illnesses that could result in structural heart defects in exposed offspring include German measles (rubella), insulin-dependent diabetes, epilepsy, lupus erythematosus, and maternal phenylketonuria. (See Chapter 22 for information on prenatal diagnosis.) Maternal alcohol abuse results in a 25 to 30 percent risk of a heart defect in the exposed offspring. Anticonvulsant medications for epilepsy have risks that range from 2 to 30 percent, depending upon the medication. Acne medications (retinoic acid) are associated with a 10 to 20 percent risk. The risk with German measles is 35 percent, and with maternal diabetes, 3 to 5 percent.

Preconception counseling should be the golden rule for avoiding cardiac complications resulting from environmental factors. Immunization against rubella is a must before conception. Tight control of diabetes is likely to diminish risks of heart defects to a rate similar to those in nondiabetics. This control, however, needs to be accomplished prior to conception. Excessive alcohol use is of course outright fetal abuse. Anticonvulsant medication may need adjustment or change in consultation with a neurologist, prior to conception, and acne medication containing retinoic acid should not be used when planning or during pregnancy.

In matters of the heart, it is unwise to choose not to know. Pay careful attention to any family history of heart problems, and do your best to determine a precise answer about the nature of the heart condition. Exact information can then be discussed with your physician and a determination made about whether you are at risk and what tests might be helpful. Ultimately, there will be many who will save their lives by being so well informed.

chapter 13

Is High (or Low)
Blood Pressure Inherited?

If you do not have high blood pressure (hypertension), then you almost certainly know someone who does: Approximately 50 million Americans (about one in every four adults) have hypertension, the majority (70–75 percent) showing only mild elevations in blood pressure. Hypertension is listed as the most common reason for visits to the doctor in the United States, as well as the most common reason for the use of prescription medications. Estimates of the total annual direct costs for antihypertensive drugs exceed $7 billion in the United States alone. Among all the known heart disease risk factors, hypertension is the most prevalent. Hypertension is the main risk factor for heart attack, stroke, kidney failure, and other disorders of the blood vessels that together account for the majority of deaths in the United States and the Western world.

What Is High Blood Pressure?

The heart and blood vessels (arteries and veins) are essentially a pump with outgoing and returning vessels, which, unlike rigid pipes, are affected by the pump pressure, the flow of blood, and the environmental and other factors that influence the tone of the smooth muscles that line these channels. It is indeed this tone in the arteries (called *peripheral resistance*) that is particularly important in raising the blood pressure. All of these factors interact to produce our blood pressure. Factors that change any part of this balanced orchestral arrangement have the potential of raising the blood pressure. For example, substances that may raise the blood pressure include some anti-inflammatory drugs, some medications for asthma, oral contraceptives, caffeine, nicotine, and cocaine and other illicit drugs.

The blood pressure is recorded by two figures, one (systolic) over the other (diastolic). Blood pressure is recorded by placing a pressure cuff around the upper arm, inflating the cuff, and then listening with a stethoscope at the site of the brachial artery—at the elbow joint just below the cuff. As the cuff is deflated, tapping sounds are heard, at which level the mercury measurement can be made, providing the systolic reading. When the sounds disappear, another reading is taken, for the diastolic pressure. The normal systolic blood pressure is less than 130 (expressed as millimeters of mercury) over 84 or less (diastolic).

Although the descriptions of *mild, moderate,* and *severe hypertension* are still used as they were decades ago, hypertension is most accurately designated by the precise levels of systolic and diastolic pressures: The words *mild* and *moderate* might convey a mistaken sense that the problem is relatively unimportant. In fact, about 70 percent of those with diastolic hypertension, and more than half of the deaths and disabilities that are attributable to hypertension, have occurred with levels of diastolic blood pressure between 90 and 104. The level of systolic blood pressure is also important, especially since it is a major contributor to complications that occur because of hypertension, which include mortality, coronary heart disease, strokes, heart failure, and kidney failure. Large differences between the systolic and diastolic pressures (so-called *wide pulse pressure*) are known to be associated with increased cardiovascular risks, especially among the elderly, because such gaps may indicate that the larger arteries are more rigid because of atherosclerosis. The serious health risks accompanying hypertension can be aggravated by the presence of diabetes, high blood fats, obesity, and smoking. In fact, 12 to 16 percent of hypertensive individuals have familial blood fat abnormalities (see Chapter 11). Many cases of early-stage hypertension can be successfully managed by weight loss, exercise, alcohol reduction, and salt restriction.

Is Hypertension Inherited?

The cause of hypertension is unknown in some 95 percent of cases—the so-called *idiopathic, primary,* or *essential hypertension.* Essential hypertension and the variations in blood pressures that are encountered represent the combined effects of multiple genes and their products interacting with some known and some unknown environmental factors. Certainly, if you have a strong family history of two or more first-degree relatives with hypertension prior to 55 years of age, your personal risks are between 2 and 5 times higher than someone without such a history. The risk is higher for those between the ages of 20 and 39, becoming progressively lower with advancing age. A familial predisposition to pregnancy-induced hypertension is well established and may herald the development of essential hypertension later in life.

The many genes that control, regulate, or modify blood pressure do so via a very large number of interactive mechanisms. Some examples include control of kidney function, with special reference to the handling of salt and the kidneys' secretion of and response to specific hormones; blood pressure and oxygen supply; the nervous system and secretion of stress hormones such as adrenaline and of hormones secreted by brain cells (so-called neurotransmitters); geometric design of the blood vessels including layout, elasticity, and contractility of the smooth muscles within the wall of these vessels, as well as the size and thickness of the vessel wall and its diameter; the permeability of blood vessels and cells, allowing the movement of salt and other molecules (ions) to and fro across cell walls; the structure, power, and function of smooth muscles within the walls of arteries and veins, and the multiple growth factors required for the growth and maintenance of these smooth muscles; molecules (such as nitric oxide) released by the internal lining cells of blood vessels that interfere with circulating platelets that cause clotting as well as effectively relax the smooth muscle within the arteries; and associated high levels of circulating insulin, which dilates blood vessels, as well as of many other hormones that, in addition to their other actions, may cause dilatation or constriction of blood vessels.

Perhaps the most frequently invoked environmental factor is salt intake. Notwithstanding superb studies that have been unable to confirm a consistent relationship between dietary salt intake and blood pressure, the most recent demonstrate significant benefit when a marked reduction in dietary salt is combined with a low-fat, vegetable and fruit diet. Researchers in Utah located a gene (called *angiotensinogen*) that may explain difficulties in the interpretation of the relationship between salt and hypertension. This gene has at least three common variations that make carriers more prone to hypertension as well as more responsive to a low-salt diet.

Very rarely, severe hypertension occurs as a consequence of a defect in a single gene. These rare cases stem from defective genes that cause dysfunction or tumors in the kidneys, adrenal glands, or tissues of the autonomic (involuntary) nervous system.

People develop a characteristic blood pressure pattern in their infancy that tends to remain with them for life. So if your blood pressure is on the high side at one month of age, it is likely to remain on the high side throughout your life (and similarly, to stay low if it was low initially). Those who begin with higher levels appear more likely to develop essential hypertension in adulthood. Since blacks tend to start with higher blood pressures, their pressures predictably rise faster as they age than do those of whites. The frequency of strokes due to hypertension is highest among people who as children had the highest blood pressures.

Hypertension commonly occurs within families. The blood pressures of related individuals clearly resemble one another more closely than those of unrelated individuals living in the same house. For example, the blood pressures of parents are much more closely related to their natural children than to their adopted children. In studies of twins, the genetic contribution to the cause of hypertension has been estimated to be between 30 and 60 percent. Attesting further to likely genetic factors in hypertension is the significant concordance found between identical twins (both affected) compared to nonidentical pairs. A study of 1,003 identical and 858 nonidentical male veteran twins with hypertension showed concordance of 62 percent and 48 percent, respectively.

Not Genetic, but Still Important

Only about 5 percent of cases of hypertension have an identifiable cause. The list of disorders that constitute this 5 percent is extensive, but the conditions are individually unusual or rare. It is, however, very important to consider and exclude the so-called secondary causes of hypertension, since their discovery may lead to a cure. For example, the discovery of a severe narrowing (stenosis) of a kidney artery can be remedied by surgery, and the hypertension thus cured. It is especially important to discover any identifiable cause, since conventional antihypertensive treatment is frequently ineffective until the cause is remedied.

Does Hypertension in the Adult Originate in the Womb?

Originally, the hypothesis that adult blood pressure might be related to fetal growth emerged from a study in England of individuals born in 1946 and followed up for 36 years. Those who had lower birth weights had higher systolic blood pressures. Subsequent studies confirmed that observation and revealed that babies who were born small for their fetal age (rather than having low birth weight due to premature delivery) had a higher risk of hypertension. Various factors were, of course, considered in these analyses. Studies concluded that blood pressure was associated with disproportionate growth of the placenta related to fetal size, indicating undernutrition of the fetus. In fact, animal experiments support the hypothesis that undernutrition in the mother leads to persistent elevation of blood pressure in the offspring. A 1999 U.S. National Institutes of Health prospective analysis of subjects in the Danish Perinatal Study concluded that 11.3 percent of pregnant women who were small for gestational age at birth developed hypertension, compared to 7.2 percent who were of normal size for their gestational age.

A number of other curious observations have been made in the context of fetal development: Fingerprint patterns (whorls and loops) also appear to be related to subsequent blood pressure levels. Fingerprints are established between the fourteenth and sixteenth weeks of pregnancy. Apparently babies with low birth weight tend to have whorls more often, as a result of swollen finger pads in early fetal development. A number of studies have concluded that people with a whorl on one or more fingers have higher blood pressure in adult life than do those without whorls. Furthermore, babies who are short at birth in relation to their head size generally have narrow palms, and adults whose palms are narrow have higher blood pressure.

Ethnic Aspects

Hypertension in blacks is especially common, 1 in 4 being affected. The reasons for this propensity are not yet precisely known, but involve the interaction of genes. On average, black women develop hypertension ten years earlier than white women. Not only is hypertension more common among blacks but it is also more severe and hence more deadly. Although the mortality rate of black hypertensives from heart attacks is lower than that of white hypertensives, blacks' risk of strokes is about double.

Screening and Prevention

Hypertension is not a disease but rather a dysfunction in normal body processes that increases the likelihood of cardiovascular complications or stroke. The Joint National Committee on Detection, Evaluation, and Treatment of High Blood Pressure has recommended that blood pressure be measured every two years if the diastolic pressure has been below 85 and the systolic blood pressure has been below 130. Annual measurements are necessary if the diastolic pressure has been between 85 and 90 or the systolic blood pressure has been between 130 and 139. Obviously, those with higher blood pressures require much more frequent measurements as well as more care. Although hypertension is rarely cured, it is almost invariably successfully treated. In combination with antihypertensive drug therapy, certain lifestyle modifications are extremely important (see Table 13.1).

Treatment of hypertension is usually successful, and often dramatically so. The importance of early and effective treatment cannot be overemphasized. By having your blood pressure checked on the schedule discussed, you will best secure

TABLE 13.1   Steps You Can Take (Besides Medication) to Reduce Your Blood Pressure

Alcohol	Limit to two or fewer drinks/day (2 glasses of wine or 2 ounces of liquor or 24 ounces of beer); total abstinence if hypertension still uncontrolled.
Obesity	Lose weight by cutting calories and exercising.
Saturated Fat	Total fat should be less than 30 percent of all calories; saturated fat should be less than 10 percent of total calories; no more than 100 milligrams of cholesterol per day.
Smoking	Stop smoking.
Exercise	Get regular, progressive exercise; avoid weight lifting.
Salt	Avoid very salty foods and reduce salt intake.
Potassium	Eat a potassium-rich diet (high in vegetables, fruits), low-fat dietary products.

your health, and you might even save your life. This is especially so since hypertension is a key cause of strokes and a major contributor to heart attacks as well as heart failure. The prime purpose of treatment and control is to prevent or limit damage to key organs (heart, kidneys, and brain) rather than trying to remedy or reverse the largely irreversible changes. A significant number of people who have hypertension (estimated at between 18 and 50 percent) are unaware that they have the condition. Even more startling is that only about 50 percent of people who know they are hypertensive pay sufficient attention to the problem to keep it under control. Be sure that you are in neither of these groups.

Is Low Blood Pressure Inherited?

For the most part, low blood pressure in an otherwise healthy person—like hypertension—is probably caused by the interaction of multiple genes and environmental factors. Dr. Clinton Baldwin and his colleagues at the Center for Human Genetics at Boston University's School of Medicine, in attempting to identify the genes that cause high blood pressure, discovered the location of a single gene that causes low blood pressure (hypotension) and that had segregated within a large family. Those who had low blood pressure all had the same genetic makeup (haplotype) for what apparently is a dominantly inherited condition in that family. Ongoing study of this mitochondrial gene for hypotension promises eventually to shed further light on the biochemical pathways regulating blood pressure.

chapter 14

Genes and Diabetes

·····························

Our bodies maintain a steady level of blood sugar whether we are at rest or on the run. After a meal, blood sugar levels rise and are slowly returned to normal within one or two hours. At least 200 genes orchestrate this fine adjustment, preventing our blood sugars from rising too high or falling too low. Any defect in one or more of these genes may interfere with the regulation of blood sugar. In addition, various environmental factors might further challenge this system. Exercise, stress, infection, illness, medications, fever, alcohol, diet, and pregnancy—to name but a few factors—may interfere with blood sugar control. Failure of the body's regulatory mechanisms to return blood sugar levels to their normal range results in persistent high concentrations of sugar in the blood, a condition known as *diabetes mellitus*.

Diagnosing Diabetes

One might think that diabetes could be diagnosed simply by determining the level of sugar in the blood during a period of fasting. In fact, the most reliable method is the oral glucose tolerance test (OGTT), which involves fasting overnight (for 10 to 14 hours), having blood drawn in the morning, and then drinking a very sweet liquid. Blood samples are then drawn for blood sugar measurements at 30- to 60-minute intervals for the next 3 to 5 hours. The body's primary response to a sugar load is to step up the production of insulin, which drives the sugar level back to normal. Failure of the blood sugar to return to normal within 2 to 5 hours reflects inadequate insulin production (or resistance to insulin), resulting in persistent high blood sugar and a diagnosis of diabetes. A normal OGTT would show a fasting level below 115 mg/dL, a peak level below 200 mg/dL, and a two-hour level below 140 mg/dL. Before a final diagnosis of diabetes is made, a repeat OGTT may be recommended. The reason for taking this precaution is that multiple factors may affect the blood sugar. For example, a low carbohydrate intake for several days before the OGTT might blunt the body's insulin response. Individuals on certain diets may fall into this category. Caffeine and nicotine should be avoided before and during the test in order to maximize the accuracy of the reading.

Interpretation of the OGTT is complicated and probably unreliable in the presence of acute or chronic illness, immobilization, prolonged bed rest, infection, or any one of a number of different medications (including diuretics, certain vitamins, anticonvulsants, steroids, aspirin, and others). The test should be performed first thing in the morning, since the blood glucose levels vary cyclically during the day. Age should be given consideration, because tolerance to glucose decreases with advancing years. Interpretive levels for the OGTT also are different during pregnancy. Hence, a mistaken conclusion could be reached if pregnancy is not mentioned (or recognized) when an OGTT test is done.

An OGTT is recommended for the following reasons:

1. Urine tests positive for sugar.
2. A fasting blood sugar level obtained during a screening test is elevated (see discussion below).
3. To identify diabetes during pregnancy (gestational diabetes).
4. Identification of diabetes in individuals who are without symptoms and who are obese, or who have a strong family history of diabetes.
5. For diagnosis in those with coronary artery disease, other problems with their circulation, eye problems (retinopathy), and unexplained symptoms and signs in their lower extremities (neuropathy) including prickling sensations, pain, and weakness.
6. For diagnosis in those with abnormal blood sugar levels found in the course of surgery, trauma, stress, or heart attack, or while on steroid medications.

The diagnosis of diabetes is usually made on the basis of the OGTT. However, other screening tests for diabetes (e.g., a fasting blood sugar level) are usually administered as part of routine annual care, and these may provide the first indication of the condition. Both the American Diabetes Association and the World Health Organization currently recommend that a blood sugar level at or above 126 mg/dL be considered indicative of possible diabetes and of the need for further testing. This lower level is recommended in order to avoid misdiagnosis. Despite these measures, however, about half of all diabetic adults (between the ages of 20 and 74) in the United States are as yet undiagnosed.

The onset of diabetes may be insidious, taking years to become obvious. Given the potential serious health complications that can result from the condition if it is left untreated (including heart attack and kidney failure), early detection and intervention are crucial. Unusual thirst or excessive urination, increasing weight, increasing appetite, and fatigue—occurring separately or together—should prompt a blood glucose test.

Types of Diabetes

Diabetes is a genetic disorder involving multiple genes (for the most part) interacting with multiple environmental factors. It involves the body's handling (metabolism) of carbohydrates, protein, and fat and is associated with relative or absolute insufficiency of insulin or with the body's biologic failure to respond normally to insulin (known as *insulin resistance*).

Four types of diabetes are recognized:

1. Type I: insulin-dependent diabetes (IDDM);
2. Type II: non-insulin-dependent diabetes (NIDDM);
3. Type III: malnutrition-related diabetes (not discussed here);
4. Type IV: diabetes associated with other disorders.

Diabetes during pregnancy may be a precursor of Type II or Type I diabetes.

Insulin-Dependent Diabetes

IDDM generally begins in childhood and adolescence, but it may begin at any age. The onset of symptoms in many cases is abrupt. Immediate insulin treatment is necessary to prevent a diabetic from lapsing into a coma and dying. The most common form of IDDM (called Type IA) appears to result from destruction of the cells in the pancreas that manufacture insulin. These so-called beta cells are progressively destroyed by antibodies (proteins) that mistake beta cells for foreign invaders. This destructive autoimmune response appears to be caused by susceptibility genes that are linked to the HLA antigen blood groups (discussed in Chapter 10). Initially, at least 80 percent of individuals diagnosed with Type IA diabetes are found to have such antibodies. Hence, it is not the diabetes that is inherited but rather the predisposition to specific environmental factors (most probably, viral infections) that lead to antibody formation and destruction of the pancreatic cells.

A second type of IDDM (called Type IB) is much less common than Type IA, but it also appears to result from a basic autoimmune disorder in which the beta cells are destroyed by self-made antibodies. Those affected often have a family history of other disorders of the glands that secrete hormones, including the thyroid, adrenal glands, ovaries, and testes. They may not only have diabetes but also thyroid disease or other so-called endocrine disorders affecting the hormone-secreting glands.

Non-Insulin-Dependent Diabetes

The majority of those with NIDDM develop symptoms in middle age. Nevertheless, this form of diabetes may also appear in children, adolescents, and young adults, in which case it is described as *maturity-onset diabetes of the young* (MODY). Dominant inheritance accounts for MODY, and there is therefore no general association of the disorder with the HLA groups.

A majority (60 to 80 percent) of those with NIDDM are overweight or obese. In these genetically predisposed individuals, insulin resistance develops gradually over time, resulting eventually in overt symptoms of diabetes. Knowing your family history of NIDDM should put you on notice not to become overweight. Many who were overweight and developed diabetes have improved or even corrected the condition by losing the excess weight. (The OGTT in such cases, however, may remain abnormal.) It is important also to keep in mind that exercise increases insulin sensitivity, whereas a sedentary existence increases the body's resistance to insulin.

Although insulin-dependent and non-insulin-dependent forms of diabetes are commonly differentiated for purposes of treatment as well as anticipation of complications and related problems, some individuals with NIDDM do eventually require insulin treatment temporarily or permanently.

Diabetes Associated with Other Disorders

Type IV diabetes may occur in association with many other disorders wherein insulin secretion is decreased or insulin resistance develops. Diabetes may also occur as a complication of certain medications, including antihypertensive drugs, diuretics, estrogen hormones, and antidepressants. Diabetes may also be part of various genetic syndromes (see below).

A Common Disorder

NIDDM is by far the most frequent form of diabetes worldwide, and it accounts for about 90 percent of diabetics in the Western world. About half of those who have NIDDM remain undiagnosed for many years. In the United States, the prevalence of NIDDM between the ages of 20 and 74 years is 6.6 percent. Some ethnic groups have higher rates of NIDDM. For example, the highest known rate in the world is found among the Pima Indians of Arizona, about 1 in 2 of whom become diabetic. The rate is also higher in Hispanic Americans, especially Mexican Americans, and higher in blacks than in whites. Japanese Americans also have a high prevalence of NIDDM. In England, NIDDM is much more frequent among South Asians than whites. Even among different ethnic groups sharing the same environment, genetic differences have influenced the prevalence of NIDDM. In Singapore, those of Chinese ancestry had a rate of 4 percent, whereas the rate among Malays and Asian Indians were 7.6 percent and 8.9 percent, respectively. Cultural and economic (dietary) factors may also influence prevalence rates. This is especially exemplified by migrations—for example of Asian Indians to England, South Africa, Malaysia, and elsewhere, where their rates of developing diabetes are much higher than in their home country of India. Similar observations have been made in Japanese who immigrated to the United States or Brazil.

IDDM ranks as one of the most common chronic childhood disorders, with at least 50,000 new cases occurring annually worldwide. In the Western world, 1 to 3 children per thousand are affected by the age of 20. The incidence of IDDM varies from country to country. Scandinavia and the Mediterranean island of Sardinia report the highest incidence rates, whereas Asians have the lowest rates. Children born in Finland are 35 times more likely to develop diabetes than children in Japan. American non-Hispanic whites are about 1.5 times more likely to develop IDDM than are African Americans or Hispanics.

During the final decade of the twentieth century, the prevalence of diabetes in the United States increased at an alarming rate. The most startling observation is the 70 percent increase in diabetes among people in their thirties. The epidemic of obesity, which is closely associated with diabetes (see Chapter 15), is regarded as the main reason for this serious situation. Lack of exercise (due to free hours being spent in front of the television set and at the computer) and easy availability of high-calorie snacks are well recognized (and preventable) factors.

Culprit Genes and Environmental Factors

Efforts to unravel the causes of diabetes have been bedeviled by a host of confusing realities. Not the least of these difficulties has been determining who is actually diabetic. Some individuals have no symptoms but show a chemical response to a sugar load that is typically diabetic as seen on the OGTT. Others with a diabetic OGTT curve who are tested after weight loss may show no indication whatsoever of symptoms or abnormal OGTT. Some women have high blood sugar and excrete sugar in their urine during pregnancy but show no symptoms of diabetes after delivery. Most of those who develop NIDDM are adults, but a few are children. Modifications of diet, weight, exercise, and the presence or absence of infection may confound efforts to determine the presence or absence of diabetes or a predisposition toward diabetes.

Diabetes may be due to the action of a single gene, to the interaction of multiple genes with one or more environmental factors, or to genetic disorders or syndromes that affect the function of the insulin-manufacturing cells (beta cells) in the pancreas. Genes predisposing individuals to diabetes are undoubtedly involved even in various nongenetic illnesses that may cause diabetes or a predisposition to it. Let's briefly examine each of these causal routes separately, even though they overlap somewhat.

Diabetes and the Single Gene

Diabetes due to a single gene is relatively rare. Mutations in the insulin gene may cause diabetes through faulty production of insulin, but this defect is responsible for fewer than 0.5 percent of all diabetes cases. A handful of other, similarly rare mutations of single genes also cause diabetes. Mitochondrial gene (see Chapter 6) mutations (not part of a genetic syndrome) may account for about 1.4 percent of all cases of NIDDM. In fact, it has been known for years that in families with NIDDM or maturity-onset diabetes of the young (MODY; see below), diabetes occurs much more frequently on the mother's side than on the father's. More mothers than fathers of MODY diabetics are also diabetic—supporting the assumption that the condition is due to mitochondrial inheritance in some of these families.

Maturity-Onset Diabetes of the Young

MODY is present if the age at onset of diabetes for at least one family member was younger than 25 years; if blood sugar levels have been controlled for at least two years without insulin; and if the diabetes has never been out of control (that is, if there is no *ketosis* indicated by excess fatty acids in the blood and urine). Thus far, three different genes (on chromosomes 20, 7, and 12) have been identified for MODY that are dominantly inherited. Hence, an affected person has a 50 percent risk of having affected offspring. Fortunately, however, MODY is very uncommon and accounts at most for only a few percent of all NIDDM cases.

TABLE 14.1   Risks for Non-Insulin-Dependent Diabetes (NIDDM)

Population risk	1 in 20
Risk for sibling if one affected	1 in 7 to 1 in 10
Risk for a child if one parent affected	1 in 7 to 1 in 10
Risk for a child if both parents affected	3 in 4
Risks for identical twin if one affected	3 in 4 to 1 in 1 (100 percent)
Risks for nonidentical twin if one affected	1 in 10

NOTE: See text for risks related to maturity-onset diabetes of the young (MODY).

Diabetes, Multiple Genes, and Environment

Those with obvious diabetes and those with abnormal OGTT have a greater number of affected relatives than do nondiabetic individuals. Identical twins, who share identical genes, are concordant (both affected) in more than 70 percent of cases. Nonidentical (fraternal) twins have a reported concordance rate of between 3 and 37 percent. This difference in concordance strongly supports the role of genes in causing or predisposing individuals to diabetes. The data on nonidentical twins are comparable to those for other siblings, which indicates that the effects of any unique, shared family environment are relatively minimal.

Genes and Non-Insulin-Dependent Diabetes (NIDDM)

A number of candidate genes have been identified for NIDDM, but this disorder is still regarded as one that evolves from the interaction of these susceptibility genes with one or more environmental factors. The risks for clinically overt NIDDM and for impaired glucose tolerance (as reflected by the OGTT) are shown in Table 14.1.

In a study conducted at the Joslin Diabetes Center in Boston, which extended over 25 years, the risk of developing NIDDM among 606 offspring of 2 parents with NIDDM was found to be 45 percent by age 65 (5 times that in the general population). In addition, the researchers noted that the onset was earlier in offspring of parents who developed NIDDM before 50 than in offspring of those who developed this disorder after age 50. In recent studies of white male twins with NIDDM, the concordance rate in identical twins (both affected) was 41 percent, dramatically higher than the 10 percent concordance rate in nonidentical twins. The twins studied ranged in age from 52 to 65 years.

TABLE 14.2   Risks for Insulin-Dependent Diabetes

Population risk	1 in 500
Risk for sibling if one affected	1 in 14
Risks for a child if mother affected	1 in 40 to 1 in 50
Risks for a child if father affected	1 in 20
Risks for a child if both parents affected	1 in 3
Risks for identical twin if one affected	1 in 3
Risks for nonidentical twin if one affected	1 in 5

Genes and Insulin-Dependent Diabetes (IDDM)

As mentioned earlier, genes that encourage susceptibility to infection or other environmental agents facilitate damage to the insulin-producing cells of the pancreas. These so-called HLA-linked genes (see Chapter 10) place the individual inheriting them at risk of subsequently developing IDDM. The risks of developing diabetes for the siblings of diabetics and the offspring of diabetics are shown in Table 14.2. These risks in fact can be modified by testing for the HLA groups. Such testing, however, is regarded as of little benefit and of no practical use, since the data gleaned from the tests do not indicate whether diabetes will emerge in the future. In fact, siblings of an individual with IDDM still have an increased risk of developing diabetes even when there is no sharing of any HLA group! Note that the risk to offspring of developing diabetes is twice as great when the father has IDDM as when the mother is affected.

Individuals and families in which IDDM occurs are at increased risk for other autoimmune disorders, including thyroid disease, myasthenia gravis (a nerve-muscle junction disease), vitiligo (white patches of skin), and celiac disease (an intestinal disorder of food malabsorption). Some studies have found that 21 percent of diabetics and 22 percent of their first-degree relatives had evidence of another autoimmune disorder. Thyroid disorder has been noted as the most common.

Fetal Origins of Diabetes

Insulin plays a key role in fetal growth. Moreover, poor growth and small size in the fetus, subject to various influences within the womb, increase the risk after birth of developing NIDDM. Infants of low birth weight later develop increased rates of NIDDM and impaired sugar tolerance. Those born at lower birth weights (thin at birth) develop less muscle and tend to develop diabetes caused by resistance to insulin, as well as hypertension and elevated levels of triglycerides (see Chapters 11 and 13). The "fetal programming" effect operating in the womb results in the offspring of diabetic mothers being more obese than those of non-diabetic mothers. It remains uncertain whether low birth weight due to a nutritional deficiency in the womb leads to impaired development of the insulin-producing cells in the pancreas and subsequently to diabetes, or whether there is selective survival of infants who are genetically predisposed to insulin resistance.

Diabetes and Genetic Syndromes

Either diabetes or glucose intolerance occurs in more than 70 different genetic disorders. Although these disorders are rare or uncommon, early diagnosis is important for prompt treatment and because of the risks of recurrence. The mechanism by which diabetes develops in these many different disorders, however, is the same: It is caused by absolute or relative insulin deficiency. This lack of insulin may come about due to destructive disorders of the pancreas (e.g., cystic fibrosis, hemochromatosis, or hereditary pancreatitis); related growth hormone deficiency; inhibition of insulin secretion (e.g., hereditary tumors of the autonomic nervous system); insulin resistance (e.g., myotonic muscular dystrophy); hereditary syndromes associated with obesity; or in association with premature aging syndromes or chromosome defects (see Chapters 3 and 4).

Many nongenetic disorders involving the pancreas, or conditions that influence or affect insulin secretion, also are known—for example, alcoholic pancreatitis. The use of certain medications, such as some diuretics (pills to lose water) or beta-blockers (for hypertension), can have the same effects. It is likely that even in cases of diabetes resulting from the so-called nongenetic causes an inherited predisposition is present: Not all alcoholics develop alcoholic pancreatitis, and relatively few individuals taking various medications develop glucose intolerance or overt diabetes.

Pregnancy and Diabetes

Women with diabetes need to be concerned about three primary issues:

1. The effects of pregnancy on diabetes.
2. The effect of diabetes on pregnancy.
3. The effect of diabetes on the developing fetus.

The effects of diabetes on pregnancy may give pause to some diabetic women contemplating childbearing. Their concerns may be well founded. Diabetic women have an increased risk of complications during pregnancy, including miscarriage, preterm labor, kidney infection, hypertension, and excess accumulation of the fluid around the fetus (amniotic fluid). Moreover, control of diabetes may become more challenging during pregnancy, with fluctuations between very high blood sugar levels and low blood sugar levels. There is also a higher cesarean section rate, which together with potential heart and kidney complications increases the risk of maternal death eight- to tenfold. Pregnancies in women who have diabetes are considered high-risk.

Fortunately, attention to fundamentals will enable the vast majority of diabetic women to have children without risking their lives. Attention to important details makes the difference: Avoid obesity; achieve tight control of the diabetes; check and treat the heart, kidneys, and blood pressure; and plan pregnancies at an earlier rather than a later age. It is critically important to seek the care of a specialist in diabetes as well as an obstetrician who specializes in high-risk pregnancies (a perinatologist).

The effects of pregnancy on diabetes reflect the stress in pregnancy in raising the blood sugar. Insulin needs steadily increase, with a significant jump in requirements between the twentieth and thirtieth weeks of pregnancy. Following delivery there is often a dramatic drop in insulin requirements that lasts several days; therefore, careful attention is necessary in order to avoid very low blood sugar levels (hypoglycemia). There is still uncertainty about whether or not the typical diabetic eye and kidney complications are aggravated by pregnancy. Prior to conception, cardiovascular, eye, and kidney function should be carefully assessed.

Diabetes also can have a negative effect on the developing fetus, resulting in abnormal growth and development. Large infants are typical in the offspring of diabetic mothers. About 30 percent of diabetic mothers have infants weighing more than 9 pounds (or 4 kg). These heavy (macrosomic) babies result from the maternal blood sugar levels, which induce high fetal blood sugar levels and high levels of insulin secretion. The insulin secreted efficiently stuffs away fats, proteins, and carbohydrates into the fetal cells. Needless to say, overweight babies are at higher risk of complications at the time of delivery, with an increased rate of death (stillbirth) and an increased likelihood of being subject to trauma during delivery. This is especially true if the mother's diabetic state is poorly controlled. If the infant is not stillborn and survives, it may still be endangered by serious complications resulting from the mother's poorly controlled diabetes, including hypoglycemia, low calcium and magnesium levels, and high concentrations of red blood cells that could lead to thrombosis in kidneys and other key organs. The offspring of mothers with uncontrolled diabetes are also subject to convulsions, respiratory distress, heart failure, liver complications, and significant birth defects (discussed in detail in Chapter 9). The offspring of fathers with diabetes are not afflicted by any of these problems.

Gestational Diabetes (GDM)

Diabetes that first makes an appearance during pregnancy is termed gestational diabetes. About 3 to 5 percent of pregnant women develop GDM. Detection of GDM is important, since many pregnancy complications and fetal consequences may result from uncontrolled diabetes, including abnormalities and subsequent learning defects. Consequently, blood sugar screening should be performed in every pregnancy. This is usually done with a sugar drink and a blood sugar level drawn one hour later. A reading above a certain value (some use 140 mg/dL) prompts a 3- to 5-hour OGTT. Clues that suggest an increased likelihood of GDM developing include obesity, a previous large baby, previous history of fetal or newborn death, a previous child with birth defects, a family history of diabetes, and hypertension during the current pregnancy.

Measures for avoiding and preventing complications and birth defects involve planning pregnancy, securing the best possible health before conceiving, attending to treatable disorders such as hypertension, tightly controlling diabetes, achieving a normal weight level, and monitoring the fetus by maternal blood screening studies at 16 weeks (see Chapter 22) as well as by ultrasound imaging.

Women who have GDM become full-fledged diabetics in up to 63 percent of cases within sixteen years after the index pregnancy. The risk of diabetes is especially pronounced for those who have high blood sugar levels during or soon after pregnancy; who are obese; and whose GDM was diagnosed prior to the twenty-fourth week of pregnancy. I cannot emphasize enough how important it is for women who now have or have had GDM to avoid becoming obese, to have regular aerobic exercise, and to avoid medications that could induce resistance to insulin. An annual blood sugar test would be prudent, as would an appointment with a physician, should there be any suspicion of the signs of typical diabetes. Effective contraception is also highly recommended following a diagnosis of GDM. This step decreases the likelihood of pregnancy occurring in the presence of unrecognized high blood sugar, and thus lessens the risk of birth defects resulting from the diabetes. The intrauterine device (IUD) is highly effective and does not influence the body chemistry. Although low-dose combination oral contraceptives do not appear to increase the risk of diabetes even after GDM, oral contraceptives containing only progestin, when used during the period of breast-feeding, result in a threefold increase in the rate of diabetes.

A Predisposition to Diabetic Complications

The individual with diabetes has to contend with a host of complications. The most important complications of diabetes include heart and blood vessel disease (atherosclerosis; see Chapter 11), eye complications with loss of vision (retinopathy), progressive kidney failure (nephropathy), an increased risk of infections, damage to nerves of the lower and upper extremities (peripheral neuropathy), and hypertension. Bear in mind as you consider these serious complications that there is much that can be done in terms of prevention, avoidance, and treatment. Certainly aggressive efforts must be made to treat high blood fat levels, to stop smoking, to control high blood pressure, and to achieve a normal weight. Diabetes may also increase the tendency to thrombosis, which, coupled with elevated blood fat levels, can lead to heart attacks and strokes. Knowledge of these potential complications should provide sufficient warning to take the critical steps necessary to achieve the best possible outcome.

Heart Disease

There is a two- to fourfold increased risk of coronary artery disease in those with NIDDM or IDDM. Increased cholesterol and other blood fat levels worsen the prognosis, as does obesity (see Chapters 11 and 15).

Eye Complications

Lining the inner layer of the eye is a thin, meshed network of nerves laid down in ten well-defined layers and interspersed by a network of tiny blood vessels (capillaries). Diabetes involves these tiny capillaries, causing blockage, dilatation, and leakage. This is especially so with IDDM, and to a much lesser extent with NIDDM. Within ten years of diagnosis, close to 90 percent have developed retinopathy of various types and degrees of severity. Proliferative retinopathy is the most serious type, the main consequence of which is impaired vision. Those with early-onset diabetes have about a 1.8 percent risk of eventual blindness, and those with NIDDM have a 4.8 percent risk. Later-onset, insulin-requiring diabetics have about a 4 percent risk. The figures for visual impairment in these three groups are 9.4 percent, 37.2 percent, and 23.9 percent respectively, according to the results of a major, ten-year study. Cataracts (cloudy to opaque lenses of the eyes) obviously impair vision, and they occur more commonly among diabetics. Glaucoma (a disorder in which there is impaired vision from damage to the mesh of nerves in the inner lining of the eye, due to elevated pressure within the eye) also occurs more commonly in persons with diabetes.

Kidney Complications

Progressive kidney failure is a long-term but common complication, mainly in IDDM. Between 30 and 40 percent of those who have had IDDM for 20 to 25 years develop kidney failure. After that lengthy period without kidney involvement, the risk that it will develop is only about 1 percent per year. The critically important message is that every effort should be made to achieve tight control of both IDDM and NIDDM long before irreversible and irremediable defects occur in the kidneys. Early, continuous, and strict control of diabetes will prevent or at least slow the progression of kidney damage as well as retinopathy.

Kidney complications do not develop in everyone with diabetes. Since only 30 to 50 percent of those with IDDM develop these complications regardless of the duration of diabetes, evidence has accumulated suggesting a genetic predisposition not only to complications of the kidneys but also to those involving the eyes. Part of the evidence has pointed to diabetics in the same family developing the same complications. Precise interpretation of these familial clusters is made more difficult by the noteworthy higher correlation between mother and child, when both have diabetes and develop the same complications. Questions therefore remain to be answered about whether the environment within the womb is important, and whether mitochondrial inheritance (see Chapter 7) is significant.

Susceptibility to Infection

Diabetics have a higher than normal risk of developing bacterial infections (such as those caused by staphylococci or streptococci) that may cause serious illness or death. This susceptibility to infection is not yet fully explainable, but it is likely to have at least some genetic component. In one major study, in almost one-third of diabetics admitted to an intensive care unit with their diabetes out of control, an infection was the precipitating factor. According to the same study, in 43 percent of those who died after being admitted, the cause of death was infection. Experts have emphasized that antibiotics are second only to insulin in increasing the life span of diabetics. Immediate attention to any infection, coupled with tight control of diabetes, can secure the health and save the life of many.

Nerve Damage

High levels of blood sugar combined with insulin deficiency over the long term can cause peripheral nerve damage *(peripheral neuropathy)*, which is extraordinarily common in diabetics. The frequency of this complication increases with the duration of diabetes. According to one study, peripheral neuropathy had occurred in as many as 50 percent of diabetics older than 30 years. Considerable health impairment typically results from this complication: The results may include recurrent infections and sores of the feet, erectile dysfunction (impotence), intestinal problems, and limb amputations due to gangrene—to mention but a few of the possibilities. A genetic predisposition may partly explain why some diabetics with poorly controlled blood sugar levels never develop this neuropathy. Once again, strong evidence shows that strict control of diabetes prevents diabetic neuropathy and, where it already exists, probably halts its progression.

Hypertension

High blood pressure and diabetes frequently coexist—possibly three times more often than predicted by chance. Moreover, glucose intolerance and diabetes appear more often in hypertensives than in those with normal blood pressure. Hypertension also increases the vascular complications of diabetes, including cardiac, eye, and kidney complications. The likelihood of serious illness and death is almost twice as great in diabetics who also suffer from hypertension. For this reason, the treatment of hypertension in diabetes is very important and should be combined with strict control of blood sugar levels. Since important protection can be achieved thereby against the progression of heart, eye, and kidney complications, aggressive control of hypertension is recommended. The treatment and management of hypertension includes antihypertensive drug therapy as well as daily exercise, reduced calorie intake, a low-salt, high-fiber diet, weight loss, a halt to smoking, and a moderation of alcohol intake. Infections should be avoided, and treated promptly when they occur.

A huge body of evidence now emphasizes that tight control of diabetes delays the onset and slows the progression of diabetic eye, kidney, and nerve damage.

The available data are especially applicable to IDDM but also important in NIDDM. Especially among non-insulin-dependent diabetics, particular care is necessary to avoid sudden, very low blood sugar levels, which could result in seizure or even in coma.

Obesity and Diabetes

High blood pressure and high blood fat levels frequently occur together with obesity and NIDDM. This combination may also frequently be a common precursor of heart disease and stroke (see Chapter 15 for further discussion).

Dietary Guidelines

Dietary guidelines for managing diabetes are available from the American Diabetes Association (ADA). The most recent studies indicate that an increase in dietary fiber (particularly of the soluble type) beyond the level recommended by the ADA helps control blood sugar, decreases high levels of insulin, and lowers cholesterol levels in those with NIDDM. In fact, the decrease in blood sugar levels achieved by increased dietary fiber intake in many cases is commensurate with that obtained by taking an additional antidiabetic pill.

The many exhortations in this chapter about the care and treatment of diabetes reflect the formidable existing challenges. New genetic technologies hold out the promise that gene therapy eventually will protect insulin-producing beta cells from damage, enhance their insulin secretion ability, and facilitate their transplantation directly into the pancreas (see Chapter 27). Beta cell transplants have already been successfully performed, and further achievements in this direction can be anticipated.

chapter 15

Obesity: Genes or Environment?

··

Obesity is epidemic—and it's worldwide. The World Health Organization and the U.S. National Institutes of Health have this problem in sharp focus, given the serious contribution of obesity to illness and mortality. The risk of dying from all causes, including cardiovascular disease and cancer, rises with increasing obesity. Recent estimates in the United States of annual deaths attributable to obesity approximate 325,000 among nonsmokers.

The millions of diet books sold each year attest to the ubiquity of the problem as well as to the desire of many overweight people to shed pounds. But the surest way to lose weight is simply to cut calories and increase exercise: Central to the reality of obesity is the fact of excess calorie intake over the number of calories expended. Keep on stuffing a closet, and it will become overfull.

This analogy has deeper biological underpinnings in the understanding of obesity. It is currently thought that the number and size of fat cells are established in the fetus toward the end of pregnancy and in early infancy. This early "programming" of fat cells essentially dictates the number and size of closets available for filling later on. In experiments with rats whose calorie intake was markedly reduced during suckling, a permanently reduced number of fat cells were found later. Obesity, however, is even more complex than these studies and analogies suggest, as will become clear to you in the course of reading this chapter.

Classifications of obesity use the body mass index (BMI) as an indicator. To calculate your BMI, divide your weight (in kilograms) by your height (in meters, squared). Adults with a BMI of 25 or more are regarded as overweight, and those with a BMI of 30 or more, as obese. Extreme obesity is at or above 35; and morbid obesity, above 40.

The Obesity Epidemic

There is a startling paradox between the image of health-conscious Americans obsessed with health, diet, and exercise and the reality that more than 50 percent of adults in the United States are overweight. Most are obviously paying lip service (pun intended) to the goal of weight loss. Particularly worrisome at the close of the twentieth century were reports in the United States that the greatest weight increases have been among those between the ages of 18 and 29 years. Substantial weight increases also have been noted in Americans with higher education and among Hispanics.

Being obese, however, is not the same as being overweight. Obesity is considered to be present when an individual is 30 percent over his or her ideal weight. About one-fifth of U.S. adults are classified as obese. The proportion of Americans who are obese has increased by more than 50 percent in the past twenty years, and the number of overweight children has doubled during the same period. The number of obese adults has increased by more than 25 percent over the past thirty years.

Obesity and Associated Disorders

Excess weight is well known to be associated with an increased incidence of cardiovascular disease, non-insulin-dependent diabetes, hypertension, stroke, high blood fat levels, osteoarthritis, some cancers, and decreased longevity. The proportions of the increases, however, are less well known—and startling. Among men under 55 years of age who are very obese, an 18-fold increased likelihood of adult-onset diabetes has been reported; among women in the same age and weight group, the risk is 4 to 5 times greater. In the same age and weight categories the likelihood of gallbladder disease among men was 21 times higher, and among women about 5 times higher, than the risks among their counterparts with normal weight. Moreover, the likelihood of having two or more serious health conditions increases with greater weights. The risks of coronary heart disease are about 14 times greater in men and 19 times greater in women. High blood cholesterol levels in both very obese men and women are about 36 times more likely than among those of normal weight.

In a major recent study of 6,987 men and 7,689 women, the most common obesity-related health condition was hypertension, which affected some 64 percent. High blood cholesterol was next, at about 36 percent. Among obese men, 14 percent developed coronary artery disease, as did 19 percent of obese women. About 11 percent of very obese men developed maturity-onset diabetes, as did almost 20 percent of very obese women. The proportion of obese women who developed gallbladder disease (24 percent) was twice that of obese men (10 percent). Obese women were also more subject to osteoarthritis than were obese men. For both sexes, all of these complications increased sharply with increasing weight.

Obesity and Diabetes

Obesity frequently precedes non-insulin-dependent diabetes mellitus (NIDDM) (see Chapter 14). Although not everyone who is obese develops NIDDM, most individuals with NIDDM are obese. This apparent inconsistency is probably due to some other factor that also leads to NIDDM. The causal factor seems to be a defect in the ability of beta cells to produce the extra insulin needed by those who are fat. Many obese individuals are insensitive (resistant) to insulin, resulting in the diabetic state. Measurements of the waist and hip circumference in such individuals show that resistance to insulin is especially closely associated with fat accumulation in the abdominal area. In fact, abdominal obesity is a more powerful predictor of insulin resistance than is total body weight.

Why and How People Get Fat

Yes, once again it's in the genes. However, in the case of fat, it is almost invariably what we do to our genetic constitution that causes the problem. Obesity is due to an excess of body fat, which in turn is directly due to an intake of calories exceeding the amount needed to produce energy (i.e., the number of calories "burned" as fuel). Although there are a host of syndromes—genetic and otherwise—that result in obesity, the overwhelming majority of persons who become obese have a chronic energy imbalance; they simply and continuously take in more calories than they use. An excess of only 2 percent of calories per day over energy expenditure results in a weight gain of about 5 lb (2.3 kg) within a year. The average American man or woman gains about 20 lb (9.1 kg) between the ages of 25 and 55 years. This substantial increase in weight is the cumulative result of only a tiny imbalance between average daily energy intake and expenditure—amounting to only 0.3 percent extra calories!

A huge body of research has revealed some of the mechanisms by which the balance between food intake and energy expenditure can be disturbed and result in obesity. As with many other bodily functions and dysfunctions, the matter of body weight is extremely complex. Without waxing technical, the following discussion provides some perspective about the complex mechanisms involved, and lends insight into how and why the treatment of obesity has been so uniformly unsatisfactory.

Signals and More Signals

Key (but not exclusive) players include the brain, all of the hormone-producing glands (such as the pituitary, thyroid, and adrenal glands, as well as the insulin-producing cells of the pancreas), the intestines, fat cells, muscles, and the peripheral nervous system. A few dozen hormones and proteins (all of them being products of individual genes) have been identified thus far that function as chemical signals to target an organ or a cell. Such signals are responded to by reciprocal signals in the form of secreted hormones or proteins, which in turn provide feedback responses to the original source as well as to other organs or cells. A continuous balancing act takes place in the body as it responds to food intake and calorie expenditure, via the exchange of signals between all organs involved.

A key role is played by a part of the brain called the *hypothalamus,* which controls hunger and satiety. The hypothalamus also controls the output of the hormones involved in depositing energy stores. It is likely that "satiety" and "hunger" centers exist within the hypothalamus. In sending a "hungry" signal, the hypothalamus may be aided and abetted by visual and olfactory stimuli (the display at a deli, or the aroma wafting from a nearby bakery), activating the appetite.

Fat cells are not inert but secrete products such as the protein leptin, which (as we know from experiments in mice) influences body weight and energy stores through actions on the brain that regulate food intake and energy expenditure. It appears likely, furthermore, that fat cells in different regions of the body (for example, the abdomen and the thighs) operate somewhat differently.

Loss of fat decreases leptin levels, thereby stimulating food intake. An increase in body fat raises leptin levels, reducing food intake. In very rare cases, mutations in the leptin gene, making it ineffective, have resulted in morbid obesity. Obesity also occasionally results from the production of too little leptin or from leptin resistance. Leptin interacts with brain cells and a whole network of chemical connections with feedback circuit control, ultimately balancing and modulating weight gain. Genetic, environmental, hormonal, psychological, and other factors have the potential to disturb the delicate balance.

Complicating this vast interplay of cell-to-cell signals is the recognized effect of programming that begins in the womb and in early infancy. It is thought that the appetite center in the hypothalamus is set in relation to body size early in pregnancy. This view is supported by a study showing that males born in the Netherlands toward the end of World War II (in the winter of 1944–1945) and exposed to famine in the first half of pregnancy became obese as adults. In contrast, those whose mothers starved later in pregnancy and in the first month after birth subsequently remained thin. It is also likely that the number of fat cells is established in late pregnancy and early infancy, and that the latter group of men were born with a permanently smaller number and size of fat cells.

In a study of the children of Pima Indian women with diabetes, a higher prevalence of obesity was found than in the offspring of nondiabetic Pima mothers, despite similar maternal weights in the two groups. Obesity can begin at any age, but when it becomes apparent in early childhood or adolescence, it tends to be lifelong. Clearly the gene machine that powers the signaling cycle for hunger, satiety, and fat storage is subject to dysregulation at many different sites. With few exceptions, the overwhelming number of those who become obese do so primarily because their caloric intake exceeds calories expended. Regardless of where dysfunction occurs in the chain of signals, the calorie intake must be controlled first, and energy expenditure through exercise must be increased. It is the dream of every pharmaceutical company to come up with a drug that can control the appetite center with no harmful side effects. And many who need to lose a few pounds also dream of finding a pill or a diet that will magically melt the weight away. But only personal resolve and lifelong discipline will make a lasting difference in achieving and maintaining a healthy body mass.

Genes and Obesity

Many genes are involved in the signaling pathways described above, but only a few have been identified thus far. The interaction of genetic and environmental factors is currently believed to be the basis for most cases of obesity. Common experience confirms that common cultural, environmental, and dietary factors operating within families come into play along with genetic programming. In combined studies of 2,002 obese children, one or both parents were obese in 72 percent of cases. Mothers were more often obese than fathers. The risk of extreme obesity is about 6 to 8 times higher in families of extremely obese individuals. Moreover, first-degree relatives of individuals who had childhood-onset obesity were twice as likely to be obese as were relatives of those who developed obesity as adults. In some 30 percent of families, both parents of obese children are obese. Up to 5 years of age, having two obese parents gives a child a 10 times greater risk of becoming obese as an adult. If the child is obese between 10 and 17 years of age, regardless of parental obesity, there is a 20-fold increased risk of that child's becoming an obese adult. The familial risk of obesity increases with the degree of obesity. Studies of adopted children in Denmark and of twins who were reared apart in Sweden showed that adopted children were more closely related to their biological parents than to their adopted parents, as far as measures of their body weight chemistry were concerned. Identical twins have similar weights when reared apart, with a concordance rate of about 70 percent for men and 66 percent for women. In all of these studies, the degree of heritability of BMI has ranged between 50 and 70 percent.

Genetic Disorders with Obesity

There are well-known but thankfully relatively rare syndromes of obesity, some already obvious through overgrowth at birth and ultimately associated with mental retardation and other features. Technical descriptions of the approximately 30 single-gene syndromes would not be appropriate here. However, if within your family there is an individual with obesity and mental retardation, with or without other abnormal features, consultation with a clinical geneticist would be appropriate—as borne out by the case history on page 199.

Another frequently encountered disorder, known as *polycystic ovary syndrome,* is characterized by multiple ovarian cysts, causing obesity, hairiness, infertility, and failure to ovulate. The cause of this disorder is uncertain. Dominant inheritance (see Chapter 7), with very variable signs even in the same family, is common. Other modes of inheritance and similar conditions are known that require specialist consultation in order to reach a precise diagnosis.

Risks of Obesity

The consequences of obesity are serious medical problems; frequently, psychological difficulties; and invariably, physical and behavioral limitations. Obesity is second only to smoking as the leading cause of preventable death in the United States. For extremely obese men the risk of dying is 12 times greater between the ages of 25 and 34 years compared to men of the same age range in the general population. Increased death rates among the obese continue throughout life. Cardiovascular disease, hypertension, diabetes, lung problems, and gallstones all occur more frequently among the obese. For men there is also an increased mortality from cancer of the colon, rectum, and prostate, whereas women suffer an increased risk of cancer of the breast, uterus, and cervix. Even moderate obesity carries with it an increased risk of hypertension, diabetes, gallstones, kidney stones, stroke, and heart disease. Osteoarthritis, varicose veins, and gout are also more frequent in the obese.

M.W. was born by cesarean section after little progress had been made in her mother's labor. At birth she was noted to be floppy, and it soon became apparent that she sucked poorly and failed to gain weight. For a while tube feeding was necessary. Her mother noted her weak cry and generally lethargic state compared to her previous child. Her motor development was slow in that she was unable to sit up unsupported until the age of 12 months and did not begin to walk until 2 years.

At about 2 years of age, excessive appetite became apparent and huge weight gains were noted, with resulting obesity. This pattern of excessive eating continued into adulthood. Schooling was a source of many problems, and M.W. attended a special class in a regular school. Her menses began at age 14, but were extremely light in volume and short in duration. Her photographs showed almond-shaped eyes, a narrow bridge to the nose, a downturned mouth, and a thin upper lip. Her hands were small and the fingers tapered and short. Her shoulders sloped, and the accumulation of fat was most obvious over her abdomen, buttocks, and thighs. She had fair skin, hair, and eye color, compared to other members of her family.

Her motor and language development were markedly delayed and she did poorly in school. Even after graduation she had slurred speech, although she was able to converse normally. Her IQ was 70. Her behavioral profile included temper tantrums, problems in behavior control, manipulative behavior, obsessive-compulsive characteristics, lying, stealing, aggressive behavior, sleep disturbances, and a marked tendency to pick at her skin. She exhibited a high pain threshold and had an unusual skill with jigsaw puzzles.

She had been seen by an endless array of physicians throughout her childhood without any definitive diagnosis. When she was 29, her supervisor at her sheltered employment location sent her home with a newspaper cutting for her mother. The newspaper carried a photograph of a woman who closely resembled M.W. The story was about *Prader-Willi syndrome* (see Chapter 7 for more about this disorder). M.W.'s mother took the photograph and her 29-year-old daughter to a clinical geneticist, who made the diagnosis.

Obesity also commonly has psychological consequences, reflected in obese individuals' feelings about themselves as well as the reactions of others. Negative self-images are commonly reported, some to the point of self-hatred. Because of the way they feel about themselves, the very obese often experience anxiety, depression, hostility, or guilt. Behavioral consequences of obesity are usually due to the effects of excessive weight, and include limited ability to exercise, shortness of breath, pain in the joints, swelling of the legs, and easy fatigue, which in many cases lead the obese person to withdraw from social contact. There is also evidence that excessively obese individuals may lose out on employment or be rejected in situations that require interviews.

Maternal Obesity and Reproduction

The consequences of obesity on women's reproductive system are many. Obesity is present about 4 times more often in women with menstrual disturbances than in those with normal cycles. Failure to ovulate is strongly associated with obesity. Teenage obesity is frequently followed years later by abnormal menstrual cycles and excessive facial and body hair. Ample published data show higher rates of infertility among obese women as well as an increased risk of miscarriage. Obese women who become pregnant have increased risks of hypertension, toxemia of pregnancy, gestational diabetes, and urinary tract infection, and more often need cesarean section.

Maternal Obesity and Birth Defects

In a study of potential prenatal influences on pregnancy outcome, one question that I and my research colleagues sought to answer concerned the association of birth defects with obesity and diabetes. This prospective study of about 24,000 women, at Boston University School of Medicine and which was supported by the National Institutes of Health, found no increase in the frequency of major birth defects among the offspring of obese women. Nonobese women with preexisting or gestational diabetes also had no excess risk overall. However, women who were both obese and diabetic were three times more likely to have children with major birth defects than were women who were not obese and not diabetic. Obesity and diabetes in combination clearly influence the development of major birth defects.

A clearer insight into the relation between maternal weight and pregnancy outcome emerged from an analysis of 56,857 children in the National Institutes of Health Collaborative Perinatal Study. Fetal and newborn deaths around the time of delivery (perinatal) increased from 3.7 percent in thin mothers to 12.1 percent in obese mothers. Most of these losses were due to preterm deliveries, but other contributing factors included diabetes, maternal age, twins, and birth defects. Maternal obesity is also associated with a higher cesarean section rate as well as other major maternal and fetal complications.

Treatment and Prevention

The treatment of obesity has to a large extent been a dismal failure. Some diets actually work—but only for a short time. Most who embark on a program of weight loss are almost certainly going to be back at their previous weight or even heavier within five years. Regardless of the special diets, behavior modification therapy, and various appetite-suppressant drugs, successful treatment of obesity has proved illusive.

What can work over the long haul is a strict limitation of calorie intake together with a change in the dietary habits of the whole family. Basic to that effort is a drastic reduction in fat intake, which is also good for the entire family. In addition, a persistent alternate-day aerobic exercise of at least 30 minutes with additional exercise on the intervening days is recommended. This approach is not a temporary program but a permanent lifestyle change.

Appetite-suppressant drugs are likely to burgeon in number following the completion of the Human Genome Project. Once genes involved in the "obesity pathway" have been identified, genetically based drugs that interrupt the pathway and still the appetite will undoubtedly emerge. Meanwhile, the morbidly obese may opt for a surgical bypass connecting the upper intestine to the lower small intestine, or a different surgical procedure that reduces stomach volume by stapling. Complications resulting from these procedures are common and can be serious or even fatal. That is why the procedures are offered only to the morbidly obese whose lives are already at serious risk from other complications.

Given the intractable nature of obesity and the dismal record of failure in its treatment, avoidance and prevention are critical. The development of a national policy aimed at the strategic prevention of obesity is long overdue. A public health initiative similar to the one that resulted in the decline of tobacco use is required. Fundamental to any such program would be a plan to increase general awareness of the many health consequences of obesity. The educational elements of such a policy should focus on effective, lifelong eating habits and exercise regimens from early childhood on. We already know that about 60 percent of overweight children between the ages of 5 and 10 years have at least one associated biochemical or clinical cardiac risk factor, such as high blood fat levels, high blood pressure, or raised levels of insulin. Indeed, 25 percent have two or more of these risk factors. By adulthood, these risk factors have been translated into chronic diseases. Close to 80 percent of the obese suffer from diabetes, high cholesterol levels, high blood pressure, gallbladder disease, osteoarthritis, and coronary heart disease. In fact, 40 percent of obese adults have two or more of these problems.

Major policy initiatives aimed at educating the public on the benefits of moderate, regular exercise are necessary, since it is clear that physical activity may prevent obesity, lessen the effect of the complications mentioned, reduce the death rate, and have other good effects on health generally.

Since we know that 8 out of 10 fat children become fat adults, we should all be ashamed that we allow our school systems to serve such high-fat offerings in their cafeterias. Equally disturbing is the move by an increasing number of school systems not to support physical education programs. We are in the midst of a worldwide epidemic that needs urgent attention. We can be thankful that genes alone need not determine the final outcome.

chapter 16

Genes and Cancer

·······························

All cancers are genetic. This is not to say that all cancers are inherited but that every cancer involves the genetic machinery of a cell; only when a sperm or egg is involved could the cancer gene be transmitted to a future generation. Fortunately, only about 10 percent of all cancers are truly inherited. The focus of this chapter is to primarily consider the hereditary types of cancer and how things go awry in the genetic machinery of the cell.

Many people have a pessimistic and fatalistic view about inherited cancer, convinced that they can do nothing about their own risk. Nothing could be further from the truth. Although predisposed individuals may indeed have a high risk of developing a specific type of inherited cancer, this knowledge should serve to stimulate initiation of detailed surveillance plans that result in early detection, preemptive treatment, and likely cure. Certainly, knowledge of specific cancers in the family should enable those who know their genes to save their lives. First, however, we should discuss some basic concepts, causes, and mechanisms of cancer.

What Is Cancer?

Cancer arises mostly from a single cell that begins to grow abnormally, invading the surrounding tissues and spreading through the bloodstream and lymphatic system to other parts of the body. Such cancers are deemed *malignant*. Tumors that grow but do not invade surrounding tissue or spread to other parts of the body are called *benign* and are easily removed. The key feature of a malignant tumor is not only the abnormal growth of certain cells but also a defect in the body mechanisms that normally set the territorial limits of cells and keep the abnormally growing cells from invading other tissues or spreading to other parts of the body.

How Widespread Is It?

The American Cancer Society expected that 1,220,100 new cancer cases would be diagnosed in 1999 in the United States alone. Since 1990, about 13 million new cancer cases have been diagnosed, not including localized, noninvasive cancer of the cervix or of the skin (basal cell and squamous cell cancers). An estimated 1,300,000 cases of basal and squamous cell skin cancers were expected to be diagnosed in 2000. The American Cancer Society also projected that about 552,200 Americans would die of cancer in 2000—more than 1,500 people a day. Although heart disease is the leading cause of death in the United States, cancer is next. The costs of cancer, including health care and loss of productivity due to illness and premature death, exceeds a staggering $107 billion annually.

Virtually all cancers caused by tobacco and alcohol use could be avoided. In 2000, the American Cancer Society estimated that about 171,000 cancer deaths would occur due to tobacco use, and an additional 19,000 due to excessive alcohol use. Available evidence indicated that up to one-third of the 552,200 cancer deaths expected in 2000 would be related to nutritional factors and thus could also theoretically have been prevented. Moreover, many of the more than 1.3 million skin cancers anticipated could have been avoided by protection from the sun. With appropriate forethought and timely action, the huge environmental contribution to cancer could be dramatically reduced. If it were, then the inherited forms of cancer with higher personal risks would become even more important.

How Does a Normal Cell Become a Cancer Cell?

We all develop from a single cell formed by the fusion of an egg and a sperm. That original cell contains the genetic blueprint in the nucleus, and related elements in the fluid that surrounds the nucleus. That original cell divides in two, the two become four, four become eight, and so on, until an entire body is formed. This process of reproducing cells is known as the *cell cycle*. For each cell that doubles, it is necessary for the chromosomes (and hence genes) to replicate, for the surrounding constituents within the cell to double, and for the chromosomes to separate precisely, as they are distributed to the daughter cells. These various processes occur in phases, each governed by genes whose function it is to guide and synchronize these complex events, together with proteins and other cell elements surrounding the nucleus that further regulate cell reproduction. Hence, at the very basis of the cell cycle are genes that control cell growth, chromosome duplication, and cell division. Only some of the genes involved in these processes have been identified, but it is expected that all will be recognized early in the new millennium.

Genes at Checkpoints

Cancer, it is thought, is a disease of the cell cycle in which one or more critical defects occur, leading the cell to go awry. Available evidence indicates that more than one defect and often multiple defects (mutations) can occur in genes controlling the cell cycle—defects that may transform a normal cell into a cancer cell.

Key genes in normal cells provide fail-safe mechanisms against the accumulation of gene mutations during cell division. Inherited or acquired mutations in these checkpoint genes, however, allow the initiation of a process that leads to genetic instability and the occurrence of additional multiple mutations, which eventually result in a failure of cell growth control and the development of cancer. Key genes (called proto-oncogenes), over 100 of which have been identified so far and which are conserved throughout evolution even in the most primitive species, produce proteins that regulate normal cell growth and the specialization of cells into particular tissues. These genes maintain an orderly progression through the cycle of cell division and specialization. When a mutation occurs in the proto-oncogene, the normal cell genes (proto-oncogenes) become cancer genes (oncogenes) whose dysfunctional protein products activate cell growth beyond normal limits. Chemical carcinogens (cancer-causing agents) may cause such mutations in proto-oncogenes.

The process by which normal cells are transformed into cancer cells is extremely complex. In addition to the cell cycle genes and their protein products, cell growth control is also influenced by proteins and other genes and their products surrounding the nucleus within the cell, as well as by the structure and function of the cell wall, the blood supply, and circulating growth factors within the body, such as hormones. Environmental poisons (tobacco, chemicals, ultraviolet radiation, and so on) may directly cause gene mutations or may only tip the balance in an individual with an existing genetic predisposition to cancer. The next seven sections provide additional insights into the complex ways in which normal cells may become cancer cells.

Cells With Brakes

Our bodies are constantly assailed by environmental toxic agents (e.g., chemicals and ultraviolet rays) that can cause mutations in our genes. Fortunately, all cells have specific genes whose protein products regulate cell growth. Well over a dozen of these genes (called *tumor suppressor genes*) have been identified. Like other genes, these genes are subject to mutation. Indeed, one relatively common mechanism by which cancer may develop involves a two-hit system. Recall that each of our genes is, by and large, in duplicate—one from each parent. In the event that one of the gene pair undergoes mutation or is transmitted with a mutation, cancer may not eventuate. However, if the other gene later undergoes mutation (the second hit), the theory is that a tumor suppressor gene is knocked out and cancer results. Many important cancers occur in this way, including specific types of breast, colon, lung, kidney, ovarian, skin, stomach, and pancreatic cancers. Essentially, the cell's brakes have failed.

Gene Silencing

A different kind of mechanism also often operates to inactivate or silence a tumor suppressor gene. Such a mechanism was discovered for von Hippel-Lindau disease or the associated kidney cancer. When no gene mutation was found in about 20 percent of cases of kidney tumors resulting from von Hippel-Lindau disease, intensive studies revealed that a biochemical change in the promoter (see explanation in Chapter 6) of the tumor suppressor gene had effectively led to its being silenced or inactivated. This process is now known to be important in various sporadic cancers (those that occur without any prior family history), including cancers of the kidneys, colon, and uterine lining, among others.

Cell Suicide

About thirty years ago, the discovery was made that cells in many species "commit suicide." Subsequently we learned that programmed cell death (called *apoptosis*) plays a critical role in normal development (see Chapter 6), including formation of complex organs such as the brain. As an illustration of the process, let us consider the fetal development of the hand and fingers: In the early weeks of fetal development, the hand and fingers form as a pad. Through the process of apoptosis, cells die in the tissues holding the fingers together. This process allows each finger and toe to be unbound from its neighboring digit. Failure of apoptosis in this process leads to webbing of the fingers or toes (commonly seen between the second and third toes). This very same process of apoptosis is controlled by specific genes that when activated produce proteins that also induce a suicide response in a cell that is becoming malignant. Indeed, this process enables the body to mount an important defense against many types of viral infection. Cell suicide following viral infection may not only put an end to viral multiplication but also prevent viruses from transforming cells into potentially malignant growths. However, even the protector mechanism is subject to mutation. Mutation in genes that control apoptosis leads to a failure in this process, which allows premalignant cells to move on and become cancerous.

It was previously thought that cancer treatments including radiation and chemotherapy function by causing chemical or physical damage to cancer cells. Now we know that these therapeutic approaches mostly work by inducing apoptosis and therefore cell suicide. Moreover, it seems that failure of chemotherapy may result from a process that interferes with or inhibits programmed cell death.

DNA Repair Systems

Our bodies are constantly challenged by numerous environmental, genetic, and other agents capable of mutating genes. Fortunately for us, normal cells also have genes that constantly monitor the genetic machinery within the cell and usually bring the cell cycle to a halt if the various orchestrated events do not occur correctly. Cell multiplication then stops, and the genetic machinery shifts into repair mode as DNA repair genes come into play.

Over twenty genes controlling the repair process have been identified so far, all of which have been conserved remarkably through evolution and are present not only in humans but also in other animal species. Even though the repair system normally works to correct induced mutations, the repair genes are also subject to mutation. The consequences of this type of mutation are uncontrolled cell growth—in other words, cancer. Defects in cell cycle gene products (proteins) lead to cancer development by allowing the cell to override or bypass controls that ordinarily restrict cell growth. These very same defects in cell cycle proteins result in the cell ignoring internal alarms that signal errors in the genetic machinery. The result is genetic instability and the steady evolution into malignant growth. Fortunately these mutations in the DNA repair system cause only rare cancer syndromes (e.g., xeroderma pigmentosum). Patients with xeroderma pigmentosum are subject to mutations in the DNA repair genes due to their extraordinary sensitivity to the sun's ultraviolet rays. As a result, they experience sunlight-induced cancer at a rate greater than a thousandfold that faced by individuals who do not have such mutations in the DNA repair genes.

Means to an End

Each of our chromosomes is capped by a complex of more than a thousand short, repeated DNA sequences and DNA-binding proteins. These caps, or *telomeres* (from the Greek *telos,* meaning end, and *meros,* a component), steadily shorten due to loss of these repetitive sequences with age, being unable to achieve repair in normal cells. With erosion of their protective caps, normal cells cannot divide, and the cells eventually die. However, cancer cells are able to execute telomere repair by switching on an enzyme (telomerase) that rejuvenates cell growth without control, resulting in a cancerous growth. Therapeutic efforts are under way to block this action of telomerase, thereby arresting tumor growth.

Imprinted Genes

Although we inherit our genes in pairs (one from each parent), one of the genes in a pair may "outperform" or "underperform" its mate. This aberrant performer may operate (or fail to operate) in the egg, the sperm, or the single cell from which we all begin. This over- or underperforming gene leads to differences in the expression or function by the pair that can result not only in developmental defects and genetic disorders (see discussion in Chapter 7) but also in certain human cancers.

Imprinted genes (see Chapter 6) have been increasingly recognized, as has their paternal or maternal origin. In certain cancers (e.g., Wilms's tumor of the kidney, acute myelocytic leukemia, and neuroblastoma of the adrenal gland), parent-of-origin imprinting has been recognized. Other disorders characterized by overgrowth and cancer (e.g., Beckwith-Wiedemann syndrome—see below) are also now recognized as evolving from an imprinted, specific parental gene.

The loss of imprinting can be similarly harmful. Tumors in which genes show a loss of imprinting include those of the kidney, liver, uterus, cervix, prostate, lung, and many others. Indeed, loss of imprinting may be one of the most common alterations causing human cancer.

Broken Threads

In Chapter 5, we discussed "broken" chromosomes—the "threads of our lives." Recall that chromosomes are made up of a string of genes. When two different chromosomes break and their broken pieces trade places and reattach, the process is called translocation. This process causes different, "unfamiliar" or incompatible genes to come into contact with one another at the site of reattachment. For the most part, such translocations occur without threat to the health of the individual. However, when at the site of interchange the genes in contact fuse, a new gene fusion product may activate proto-oncogenes, resulting in a malignant disorder. Such is the case in chronic myeloid leukemia, characterized by a translocation between chromosomes 9 and 22. A therapeutic strategy to inhibit the fusion product with a pill taken by mouth has had remarkable success. Up to 78 percent of those with this form of leukemia who had failed all available treatment, when given this new inhibitor drug, promptly went into remission. Time will tell whether this remarkable response lasts and how permanent it is.

The unexpected juxtaposition of two chromosomal segments may also lead to the activation of a cancer gene by introducing a gene enhancer or promoter inappropriate to the new location. Indeed, these mechanisms of gene dysregulation cause well over half of all leukemias and a significant proportion of lymphomas.

Clearly there are many other factors that are involved in the initiation of cancer, including inherited susceptibility to infection and certain carcinogens; genetic influences on the immune system; genes whose products may contribute to the spread of cancer within the body (metastases); and genes that regulate the formation of blood vessels that feed tumors. Carriers of certain gene mutations (e.g., in ataxia telangiectasia) are thought to have an increased risk of developing several common cancers. Suffice it to say that all of these and other factors are regulated by genes and influenced by many environmental factors.

Causes of Cancer

All cancer involves the genetic machinery of cells; but about 90 percent of cases result primarily from environmental factors. For this reason, although the focus of this chapter is on the genetic basis of cancer, it is important to note the established environmental causes (see Table 16.1). It is important to keep in mind that some individuals are genetically predisposed to developing certain "environmental" cancers.

Genetic Causes

About 10 percent of all cancers are directly inherited, mostly via single gene mutations, usually from one parent but occasionally from both. Recognition of

TABLE 16.1   Examples of Established Environmental Causes of Cancer

Cancer-causing Agent	Cancers
Personal habits	
Smoking; chewing tobacco	Lung; mouth, tongue, larynx, esophagus, pancreas, bladder, kidney
Alcohol	Liver, esophagus, mouth, pharynx, larynx
Radiation	
Sun exposure	Skin; eye (melanoma)
Ionizing X-rays (therapeutic or accidental)	Most organs
Viral infection	
HIV (AIDS)	Lymphomas
Hepatitis B	Liver
Papillomavirus	Cervix
Drugs	
Synthetic estrogens	Uterine lining
Male hormone (e.g., steroids for body-building)	Liver
Therapeutic anticancer drugs	Leukemia, bladder, bone, lymphoma
Phenacetin	Bladder
Prenatal Diethylstilbestrol (DES)	Vagina

the genetic causes of cancer is very important, and predisposed individuals who know their genetic endowment can take preemptive actions leading to prevention or early diagnosis, treatment, and cure. This requires an in-depth discussion, and it is thus the subject of the next chapter.

chapter 17

Is There a History of Cancer in Your Family?

·····································

Genetic Causes

The importance of your family history to your health and the health of other family members has been emphasized throughout this book. Knowledge of cancers, mental retardation, birth defects, genetic disorders, stillbirths, and recurrent miscarriages is vital to all family members and could enable them to avoid catastrophic illness and even death. The construction of a family tree (pedigree chart) was described in Chapter 2. Every family should draw up such a chart, maintain it, and regularly update it. As you create your own similar chart, you should pay close attention to a family history of any cancer—especially to those that occurred before age 50. The earlier the onset of cancer, the more likely it is to be of genetic origin. The type of cancer and the pattern with which it occurs in the family may also yield critically important information. For example, most of the inherited cancers are due to mutations in dominant genes (see Chapter 7 for a detailed discussion of dominant, recessive, and X-linked modes of inheritance). In families showing a pattern of dominant inheritance, each child of an affected parent has a 50 percent risk of inheriting the cancer gene. However, variations occur in the proportion of affected offspring even when a dominant inheritance is present.

Further complicating interpretations of the family tree is the fact that there may be different types of cancer within the same family, but all due to the same defective gene. These important inherited cancer syndromes are discussed below. Hence, a family might have relatively young members who developed breast, ovarian, prostate, or intestinal cancer, without anyone realizing that all of these cancers resulted from a single mutation in the same breast cancer gene (BRCA-1 gene). Concerns about various cancers in the family are best addressed by a clinical geneticist. Consultations with such specialists can prove lifesaving, as exemplified in the case of P.W. (page 213).

In interpreting a family history, certain important considerations should always be borne in mind. The observation, for example, that one generation appears to have been skipped and is free of cancer may be explained in a number of ways. The individual destined to have been affected may have died before a tumor became manifest. For some cancers, there is virtually 100 percent certainty that the individual inheriting the specific gene will develop that cancer in his or her lifetime. However, there are cancers in which the *penetrance* is not absolute, despite the presence of the mutated gene. It is also possible that tiny, undetectable cancers may be present in so-called unaffected, but clearly transmitting, individuals. In addition, the same defective gene causing cancer in the family may have several different effects, including cancers in different organs and even birth defects. There are also families that exhibit the phenomenon of *anticipation,* in which the same mutated gene results in the same or a related cancer occurring progressively earlier with each successive generation.

Although inherited cancers fit patterns of dominant inheritance, a few occur as a consequence of mutations in recessive genes. Examples include ataxia telangiectasia, Fanconi anemia, and Bloom syndrome (all may lead to acute leukemias and non-Hodgkin's lymphoma); and xeroderma pigmentosum (causing skin cancer). Rarely, defective genes on the X (female) chromosome may result in lymphomas.

Prenatal Risk Factors for Cancer

Do some breast cancers originate in the womb, during early fetal development? Since this hypothesis was first advanced, evidence has surfaced linking high birth weights to subsequent increased rates of breast cancer. It remains uncertain whether estrogen hormones, susceptibility genes, or other factors are singly or jointly culpable. Similar associations have been made for cancer of the testes and the prostate, and the same array of causal factors has been invoked. Altered patterns of hormone release established in the womb may explain the increased risk of ovarian cancer among women who had a high rate of weight gain in infancy.

Cancer and Ethnic Origin

Beyond smoking (which accounts for about 30 percent of deaths due to cancer) and dietary habits (especially excess fat intake), and other environmental carcinogens, are cancer susceptibility genes. Certain mutations in cancer susceptibility genes are predominantly found in particular ethnic groups—for example,

P.W. was 21 years of age when his mother was found to have a brain tumor. The family had known that his maternal grandmother had died as a consequence of kidney cancer. P.W.'s maternal aunt had had an eye tumor, but before treatment could be initiated she had been killed in an automobile accident.

When P.W. planned to marry, his fiancée, who was a nurse, insisted that they see a clinical geneticist to determine whether there was any increased risk of cancer for P.W. or their future children.

During this consultation, I discussed the likely connection of the three cancers mentioned and described the condition called von Hippel-Lindau disease. At this first consultation, I indicated that the gene for this condition had been cloned and that analysis of the gene for mutations was now available in our laboratories. I recommended that they speak to P.W.'s mother with the aim of obtaining a blood sample from her for DNA mutation analysis. This they did, and sure enough, we detected a specific mutation in P.W.'s mother's von Hippel-Lindau gene. Simultaneously I had recommended an abdominal ultrasound study and eye consultation for P.W. The kidney ultrasound revealed a small kidney tumor, which later proved cancerous. Fortunately, immediate surgery removed the kidney cancer, which had not spread. Subsequent imaging studies of P.W.'s brain yielded normal results, as did the eye consultation.

At another consultation we arranged for a system of surveillance with particular but not exclusive focus on P.W.'s remaining kidney and his brain and eyes. Mutation analysis confirmed that P.W. had inherited the identical mutation from his mother. P.W. and his fiancée also learned that there was a 50 percent risk of transmitting this gene to each of their future offspring. Moreover, they had the option of selecting prenatal diagnosis in any future pregnancy and would be able to determine whether or not the fetus had the culprit gene. They would then have the option of terminating the pregnancy. A final option included artificial insemination by donor, should they decide to avoid the 50 percent risk.

Clearly, P.W.'s fiancée had saved his life because of her awareness of the importance of the family history and her insistence on genetic consultation.

three mutations in the two common breast cancer genes, BRCA-1 and BRCA-2. In Ashkenazic Jewish women, two specific mutations in BRCA-1 and one mutation in BRCA-2 occur frequently. Taken together, about 1 in 50 Jewish women have one of these three mutations, each of which yields an approximate 51 percent risk of breast cancer by age 50 and an 87 percent lifetime risk. The same number also have a lifetime risk of between 42 and 64 percent for ovarian cancer.

About 6 percent of Ashkenazic Jews also carry a specific susceptibility gene for colon cancer. African Americans have a prostate cancer rate that is among the highest in the world, even though this cancer is rare in Africa. Evidence indicates that an inherited predisposition may be responsible for up to 10 percent of prostate cancers. Dietary fat, however, is also considered a major direct or precipitating factor. In contrast, the dreaded skin cancer melanoma occurs some 10 times more frequently among whites than in African Americans. Although family history is important in melanoma, excessive exposure to ultraviolet light (sun) is more critical.

Cancer rates differ throughout the world. Studies of immigrants have shed light on key environmental factors in the development of cancer. For example, there is a higher rate of stomach cancer among the Japanese. However, Japanese immigrants to the United States steadily acquire the same stomach cancer rates of whites in the United States over time. Certain cancers occur more frequently in one or another nonwhite U.S. population group than in whites. Hispanic Americans and Native Americans have higher rates of gallbladder, stomach, and cervical cancers. Cancers of the throat and liver are more frequent in Chinese Americans, whereas lung, stomach, and cervical cancers are more frequent among Hawaiians and liver cancer is more frequent among Filipino Americans.

Birth Defects and Cancer

A number of well-known but mostly rare birth defects are associated with a significantly increased risk of cancer. The most common is Down syndrome (see Chapter 3), in which there is a 20-fold increased risk of leukemia. In the small subgroup of women with Turner syndrome (see Chapter 5) who not only have a missing X chromosome but who also retain a tiny fragment or more of a Y chromosome, there is a 20 to 30 times greater likelihood of a tumor arising in what would have been ovarian tissue. This realization necessitates a special study to determine the presence of any Y chromosome material in all such cases. Much rarer are cases of hemihypertrophy, in which one side of the body is larger than the other. This feature may be part of a syndrome (Beckwith-Wiedemann) in which various other features also may be present at birth, including a larger-than-normal baby; a large tongue; a crease in the lobe of each ear; a possible opening at the navel, allowing the bowel to protrude (omphalocele); enlarged liver, spleen, and kidneys; and low blood sugar. Infants with this syndrome are at risk for developing cancers of the liver and/or kidney, particularly on the larger side of the body. Imaging and blood surveillance studies are necessary for diagnosis at birth; at three-month intervals thereafter, for the first five years; and every six months between the ages of 5 and 12 years. An even more sensitive indicator of liver cancer is a blood test measuring alpha-fetoprotein, which should be done at birth and every six months thereafter, until age 6. A rapid diagnosis of this syndrome is necessary in the newborn in order to detect and treat low blood sugar and thereby avoid mental retardation.

Another disorder, Fanconi anemia, may first come to attention at birth following observation of abnormalities of the thumbs and forearms. Later, children with this disorder are at greater risk for leukemias and lymphomas. An even rarer birth defect syndrome causing deformity—usually of one arm—and absence of the large muscle over the upper chest may also be associated with an increased risk of leukemia.

Boys born with undescended testes that have to be surgically brought down into the scrotum have an increased risk of testicular cancer. This is especially the case if a biopsy of the testis was taken during the procedure. Either way, long-term annual examinations (and personal expertise) are recommended.

Cancers of the Breast, Ovary, Colon, and Prostate

The number of cancers that may occur in all organs of the body is too great to allow each one to be addressed here. However, the following discussion of genetic aspects of four common cancers should be helpful to many families, together with some additional considerations.

Breast Cancer

More than one million women worldwide are diagnosed with breast cancer every year. About 1 in 8 American women who reach the age of 85 years will develop breast cancer. The American Cancer Society estimated that over 182,800 new cases of breast cancer would be diagnosed in 2000. This number reflects the steady increase that has occurred in the rate of breast cancer over the past 50 years. Currently it is estimated that there are more than 2 million breast cancer survivors in the United States. Up to 10 percent of all breast cancers are genetic. Two genes (BRCA-1 and BRCA-2) account for 40 to 50 percent and 30 to 40 percent of inherited breast cancers, respectively. Note that a mutation in either of these genes occurs in only 1 in 800 women in the general population.

Environmental factors play a major (but not exclusive) role in the vast majority of breast cancers. In most cases, internal hormone (estrogen and progesterone) production is a root cause. This is why the rate of breast cancer increases with age and in women who experienced early onset of menstrual periods, a late age at first pregnancy, no pregnancies whatsoever, and/or a late menopause. Women having their first child before the age of 18 years have about one-third the lifetime breast cancer risk of women who have their first child after age 30. Women who give birth to three or more children also have a decreased risk. Although no clear association between dietary fat and the risk for breast cancer has emerged, obese women have an increased risk after menopause. In contrast, women who exercise at least four hours a week experience a 40 percent reduction in breast cancer risk. Exercise decreases estrogen and progesterone secretion. Estrogen taken for 15 to 20 years after menopause is known to increase the breast cancer risk by as much as 40 to 50 percent. Use of oral contraceptives for at least 10 years appears to increase the risk of breast cancer, possibly by as much as 36 percent for women under age 45. High alcohol intake (three or more drinks per day) approximately doubles the risk of breast cancer. Alcohol increases the levels of estrogen in women prior to menopause.

The family history and ethnic origin, however, weigh most heavily in assessing the risk of breast or ovarian cancer. Both of these cancers may be caused by a mutation in one of two key genes (BRCA-1 and BRCA-2). The three mutations noted earlier account for about 25 percent of early-onset breast cancer in Ashkenazic Jewish women. For any woman carrying one of these mutations, there is about a 56 percent lifetime risk of breast cancer and a 16 percent risk of ovarian cancer, based on population studies. Published lifetime risk figures as high as 87 percent and 60 percent, respectively, have been based on studies of high-risk families (e.g., those with four or more affected family members). These figures are likely overestimates of risk. Much more prospective research, especially after mutation identification, is needed before thoroughly reliable risk figures can be provided. However, we do already know that Ashkenazic Jewish women with an identified mutation in BRCA-1 or BRCA-2 have a worse prognosis than Jewish women without such a mutation. This knowledge would influence decisions about treatment options (lumpectomy; mastectomy, and whether it is unilateral or bilateral; radiation; and chemotherapy). About 20 percent of female BRCA-1 mutation carriers will develop breast cancer by age 40; 51 percent, by age 50; and 87 percent, within their lifetimes.

The risk of developing a second breast cancer in the opposite breast in BRCA-1 mutation carriers is estimated to be 60 percent. Factors that escalate the risk to the higher percentage include a family history of one or more first-degree relatives affected (mother, sister, grandmother); ovarian cancer in one or more first-degree relatives; and cancer in both breasts in a first-degree relative. Estimated risks of ovarian cancer for a woman already affected by breast cancer are about 29 percent by age 50, and 44 percent by age 70. Further research will lead to some refinement of all of the risk figures discussed but will not change their general import.

Although three particular mutations in BRCA-1 and BRCA-2 predominate in Ashkenazic Jewish women, between 10 and 40 percent of those affected who have a positive family history of breast or ovarian cancer do not have one of these three mutations. Nevertheless, given the frequency of the three common mutations, it is not rare for a woman in these families to inherit two different breast cancer gene mutations—one from each parent. The consequences seen thus far in such cases have been more severe, with onset of breast and/or ovarian cancer at an earlier age. Mutations in BRCA-1 and BRCA-2 also increase the risk of their family members' developing cancers of the prostate, pancreas, gallbladder, and stomach, as well as melanoma. Consequently, family inquiries should cover *both* maternal and paternal sides for any of these cancers. Special attention should be paid to age at onset, bilateral involvement (especially of the breasts), and the occurrence of two cancers (e.g., breast and ovary) in the same relative.

Since mutations occur in genes continuously, it is not surprising that founder mutations leading to breast cancer occur in other populations besides Ashkenazic Jews. Indeed, such founder mutations have been discovered in young women with breast cancer in Iceland and the very same mutation has been found in Finland. The founder is considered to have been a common Viking ancestor somewhere between A.D. 800 and 1050. Similar observations have been made with different mutations in Norway, Sweden, Russia, and the Netherlands.

Notwithstanding the increased risks for a woman with one of these mutations, the reality is that many such women never develop breast or ovarian cancer. The precise reason for their immunity remains obscure, but it is possible that modifying genes (perhaps from the other parent) might in some way provide a protective mechanism.

Male breast cancer occurs more frequently due to mutations in the BRCA-2 gene, which accounts for up to 40 percent of cases. Men with a BRCA-2 mutation are estimated to have an approximate 6 percent risk of developing breast cancer by 70 years of age as well as a threefold increased risk of prostate cancer and a fourfold greater risk of colon cancer—in total, a risk of about a hundredfold that of the general population. Most men who develop breast cancer have a family history of this malignancy (or a sex chromosome defect called Klinefelter syndrome; see Chapter 4 for more detail). In a Danish study, the daughters of men with breast cancer (whose mothers did not have breast cancer) had a 16 times greater risk of developing breast cancer. Such women should have routine breast cancer screenings early in adulthood and institute systematic, lifelong surveillance.

Gene Tests and Treatment Options.   Analysis of a gene cannot identify whether an individual has cancer but only the predisposition to develop cancer. The rational approach, therefore, is for gene analysis to first be conducted on a family member already known to have cancer. Only if a cancer-causing mutation is recognized is it possible to test other family members specifically for this mutation. Before proceeding with any gene analysis, it should be understood that a negative result can have at least four possible explanations. First, the result is truly negative and the individual at risk did not inherit the known family mutation. Second, another (different) cancer gene is involved. Third, the laboratory method used was unable to detect the family mutation. Fourth, an error was made in collecting or analyzing the blood sample.

The decision to have genetic testing for cancer is an intensely personal one. Before seeking such tests, you should be aware of the possible implications for yourself and your family. The specific aim of gene testing for cancer is to avoid death through early detection and/or elective, preemptive surgery; but reproductive plans also may be influenced by the results of testing, if the individuals involved discover that they carry a cancer-causing mutation. Such persons may possibly be unwilling to take the risk of transmitting the harmful gene to future offspring.

Who Needs Testing?   A genetic counselor can advise, based on your risks, whether an analysis of the three specific Ashkenazic mutations or a DNA-sequence study of the BRCA-1 and BRCA-2 genes is indicated. You should consider testing if (1) you have/had breast or ovarian cancer (primarily for the sake of your children); (2) you have a first-degree relative who has tested positive for a specific mutation; (3) you have a family history of at least two first- or second-degree relatives with breast or ovarian cancer occurring at a younger age than 50 years. Multiple other permutations of affected relatives would be discussed by a genetic counselor, who could also make you aware of other possible scenarios in which testing would be recommended, as well as of the limitations of such analyses. Informed consent is needed before testing, and the testing of children is proscribed.

Recognition of a specific mutation during laboratory analysis might result in the diagnosis of an inherited cancer syndrome in your family. Such a finding should prompt immediate testing of other family members at risk. Even an apparently sporadic cancer (that is, a cancer occurring "out of the blue," without any prior family history) may necessitate a gene analysis. This would especially be the case if the cancer occurred at a young age or in the presence of other features suggesting an inherited syndrome. In these instances, it is possible that a new mutation occurred in this family member for the very first time. Such a finding for an individual with a dominant gene mutation would mean that his or her children would have a 50 percent risk of inheriting the disorder in question. Gene tests are also important when trying to determine the exact cause of a family cancer syndrome in which a number of different genes are implicated. In some instances, more than one gene may need analysis, as is the case for BRCA-1 and BRCA-2 in families with breast and ovarian cancer. Absent any family history of cancer, cancer gene testing is discouraged. Testing may be considered by some on the basis of ethnicity (e.g., Ashkenazic Jewish women); but until more risk information is available, it is not generally recommended. The primary goal is to recognize high risk and then to consider a DNA test, but only after genetic counseling.

How to Proceed. The decision to seek cancer gene testing should be made only after careful consideration and genetic counseling with a clinical geneticist. This consultation will assist you in identifying any appropriate tests that are currently available, as well as in answering the following key questions that may affect your decision on testing:

1. Whether the mode of inheritance is dominant or recessive.
2. Whether the diagnosis in question is accurate.
3. Whether more than one gene is causing the cancer.
4. Whether or not an accurate test result can be obtained.
5. How likely it is that a cancer will develop.
6. What alternative surveillance tests are available.
7. How efficacious the surveillance tests are.
8. Where such tests are best done.
9. What psychological implications will have to be addressed.
10. What reproductive consequences may follow.
11. Whether prenatal diagnosis of a specific mutation is possible in case of future pregnancy.
12. Which other family members, including children, might need testing.

13. What the implications might be for life and health insurance planning.
14. What costs are involved.
15. What effects this decision might have on a partner.
16. What the options are for preemptive treatment.
17. How successful such preventive interventions have been.
18. Which and how many other family members would need simultaneous testing if direct mutation analysis is not possible and only gene linkage studies can be done.
19. How private and confidential the results will be.
20. Who will provide psychological support, if needed.

If you do opt for gene testing, you also will want to schedule a follow-up consultation with a clinical geneticist. The veritable explosion of knowledge in human genetics over the past two decades has made it practically impossible for family physicians to be fully informed about the latest advances in gene testing. Moreover, given the increasing complexities of DNA analysis, most practitioners are not well prepared to interpret the results of such laboratory tests. The importance of seeing a clinical geneticist cannot be overemphasized.

Psychological Considerations.   All of us would experience some anxiety upon realizing that we have an inherited cancer risk. Knowledge that a specific cancer gene test is available would automatically spawn thoughts about how an unfavorable result might affect our lives. Studies of people at risk for cancer show that many anticipate the depression they will feel if they receive news that they carry a specific cancer gene mutation. Such information has resulted in a number of individuals actually becoming less careful of their health and less attentive to medically indicated treatment or surveillance. On the positive side, however, most persons found to have a cancer gene mutation have felt much better after testing and the removal of uncertainty. With a positive attitude, they have instituted careful surveillance or undertaken prophylactic surgery. Obviously, those receiving the good news that they do not have the suspected gene mutation invariably experience relief, not only for their own sakes but also for their children's. Yet as many as 1 in 4 persons testing negative for a colon cancer gene at the same time that a close family member was affected by the gene were found in one study to experience "survivor guilt," and about half continued to worry about cancer. Such experiences may reflect the fact that most individuals in these circumstances have not seen and been counseled by a clinical geneticist. Predictive genetic testing is explored further in Chapter 24.

Difficult Decisions. The revelation that a person carries a breast cancer gene mutation should be made only when that person is under the care of a team of physicians. Optimally this would include the family doctor, a clinical geneticist, and a surgeon (certainly, a general surgeon; and possibly a plastic surgeon as well). A woman's choice to surgically remove both breasts as a preventive maneuver is extraordinarily difficult and very personal. Such decisions should not be made in haste but only after extended discussions with the physicians involved. Many would also recommend consultation with a mental health professional. A thorough exploration of the various options and recommendations for carriers of known breast cancer gene mutations is necessary and is summarized in Table 17.1.

Considerable reassurance about the potentially successful avoidance of breast cancer by obtaining bilateral mastectomy comes from a major Mayo Clinic study of 639 women with a family history of breast cancer who underwent the procedure. This important study revealed that the frequency of breast cancer among these women was 90 percent lower than was statistically predicted had they not undergone the radical surgical procedure. (Note that even total mastectomy inevitably leaves a tiny amount of breast tissue attached to the chest wall, in which cancer could develop.)

The breast cancer gene carrier faces many difficult, complicated decisions. For example, there is some evidence that long-term oral contraceptive use may increase breast cancer risk in gene mutation carriers. Therefore, oral contraceptives are not recommended for these carriers. However, there is also a decreased ovarian cancer risk from using oral contraceptives in BRCA-1 and BRCA-2 mutation carriers. Anti-estrogen medications (such as Tamoxifen) have become important in preventing breast cancer. A 49 percent decrease in breast cancer risk has been achieved by taking Tamoxifen, which incidentally also lowers cholesterol and low-density lipoproteins (see Chapter 11), helping prevent coronary artery disease. Women with a mutation in the BRCA-1 or BRCA-2 gene who already have cancer in one breast have a very high risk of developing malignancy in the other breast. Tamoxifen taken for two to four years reduces their risk by 75 percent. Newer anti-estrogens such as Raloxifene may also offer estrogen-like protection of the heart and diminish the likelihood of osteoporosis in women at the same time as they decrease breast cancer risk.

TABLE 17.1　Guidelines For BRCA–1 Mutation Carriers for Avoidance and Early Detection of Cancer

Recommended Action	Timing
Breast-self-examination	Monthly from age 18
Breast examination by your physician	Twice yearly from age 25
Mammogram	Annually from age 25
Internal (pelvic) examination of ovaries	Twice yearly from age 25
Blood test for CA–125 (for ovarian cancer)	Twice yearly from age 25
Ultrasound via vagina (color Doppler)	Biannual from age 25
Prophylactic bilateral mastectomy and reconstruction	Optional
Prophylactic removal of both ovaries	Optional. Consult cancer specialist
Visual examination of the colon (colonoscopy)	Every 3 to 5 years, starting at age 50
Men: – Rectal examination	Annually starting at age 50
– Blood test for prostatic-specific antigen	Annually starting at age 50
– Rectal ultrasound	Depends on physical exam and blood test results

Ovarian Cancer

Ovarian cancer is much less common than breast cancer but much more insidious in its presentation. About 1 in 70 women will be diagnosed with this cancer during their lifetimes. Over 23,000 cases occur each year in the United States. A decreased rate of ovarian cancer is associated with factors that prevent ovulation, such as oral contraceptives, frequent periods of breast feeding, and of course pregnancy. There is a remarkable 40 to 50 percent reduction in risk among women who have used oral contraceptives for 10 to 15 years. The most important indicator of risk for ovarian cancer is a family history of this disorder.

Ovarian cancer is seen in three different settings:

1. Lone, sporadic ovarian cancer.
2. Inherited breast and ovarian cancer in the same family.
3. Ovarian cancer as part of a familial cancer syndrome such as hereditary nonpolyposis colon cancer, a syndrome that also includes cancers of the uterus, colon cancer, and other tumors.

Mutations in BRCA-1 and BRCA-2, described earlier in this chapter, occur in only a few percent of women in the first group. In the second group (breast/ovarian cancer families), a mutation in one of these two genes occurs more frequently, reportedly ranging between 10 and 80 percent of cases. Women with a BRCA-1 mutation have a 40 to 60 percent lifetime risk of developing ovarian cancer. A mutation in BRCA-2 conveys a lifetime ovarian cancer risk of 10 to 27 percent. These risks markedly exceed the population risk of 1.4 percent by age 70. Ovarian cancers in the third group appear to arise as a consequence of mutations in other genes.

Early detection of this cancer is highly desirable, but in many cases it goes unnoticed until an advanced stage. Beyond routine annual pelvic examinations, two additional tests are recommended for women at high risk. The first is ultrasound examination performed via the vagina. The other is a blood test that measures a particular tumor marker (CA-125) that reflects a protein leaking into the bloodstream from the ovarian cancer. Unfortunately, this tumor marker is nonspecific and is also found in endometriosis, pelvic infections, and even early normal pregnancy, among other conditions. One serious problem experienced in using both of these diagnostic methods together (or separately) is the unfortunate high false positive rate. This means that quite often a positive test turns out to be a false alarm. Nevertheless, in women at high risk, twice yearly testing with both modalities is recommended, beginning at age 25. Of course, BRCA-1 and BRCA-2 gene analyses are the most important options if there is a family history of breast or ovarian cancer.

Gene Tests and Treatment Options

BRCA-1 and BRCA-2 gene analyses are options for all women, with or without a family history of breast or ovarian cancer. If a mutation is detected, they have the option of prophylactic removal of their ovaries. A Canadian-U.S. collaborative study of 208 Ashkenazic Jewish women with ovarian cancer noted a BRCA-1 or BRCA-2 mutation in 28 percent of those without a family history of breast or ovarian cancer, and in 64 percent with a family history of ovarian cancer. Because of these high risks and the limited ability to detect ovarian cancer early, prophylactic removal of the ovaries was recommended in these cases. An earlier study found a 45 percent reduction in the risk of ovarian cancer following removal of the ovaries.

In a small but important study at the University of Pennsylvania School of Medicine, women found with a BRCA-1 mutation were told of the option of surgical removal of their ovaries. Some 47 had the surgery; the other 79 did not.

After ten years, a 67 percent reduction in the incidence of *breast* cancer was observed among those who had this surgery. Data from a National Cancer Institute study noted that 18 percent of women who had their ovaries removed later developed "ovarian" cancer. The explanation for both of these apparent surgical failures is either that microscopic cancer cells had already spread before surgery or that these tumors developed from adjacent tissues that had the same embryonic origin as the ovary.

Another complication accompanying the surgical removal of the ovaries is an increased subsequent risk of heart disease and osteoporosis due to the lapse in estrogen production with the induction of early menopause; yet estrogen replacement therapy can increase the risk of breast cancer. The answer to this dilemma appears to be oral contraceptives containing both estrogen and high dose progestogen, which seem to provide safer hormonal protection to women with an inherited risk of ovarian cancer. Twice-yearly consultation with an oncologist is recommended in these circumstances.

Where ovarian cancer has occurred as part of a hereditary cancer syndrome, analyses of other genes also are indicated (see the following section).

Colon Cancer

Colon cancer is the most common of the hereditary tumors that affect both men and women and is the second leading cause of cancer deaths in the United States. About 98,400 new cases are expected to be diagnosed in 2000. There is about a 1 in 25 to 1 in 50 lifetime risk of developing this malignancy. Environmental factors that increase individual risk include chronic inflammatory diseases of the bowel and high-fat, low-fiber diets.

Most colon cancers are not directly inherited; only about 3 to 5 percent of cases are thought to have a direct genetic basis. Yet genes may well be involved even in seemingly sporadic instances of colon cancer, given that relatives may have a somewhat increased susceptibility not only to colon cancer but also to other, associated tumors. Those with an affected first-degree relative over 45 years of age have about a 1 in 17 risk of colon cancer, a 1 in 10 risk if the relative is under 45, and a 1 in 6 risk if two first-degree relatives are affected. The search continues for other hereditary factors, with evidence pointing increasingly to a combination of multiple genes.

The American Society of Colon and Rectal Surgeons has made the following recommendations for colon cancer screening in first-degree relatives of patients with colorectal cancer:

1. An annual test of the stool for nonvisible (occult) blood, beginning at the age of 35 years.
2. Colonoscopy (visual examination of the colon through a flexible "telescope") every 3 to 5 years, beginning at age 40.

Inherited colon cancers, which are much less common, occur either as tumors growing in the wall of the colon or arising as polyps (lollipop-shaped growths in the inner lining of the bowel). The first, most common type is known as *hereditary nonpolyposis colon cancer syndrome* (HNPCC); the second, rarer type is called *familial adenomatous polyposis* (FAP).

Hereditary Nonpolyposis Colon Cancer (HNPCC). HNPCC (also known as Lynch syndrome) is the most common hereditary colon cancer syndrome, and it typically occurs in families with other related cancers. These include cancers of the uterus, stomach, ovary, small intestine, ureter, and kidney. The lifetime risk of colon cancer for genetic carriers of this syndrome is 70 to 80 percent, and their risk for cancer of the uterus is 30 to 60 percent. HNPCC is dominantly inherited, and the affected person has a 50 percent risk of transmitting this disorder to each of his or her offspring. Thus far, five different genes have been recognized as causing HNPCC. Diagnosis of HNPCC syndrome generally depends upon fulfillment of the following criteria:

1. Colon cancer in three or more relatives, one of whom is a first-degree relative of the other two;
2. Colon cancer involving at least two generations;
3. One or more family members in whom colon cancer has occurred before age 40, associated with the occurrence of cancers of other organs as noted above.

Given the interplay between genes and environmental factors, however, these diagnostic criteria do not describe all cases. For example, colon cancer in some cases might be associated with certain skin disorders. Fortunately, conditions that do not fit these criteria are relatively rare.

Early Detection and Treatment Options for HNPCC. Analysis of the five HNPCC genes already recognized requires complex methods, which individually may provide only incomplete answers. Continuing technical advances will undoubtedly lead to better and less costly diagnostic methods. Individuals at risk are advised to consult with clinical geneticists for the latest diagnostic develop-

TABLE 17.2   Guidelines for Colon Cancer Gene Carriers for Avoidance and Early Detection of Cancer

Hereditary Non-Polyposis Colon Cancer

Recommended Action	Timing
Visual examination of the colon (colonoscopy)	Every 1–3 years from age 20
Surgical removal of colon (colectomy)	If early cancer or polyp detected
Women: Internal (pelvic) examination	Annually from age 25
Mammogram	Annually from age 40
Ultrasound of the uterus via vagina (color Doppler)	Annually beginning between ages 18 and 25
Surgical removal of the uterus (hysterectomy) and ovaries plus tubes	Optional—consult cancer specialist

Familial Adenomatous Polyposis

Colonoscopy	Annually from age 10
Prophylactic colectomy	As soon as colon polyps detected, preceded by in-depth counseling with both a geneticist and a cancer specialist
Establish surveillance for upper intestinal	Annual checkups beginning in the teens and other possible cancers

ments. For those found to have an HNPCC gene mutation, a clear set of recommendations has emerged (see Table 17.2.). Colonoscopy screening in such cases is best begun by 20 years of age, especially if family members have been developing colon cancer at increasingly younger ages. Most oncologists now recommend surgical removal of the colon when cancer of the HNPCC type is diagnosed. The reason is the high rate of development of colon tumors at other sites. For example, in one study, 15 of 37 patients followed for seven years developed a second colon cancer. Women at risk because of their family history of HNPCC should also be screened annually for uterine cancer by means of transvaginal ultrasound, beginning between the ages of 18 and 25. Ultrasound studies of kidneys and ovaries can be achieved at the same time. Thus far, DNA analysis of the two most common genes may enable mutation recognition in up to 60 percent of affected individuals. Further technical improvements in detection of these and other gene mutations involved in the syndrome can be anticipated.

Familial Adenomatous Polyposis.   In this dominantly inherited disorder, hundreds and even thousands of polyps are seen in the colon, appearing as early as the late teens and early twenties. Progression to cancer is virtually inevitable in a lifetime. About a third of carriers exhibit cancer by age 37, and the vast majority, by age 50. There is also a much greater risk of cancer occurring higher in the intestine (e.g., the duodenum). Curiously, a valuable diagnostic sign is seen in the back of the eye (called congenital hypertrophy of the retinal pigment epithelium). Bone, skin, and other tumors may also occur in affected individuals, requiring additional surveillance.

With FAP, there is a 50 percent risk that an affected person will pass the gene to each of his or her children. Failure to pay attention to the family history may result in individuals having children prior to their own diagnosis being made. However, up to a third of those diagnosed with FAP may have a new mutation in the culprit gene. One particular mutation in the FAP gene actually causes colon cancer of the nonpolyposis type in about 6 percent of Ashkenazic Jews. Fortunately, in this group only 9 to 10 percent of carriers will develop colorectal cancer.

Early Detection and Treatment Options for FAP.   If a mutation causing FAP is recognized through gene analysis, immediate surveillance and treatment are recommended, with the particulars dictated by the individual's age (see Table 17.2 and the section on "Repair and Reconstruction" in Chapter 27). Even young children should be tested for the FAP gene. Those found to have the mutated gene should undergo annual colonoscopies, beginning at age 10. When polyps are detected, which usually occurs in the teens, serious consideration of prophylactic removal of the colon is required. Simultaneously, screening surveillance must be initiated for tumors in the upper intestinal tract.

Prostate Cancer

The frequency of prostate cancer in men rivals that of breast cancer in women. An estimated 180,400 new cases were expected to be diagnosed in the United States in 2000. This is typically a disease of older men, and the average man has about a 10 percent lifetime risk of developing clinically obvious prostate cancer. Prostate cancer is remarkably silent and is frequently discovered incidentally. Major autopsy studies have reported that 30 to 40 percent of men older than 50 who died of other causes were found to have prostate cancer. The mechanism by which such cancers are curtailed in their growth and kept prisoner within the prostate gland remains to be discovered. This disease has its highest incidence among African Americans (twice as high as among American whites), occurs infrequently in Scandinavian countries, and is very uncommon in Asia. Clearly, environmental factors play a major role in the evolution of this cancer. Other than age and race, family history is the other recognized significant risk factor. This is especially the case if a brother or close family relative(s) has developed prostate cancer at an early age.

Up to 10 percent of prostate cancers may occur as a consequence of an inherited dominant gene mutation. A male with such a mutation has a 50 percent risk of transmitting the gene to each of his offspring. It is anticipated that the specific gene or genes will soon be cloned. The vast majority of prostate cancers are due to susceptibility genes (some already identified) interacting with environmental factors such as obesity, red meat, fat, and certain toxins (e.g., cadmium). Meanwhile, particular attention is necessary for families in which there may be both breast and prostate cancer. Prostate cancer also may rarely occur as part of inherited syndromes associated with various other cancers (see "Family Cancer Syndromes," below).

Early Detection and Treatment Options for Prostate Cancer.   Annual physical examinations and blood screening tests for prostate-specific antigen are the basic tools for timely detection. The blood test, however, will miss at least 25 percent of prostate cancers. Any suspicion aroused by a physical examination or a blood test should lead to an ultrasound study via the rectum, which may include a needle biopsy at the same time. For those diagnosed with prostate cancer, good therapeutic options are available, including curative treatment by total removal of the prostate, with or without radiation therapy. For those with increased genetic risks, the prophylactic use of female hormones could be valuable. Prophylactic removal of the prostate is yet another choice. A number of drugs are currently in clinical trials to determine their chemopreventive ability.

Family Cancer Syndromes

A particular cancer can either originate spontaneously or be directly inherited via a mutated gene. For most cancers, both roots of origin exist. This section, however, focuses on the very uncommon or rare families in which a specific gene causes either a single cancer type or is the basis for the development of various cancers within the same family, all due to the same gene. I cannot describe all of these many cancers in detail here, but I have summarized the most important ones in Table 17.3. For each type, a consultation with a clinical geneticist is recommended, and where treatment questions exist, an appointment with a cancer specialist should be sought. On occasion, for some disorders like the Li-Fraumeni syndrome, in which a large number of various cancers is possible in the same family, a child may develop a particular cancer even before a parent has been diagnosed. All of the cancers in the table, when inherited directly, are dominant. Hence, the affected individual has a 50 percent risk of transmitting the mutated gene to each of his or her offspring. The table indicates whether a gene test is available in each case. Such a test could be used to assist in diagnosis and would also be available for predictive diagnosis in family members at risk. However, a clinical geneticist should be consulted prior to any predictive testing, given the potentially serious medical, ethical, and legal implications of positive test results. The key guiding principle for deciding whether or not to have a predictive test should be the lifesaving opportunity that this knowledge would facilitate. Even then, strong psychological support would be important both before and after such testing.

Further details about some of the syndromes listed in Table 17.3 follow below, in the same order in which they are presented in the table, or may be found elsewhere in the book.

Basal cell cancers are the most common of all cancers, with upward of three-quarters of a million new cases being reported in the United States each year. For the most part, these cancers occur spontaneously in areas of skin that have been exposed to the sun. This common condition is not directly inherited. However, in a very few cases, these cancers are inherited, occur soon after puberty, and are eventually associated with cysts in the jaw, bony abnormalities, deposits of calcium within the skull, and characteristic "pits" in the skin of the palms and soles. Various birth defects may also be present, further pointing to this diagnosis. The causal gene, conserved through evolution, is also found in the fruit fly. Affected individuals are extremely susceptible to developing skin cancer from radiation therapy.

The *Beckwith-Wiedemann syndrome* was discussed earlier (see "Birth Defects and Cancer"), and the associated cancer risks outlined.

Between 10 to 15 percent of individuals with *melanoma* have a positive family history of this disease. At least three different genes have been implicated, but thus far analysis for only one is available. In multiple endocrine neoplasia, critical attention is necessary, especially for Types 2A and 2B. If gene analysis reveals

TABLE 17.3 Family Cancer Syndromes and Steps to Prevention and Early Detection

Disorder	Cancers	Prevention and Early Detection	Gene to Be Analyzed
Basal cell nevus Syndrome of Gorlin	Skin, brain	• Know your family history • Avoid excess sun • Consult a clinical geneticist • Avoid radiation therapy	NBCC
Beckwith-Weidemann syndrome	Liver, kidney, adrenal	• Consult a clinical geneticist • Diagnose low blood sugar promptly • Monitor liver and kidney by imaging and blood studies • Body asymmetry from birth is an early clue, as are other birth defects	Not yet available
Breast/ovarian cancer (see text and table 17.1)			
Familial adenomatous polyposis of the colon (see text and table 17.2)			
Hereditary nonpolyposis colon cancer (see text and table 17.2)			
Li-Fraumeni syndrome	Brain, bone, and fibrous tissues, adrenal, leukemia, lymphoma, intestinal, breast, lung, prostate, cervix, melanoma	• Know your family history • Early surveillance • Early detection and treatment	
Melanoma	Skin and eye, rarely pancreas	• Know your family history • Avoid blistering sunburn • Report any change in a mole • Limit sun exposure in childhood • Blondes and redheads at greater risk • Early detection enables cure	MLM

Table 17.3 (continued)

Disorder	Cancers	Prevention and Early Detection	Gene to Be Analyzed
Multiple endocrine neoplasia			
Type I	Parathyroids, pancreas, pituitary, thymus, stomach, facial tumors, lipomas	• Annual skin exam from age 10 • Observe skin closely during pregnancy • Sunscreen is not sufficiently reliable	MEN–1
Type 2A	Thyroid, adrenal, parathyroid	• Know your family history • Gene analysis by age 5 • Early diagnosis and surgery • Gene analysis by age 1 year • Congenital bowel obstruction may occur (Hirschsprung's disease) • Surgical removal of thyroid gland by age 2 • Early diagnosis facilitates preemptive surgery and is lifesaving	RET
Type 2B	Mainly thyroid and adrenal	• Gene analysis in the first 3 months of life • Watch for small growths on lips and tongue and in mouth and conjunctiva • Thin build, long arms and fingers • Surgical removal of thyroid gland during first year of life • Early diagnosis facilitates preemptive surgery and is lifesaving	RET
Neurofibromatosis			
Type I	Brain, nerves, sarcomas, adrenal, leukemia	• Know your family history • Consult your doctor if a noticeable change occurs in a skin "lump"	NFI

(continues)

TABLE 17.3 (continued)

Disorder	Cancers	Prevention and Early Detection	Gene to Be Analyzed
Neurofibromatosis			
Type 2	Inner-ear tumors, brain, nerves, spinal cord	• Know your family history • Gene analysis by age 10 • Early diagnosis by imaging (MRI)	NF2
Paraganglioma (Chemodectoma or glomus tumor)	Autonomic nervous system	• Know your family history • Consult a clinical geneticist • Initiate early imaging surveillance • Predictive gene analysis can be lifesaving	SDHD
Retinoblastoma	Eye, bone, brain, skin, breast, leukemia, fat (lipoma)	• Know your family history • Eye exam under anesthesia (see text) • Consult a clinical geneticist	RB1
Tuberous sclerosis	Brain, kidney, heart, eye	• Know your family history • Consult a clinical geneticist • Note many other features (see text) • Initiate surveillance program	TS1 and TS2
von Hippel-Lindau disease	Kidney, brain, eye, spinal, adrenal, pancreas, others	• Know your family history • Early gene analysis • Consult a clinical geneticist • Establish solid surveillance program (see text)	VHL
Wilms's tumor	Kidney	• Consult a clinical geneticist • Other birth defects may be an early clue • Early diagnosis enables a high cure rate	WT1

the diagnosis, prophylactic removal of the thyroid gland is regarded as the primary preventive treatment. Prophylactic surgery may be performed as early as one year of age for Type 2B, since thyroid cancers have occurred that early, and at 2 years of age for Type 2A. Although neurofibromatosis Type 1 is common (about 1 in 3,000), the risk of cancer developing is less than 10 percent. A lump in the skin (there may be many, but they are usually benign, nerve-derived tissue) or inside the body may rarely become malignant. Any sudden change in the size and appearance of such a lump warrants consultation with a physician. For neurofibromatosis Type 2, early diagnosis by magnetic resonance imaging (MRI) of the inner ear in those at risk would facilitate timely surgery.

Paraganglioma tumors occur mostly in the neck, along the course of nerves in the autonomic nervous system. About half occur on both sides of the neck and about 10 percent are directly inherited. A striking feature of the mode of inheritance is that children of affected fathers but not of affected mothers develop these tumors. This occurs through a process of gene imprinting, as discussed earlier (see also the end of Chapter 6). Gene analysis makes early and predictive diagnosis possible, which turns out to be extremely important, since late diagnosis causes considerable morbidity as well as other forms of suffering and aggravation from complications, even in the face of a benign tumor. Of course, awareness of the family history, and timely consultation, will make early surgical treatment possible and save the life of the 10 percent who have a malignant tumor.

Retinoblastoma is the most common eye cancer in children. These cancers are inherited in about one-third of cases, and the remainder occur spontaneously. Remember, however, that in the fullness of time, even in the latter group of cases, the affected children will have a 50 percent risk of transmitting this cancer to each of their offspring. When tumors occur in both eyes, the retinoblastoma is invariably genetic. Late diagnosis may result in the removal of the child's eye or eyes and possibly even in death. Early diagnosis is critical, and prompt treatment may preserve the eyes and the quality of vision. Parents are cautioned that any child whose eyes are still "crossed" by the age of three months should have a consultation with an eye specialist. Children at risk because of their family history (for example, if their mother or father was affected as a child) require eye examinations at birth and monthly for three months thereafter, and then at three-month intervals until age 2, and four-month intervals until age 3—all under anesthesia. Thereafter, adequate examination is probably achievable without anesthesia every six months until age 5, and every twelve months until age 11. Among those found to have a mutation in the retinoblastoma gene, about 1 in 4 are likely to develop a second cancer.

In *tuberous sclerosis,* about 50 percent of cases are directly inherited, the remainder being due to spontaneous mutations. Initial and annual visits to a clinical geneticist should help secure a diagnosis and establish a surveillance program including imaging of brain and kidneys. Two different genes may cause this disorder, which is manifested in variable features involving the skin and brain, including mental retardation.

In the relatively rare von Hippel-Lindau disease, cancer arising in both kidneys is the most frequent manifestation. For those at 50 percent risk, the following surveillance program is recommended: annual physical examination; eye specialist examination from age 5, and special eye studies (angiography) from age 10 until age 60; magnetic resonance imaging of the brain every three years from age 15 to 40, and then every five years until age 60; annual kidney ultrasound and abdominal CAT scans every three years; and annual twenty-four-hour urine collection for biochemical studies. Individuals who present with a kidney cancer and no family history could be carrying a new mutation, and consideration must be given to gene analysis to determine whether future surveillance is necessary and whether a reproductive risk exists. As rigorous and demanding as the surveillance program is for von Hippel-Lindau disease, quality of life—and life itself—depends upon early diagnosis and appropriate clinical genetic assessment and management.

Wilms's tumor is only occasionally directly inherited. When this kidney tumor occurs bilaterally, it is commonly associated with defects of the iris (the colored part of the eye surrounding the pupil), abnormalities of the genitals or urinary tract, and mental retardation. At least four different genes may cause Wilms's tumor, with or without other features. In children at risk, abdominal imaging studies of the kidneys are recommended frequently through the first eight years of life.

Other Cancers

Cancer may of course develop in any tissue. Some of the cancers we have discussed may occur spontaneously and not be inherited. These include cancers of the bladder, brain, stomach, lung, pancreas, and testis, and the leukemias and lymphomas. For all of these, there are families in which more than one member has been affected. In some of these families the cancers may be due either to a chromosomal disorder or to a single, segregating cancer gene.

Cancer of the cervix is second only to breast cancer in its frequency and in the associated mortality in women. As in every other cancer, chromosome abnormalities and cancer gene activation in these malignancies are invariable. Fortunately, however, cancer of the cervix is largely not an inherited disorder but rather almost exclusively one that arises from environmental factors. Indeed, some view cancer of the cervix as an infectious disease, given that more than 90 percent of these cancers contain papilloma viruses. These cancers are curable through early detection, for example, biannual examination, "PAP smears," and attention to any abnormal symptoms or signs, such as vaginal bleeding or discharge.

Table 17.3 indicates the specific gene to be tested in cases where the dominantly inherited cancer syndrome is suspected. In some cases, analysis may proceed by direct mutation detection. If the gene is too large, however, it may prove too costly and thus impractical to perform such studies, which may only be achievable in research laboratories. An alternative in such cases is DNA linkage analysis (see Chapter 6), which requires blood samples from multiple, directly related family members. Surprisingly, such studies may well be remarkably accurate, reaching 99 percent certainty. Those at risk are advised to see a clinical geneticist to discuss these options.

Cancer and Childbearing

Advances in genetics have provided new options for women and men with a cancer diagnosis (or a mutation destined to cause inherited cancer). Mindful of the loss of a parent and other close relatives from a particular cancer, and knowing that they themselves carry a cancer-causing mutation, both men and women may choose not to transmit this genetic defect to their future children. Their options are either to avoid the use of their gametes (egg or sperm), through one of the methods of assisted reproduction (e.g., artificial insemination by donor; receiving a donor egg); or to avail themselves of prenatal diagnosis, with the option of terminating a pregnancy in which the cancer mutation is found. Decisionmaking in these situations is complex and should be preceded by in-depth counseling with a clinical geneticist.

Facing chemotherapy or radiation therapy, both men and women with cancer may elect to collect sperm and eggs for deep-freeze storage. The concern has been that these powerful therapies might cause gene mutations later resulting in birth defects or cancer in their offspring. The good news is, major studies of individuals treated with both chemotherapy and radiation have shown that those who are properly protected have not appeared to experience any increased risk in birth defects in their offspring. Another concern for males is the effect of certain chemotherapeutic anticancer agents, which may cause sterility. Pregnant women who have cancer and require immediate chemotherapy treatment face difficult decisions: Exposure of the fetus in the first three to four months of pregnancy to either radiation or chemotherapy may cause serious birth defects and/or mental retardation. Chemotherapy after four months of pregnancy has not resulted in an increased frequency of such complications.

Should Children Be Tested?

Cancer gene testing of children is absolutely indicated if such tests might save their lives by enabling the diagnosis of a disorder that can be either avoided or cured by timely intervention. Among the justifiable reasons for undertaking such studies would be evidence of hereditary cancer syndromes, including familial polyps of the colon; retinoblastoma (eye cancer); multiple endocrine neoplasia type 2 (tumors of the thyroid, adrenal gland, pancreas, and associated tissues); and von Hippel-Lindau disease (tumors of brain, kidney, and eye).

Privacy and Confidentiality

All genetic information is considered highly confidential, and no health care provider is at liberty to disclose such information to family members or to anyone else, unless instructed to do so by the individual receiving medical care. However, information placed in medical records *can* be accessed by insurers or other involved third parties (e.g., employers). Consequently, issues of privacy must be considered prior to testing, until federal or state laws enshrine the privacy of this vital information (see the full discussion of legal issues in Chapter 26).

What You Should Know Before Having a Cancer Gene Test

The primary reasons for having a cancer gene test are to determine personal risk and therapeutic options, to identify the risks to your present and/or future children, and on occasion to assist very close relatives in determining their own risk. An accurate diagnosis of the cancer is necessary, as is a clear understanding of the pattern of inheritance, before the appropriate tests can be performed. The entire process may depend upon the information you can glean from autopsy reports, photographs, X-rays, and laboratory reports, as well as from medical records. Care should be taken to see a physician geneticist who will spend the necessary time (at least 45 minutes to one hour) in genetic counseling in order to cover this complex territory. Exploration of all of the issues is indispensable, including any new DNA tests and surveillance and treatment options; and one cannot reasonably expect this to be provided during routine visits to local family physicians. You also should inquire about the laboratory to be used for analysis and about its reputation. Detailed discussions are necessary about the potential accuracy of the test. For example, if mutation detection is planned, results often return with a certain detection rate—say, 98 percent. This result implies that 98 percent of the known mutations have been examined, but there is a 2 percent possibility that another mutation might be present. Prior discussion is also necessary about those disorders that could result from different genes. Some commercial laboratories provide results without interpretation, leaving it to the physician to draw the necessary conclusions. One major study revealed that in about one-third of cases, complex DNA reports had been misinterpreted by medical practitioners.

Prior to any test, you should have a clear understanding of the efficacy of surveillance if a test is positive as well as of specific treatment options and their success rates. Time should also be spent discussing the anticipation of positive or negative results, the possible consequences of psychological distress, and the possible effect on other family members. Confidentiality and privacy of results must be assured, and those tested must be provided with full assurance that their results will not be communicated to any other individual. Issues related to employment and/or insurance discrimination still remain a problem in some states that have not yet enacted protective legislation, so these matters need careful discussion. The various costs—of consultation, testing, and surveillance—need to be explored, including the extent to which they are covered by third parties.

Some very personal issues also need exploration in the genetic counseling that should precede any cancer gene testing. The decision to be tested might be made by a parent already affected by cancer. Although parents may be undertaking this test for the sake of their children (aiming to define a mutation so as to better assess risk and options), the result may unexpectedly shed light on additional personal risk—for example, of other new cancers. Since both depression and suicide attempts have followed communication of results about mutations that cause cancer, very careful consideration should be given to a personal decision to be tested. No family member should pressure an individual to have a test that has such serious personal implications. DNA linkage analysis may inadvertently reveal nonpaternity. Those being tested are best warned about this before testing. These "theoretical" possibilities occur more frequently than most people realize.

Finally, those who receive an adverse genetic test result must realize that their having a mutation does not necessarily mean they will develop cancer and is very different from their potential risk of succumbing to cancer.

Gene Therapy for Cancer

The medical treatment of cancer, despite the remarkable progress seen, has been bedeviled by devastating side effects and sometimes by the destruction not only of cancer cells but also of normal, healthy cells. Nevertheless, medical treatment, surgery, and radiation therapy have enabled major advances in the treatment of cancer. The advent of the "new genetics" has introduced an exciting era that holds great promise for the increasingly effective treatment of various cancers (see Chapter 27). Specific advances in biotechnology now allow remarkable opportunities to introduce specific genes into the body or directly into tumors. Introduction of normal genes that control the growth of cells (as discussed earlier) would literally focus on the root of the problem. Experimental success has already been seen in introducing cell cycle genes to suppress a cancer. Efforts to introduce "suicide" genes into cancer cells, leading to tumor cell death, are also promising.

In 1960, only about 40 percent of transplanted organs lasted one year, whereas today some 90 percent survive at least a year. However, long-term survival remains problematic; the longest-surviving transplants have lasted only 10 to 15 years. This long-term failure has been due to the body's rejection of the transplanted organ. Efforts are now well under way to vector in genes that aim to foil the body's rejection of transplanted organs. Other efforts are being made to manipulate the body's immune protective system; for example, immune "killer cells" are being engineered with the mission of destroying cancer cells (in kidney cancer and melanoma). DNA vaccines are also under development that could one day provide an immune attack targeted specifically on the patient's particular type of tumor. Remarkable progress also has been made in the use of molecular techniques to prevent the development of blood vessels necessary to feed the growth of tumors. In one variation, genes have been used to destroy developing blood vessels that are growing in to supply a tumor (see the gene therapy discussion in Chapter 27 for more details).

Cells harvested from human embryos aborted in a very early stage of development were successfully cultivated in 1998, opening up an era of potential miracles in medicine. These so-called stem cells possess the blueprints to grow into various types of tissues and organs. Examples include liver, kidney, spinal cord, and bone marrow. Embryo stem-cell therapies for cancer, certain birth defects, spinal cord injuries, and arthritis—to mention but a few uses—present phenomenally exciting vistas.

These and other exciting developments can be anticipated to significantly improve the treatment of cancer and to achieve more frequent cures. Notwithstanding these impressive developments in the "new genetics," however, the best guarantee of freedom from cancer is outright avoidance by pursuing healthy dietary and exercise habits, excluding known carcinogens from your environment (e.g., quitting smoking), and being fully aware of the warning signs present in your family's medical history. Such an awareness will enable you to prevent cancer, at best—or at worst, to detect it in the early stages when it is most effectively treated and cured. Knowing your genes allows you to exercise some control over your genetic destiny and could save your life as well as the lives of those you love.

chapter 18

Mental Illness in the Family
··
Schizophrenia

Is Schizophrenia Inherited?

Contemplate for a moment the tragedy of having a son or daughter develop a mental illness that is lifelong and that interferes with or totally prevents school or college attendance, a useful occupation, close friendships, marrying, and having children. Such is the frequent burden for those affected with schizophrenia and the devastating reality for their parents. Schizophrenia and related mental illnesses are known in all cultures and well recognized down through the ages.

The search for the cause(s) was for many years impeded by the lack of accurate diagnostic criteria. Even with more precise criteria, some diagnostic problems remain. This is due to the variable age of onset, the marked variation and intermittent expression of symptoms over time, and the apparent influence of precipitating environmental factors. In addition, mental illness may be mild or intermediate, and genetic studies may therefore be confounded by an inability to recognize a disorder that has yet to manifest obvious outward signs. Particularly problematic is the fact that none of the features of schizophrenia alone are absolutely diagnostic. Since there are no reliable diagnostic laboratory tests, reliance for the diagnosis is based on the individuals' medical history and on an assessment of their mental status.

The Diagnosis

An extensive array of symptoms characterizes schizophrenia. Hallucinations, delusions, bizarre behavior, emotional withdrawal, and thought disorders with incoherence (and other features) are typical. Hallucinations may vary, most individuals indicating that they "hear voices." Some hear voices that offer constant commentary; others converse with the voices they hear. Others report that they can smell particular odors or see various nonexistent scenes. Paranoid delusions and feelings of persecution are also common. Grandiose or religious delusions are not unusual, as are inappropriate delusions of guilt or sin. Behavior may be bizarre in a social, sexual, or aggressive form. Schizophrenics may also engage in uncontrolled, repetitive activity. Affected individuals may be incoherent and illogical. They may show many different types of speech difficulties, including poverty of speech, speech without content, repetitive words or sentences, or disorganized or incoherent speech.

The vast majority of schizophrenics show a lack of normal emotion, an unchanging facial expression, decreased movements and gestures, an unwillingness to make eye contact, and the tendency to speak in a monotone. Almost all fail to take care of themselves, would remain unwashed and unkempt if not helped by others, lose interest in work or school, and/or appear to have no energy or interests. Although sexual behavior in some may be bizarre, much more frequently there is little interest in sex. Intimacy and closeness are absent and relationships inevitably fail.

Among the most startling symptoms are the various delusions that schizophrenics experience. Schizophrenics with grandiose delusions will inform others about their great wealth, beauty, or special abilities (such as extrasensory perception), or they may brag about their famous friends (e.g., General Patton or Napoleon). Those with paranoid delusions complain of being persecuted, monitored, followed, or hunted. Those with religious delusions may believe that they are God's special messengers or that they are the Devil. Sexual delusions include ideas that they are prostitutes, pedophiles, or rapists. Some suffer from somatic delusions, believing that one of their vital organs, such as the heart, is no longer functioning or that their organs are rotting away. In the more extreme cases, nihilistic delusions include ideas that they are already dead or dying, that they do not exist, or that the world does not exist.

Family Studies

Efforts to determine whether schizophrenia is inherited have depended upon studies of families, twins, and adopted individuals. Almost ninety years of such studies have confirmed that schizophrenia is familial. These studies have not, however, precisely identified or defined the specific genetic elements (as opposed to environmental factors) that jointly lead to this form of psychosis.

The increased risks for first-degree relatives of schizophrenic individuals are shown in Table 18.1. The parent of a schizophrenic is affected about 5.6 percent of the time, whereas a brother or sister of a schizophrenic has about a 10.1 percent risk of developing schizophrenia. The likely concordance rate for identical twins averages 46 percent (both twins affected), as compared with 14 to 17 percent for nonidentical twins. Studies of identical twins reared apart also have shown high concordance rates for schizophrenia. The heritability of schizophrenia is especially notable in the adopted-away child of an affected mother, with the risk averaging about 17 percent. In situations where both parents have schizophrenia, the risk to a child may be as high as 46 percent. Family studies have helped identify a spectrum of related problems associated with the presence of schizophrenia in the family bloodline. Various, highly variable disorders, seen in the close relatives of schizophrenic individuals, have included schizoaffective disorder, paranoid disorder, atypical psychosis, and schizoid personality disorder. The diagnostic boundaries between any of these disorders and schizophrenia are blurred.

It is important to note that the risk figures shown in Table 18.1 refer to *lifetime risk,* assessed from birth. If by the age of 30 years no signs of schizophrenia have surfaced, then the risk is reduced by about one-half; and if by the age of 50, then the residual risk is very small indeed. However, risk increases with the number and closeness of schizophrenic relatives. Particular care is taken in the genetic counseling provided in such families, since misunderstanding of genetic information, and feelings of fear, guilt, and shame, all tend to be exacerbated. This carefulness is especially important because schizophrenia is characterized by illogical thinking and distortion of reality, and because of the potential stigma and associated problems that might be causing great emotional distress in the family.

Family studies also have helped elucidate and clarify the fact that certain environmental factors, structural brain abnormalities, and/or the age of onset may help predict familial rather than sporadic forms of schizophrenia. For example, schizophrenics with a history of obstetrical complications at birth are thought to have a lower familial risk than those who had no such complications. Schizophrenic males, especially those with some enlargement of the spaces inside the brain (ventricles), may have lesser risks than females. However, both males and females with very early onset schizophrenia likely reflect a genetic basis.

Vulnerability or susceptibility to schizophrenia is illustrated by the finding of a greater incidence of schizophrenia (by four to fifteen times) among the siblings of British-born, second-generation, African-Caribbean schizophrenics than among the siblings of their white counterparts. These observations suggest the

TABLE 18.1   The Risk of Schizophrenia Among Relatives of Schizophrenics

Relationship to Schizophrenic	Estimated Risk of Having Schizophrenia (percent)
Parent	5.6
Brother or sister	10.1
Brother or sister (one sibling and one parent affected)	17.0
Identical twin	46.0
Nonidentical twin	14.0–17.0
Child (one parent affected)	12.0–17.0
Child (both parents affected)	39.0–46.0
Uncle or aunt	2.0
Nephew or niece	2.2
First cousin	2.9
Grandchild	5.0
Nephews/nieces	4.0
Adopted child of an affected mother	17.0
Unrelated (general population)	0.9

operation of environmental factors acting on a group of individuals with genetic susceptibility.

Women tend to develop schizophrenia later than men. One major study reported an average age at onset of 21.4 years for men and 26.8 years for women. The onset of schizophrenia after the age of 35 was noted to have occurred in 17 percent of women, but in only 2 percent of men. The precise reason why sex differences occur in the age at onset remains uncertain, evidence pointing to environmental factors or early influences on brain development.

Researchers in Iceland, in a study of mathematically gifted individuals, demonstrated an increased risk of psychosis among them. What prior selection or what precise mechanism accounts for this increased frequency of psychosis among mathematicians and their first-degree relatives is unknown.

Twins and Schizophrenia

The more frequent occurrence of a disorder among identical twins compared with nonidentical twins highlights the importance of genetic factors for that disorder. Knowledge about the excess occurrence of schizophrenia in identical twins compared with nonidentical twins dates back some sixty years. A 1986 study reviewed concordance rates (both twins affected) in 817 identical and 1,016 same-sex, nonidentical twin pairs. The concordance rates for identical and for nonidentical twins in that study were 59.2 percent and 15.2 percent, respectively. On average, about 46 percent of identical twin pairs show concordance for schizophrenia. Explanations for the other twin not being affected point directly to an environmental factor or factors functioning in the presence of genetic identity. Our understanding is further complicated when one considers identical twins, one of whom is unaffected: The offspring of the unaffected twin have an increased risk for developing schizophrenia equal to the increased risk borne by the offspring of the affected twin. The clear implication is that the apparently normal twin does in fact carry and transmit the schizophrenia susceptibility genes. It is these susceptibility genes that interact with heretofore unrecognized environmental factors, leading to brain alterations that cause schizophrenia.

A few key observations signal the involvement of extraneous factors. Fingerprint patterns, which are established in early pregnancy (at fourteen to sixteen weeks after conception), may be different between identical twins when only one is affected. The affected twin in a discordant identical pair more often shows abnormalities on the left side of the brain (hypodensity), as demonstrated by brain imaging; a reduced area of one portion of the brain (hippocampus); and diminished function of one area of the cortex.

Adoption and Schizophrenia

The rates of occurrence of schizophrenia in adoptees, their adopting families, and their biological parents have clearly implicated a key role of genes in causation. The evidence is compelling. A greater risk for schizophrenia was noted among adoptees who were offspring of schizophrenic mothers and who had been separated from their mothers at birth than among those adopted from unaffected mothers. Similarly, lesser rates of psychosis were found among adoptees who were the offspring of healthy biological parents, whether or not these children were reared by adoptive parents with or without schizophrenia. The rates of schizophrenia in both adoptee groups were clearly lower than those found in adoptees from biological parents with schizophrenia. Schizophrenia and related disorders also have been reported four times more frequently among biological relatives of schizophrenic adoptees. This observed increased frequency among biological relatives of schizophrenic adoptees includes schizoid and paranoid personality disorders, suggesting gradations of schizophrenia or a variable spectrum of schizophrenic manifestations. This risk for schizophrenia and this variable spectrum of manifestations occur some ten times more frequently among the biological relatives of schizophrenic adoptees than among the biological relatives of unaffected individuals.

Early Signs

Looking back at the developmental and medical history of schizophrenics has yielded observations of signs that retrospectively enable recognition of higher-risk infants or children. For example, an increased risk of developing schizophrenia has been noted in those who had serious anxiety and attachment behavior difficulties in infancy. At two years of age these infants were noted to be more passive and less attentive, and later still had fewer social connections than their peers. A few years later, these children had difficulty making emotional connections as well as in learning and attention, and tended to be loners. About half of those children predestined to become schizophrenic were noted to be socially deviant in early adolescence. Boys tended toward aggression; girls, toward emotional hypersensitivity and excessive conformity. An important study of children at higher risk by virtue of having a schizophrenic mother recognized eccentric behavior and paranoid tendencies as characteristic of such children, rather than the expected introversion or social withdrawal. Behavior that caused school class disruption was also typically noted among those who later became schizophrenic.

Unfortunately, the sheer nonspecificity of all of these findings, coupled with uncertainties about the absolute definitions of schizophrenia used in some of these studies, makes interpretation of these early signs problematic at the very least. For example, obstetrical complications associated with hypoxia (insufficient oxygen to the brain during labor) and other difficulties in labor and delivery may subsequently also manifest with these various behavioral, personality, and cognitive signs. Moreover, many of these signs, considered in retrospect, appear to have been present in less than 25 percent of those who later became schizophrenic.

Other observations have been based on direct examination. Imaging studies of the brain (CT scans) have shown an increased frequency of enlarged spaces (ventricles) within the brain, in particular of the third ventricle. Once again, such a finding may well be nonspecific and could even reflect a lack of oxygen during the labor and delivery process. Either way, it has been known for many years that abnormal neurological signs are more common among schizophrenics than among those who are not schizophrenic, including the relatives or unaffected twins of schizophrenics.

Eye Movements and Susceptibility to Schizophrenia

Almost 100 years ago, researchers in New England noted that patients with schizophrenia had difficulty following a moving pendulum. Only in the past quarter century have efforts been made to further investigate this observation. Eye movements have been followed while tracking a moving pendulum. Electrical signals through tiny electrodes placed near each eye have recorded how smoothly the eyes move in response to a slow-moving target. Significantly, different deviations have been noted among schizophrenics compared with healthy individuals. In particular, "abnormal" eye movements have been noted in at least one parent in 55 percent of couples with a schizophrenic child, compared to 17 percent of parents with a manic-depressive child. Studies of eye movement dysfunction have become quite complex and many have reported that schizophrenics have abnormally high rates of this disordered function. Another observation has been made that nonpsychotic close relatives of schizophrenics also consistently exhibit eye movement dysfunction. Thus far, it seems that this difficulty reflects dysfunction in a specific area of the brain (prefrontal cortex) and that one or more genes are directly involved. How closely such genes relate to schizophrenia remains a subject of intensive research.

Genes and Schizophrenia

There is no longer any question about the important genetic influences of genes as a causal factor for schizophrenia. A huge number of studies have been done to determine the mode of inheritance. It appears all but certain that a number of genes make an individual susceptible to develop schizophrenia, with the interaction of one or more extraneous environmental factors. This *polygenic* or *multifactorial mode of inheritance* (see Chapter 7) does not exclude the possibility in some families of a dominant or X-linked gene playing a major or significant role.

Despite enormous efforts to find one or more genes that indisputably cause or contribute to schizophrenia, and many false leads, results have been largely disappointing. Candidate genes on chromosomes 5, 6, 8, 13, and 22 remain under the highest suspicion. Chromosome 22 harbors a gene that when mutated (frequently by a deletion) results in a condition called velocardiofacial syndrome (see Chapters 7 and 2 for a description). In this disorder, there is also a high rate of schizophrenia. In fact, many schizophrenics have a high rate of the abnormal facial features and other characteristics of this velocardiofacial syndrome.

Some researchers have pointed to the possibility that the X chromosome harbors a schizophrenia susceptibility gene: My colleagues and I reported two unrelated young females with paranoid schizophrenia both of whom were found to have a tiny chromosome deletion at the same location on the short arm of the X chromosome. It thus seems very likely that this region harbors a schizophrenia susceptibility gene.

Schizophrenia within families may show a pattern of increasingly earlier onset and more severe illness with each successive generation. This phenomenon, called *anticipation,* is discussed in more detail in Chapter 7.

Treatment

Important advances have been made and continue to be made in the development of antipsychotic drugs. Discussion about these medications is beyond the purpose of this book. It is important, however, to recognize that a delay in initiating treatment appears to reduce the likelihood of a good response. Prompt treatment, initiated as soon as possible after the recognition of symptoms, is linked to a better prognosis. Precisely why delays in early treatment decrease the rate and degree of remission remains unclear. Some evidence suggests that there may be an ongoing process destroying the internal function of certain brain cells. One study of 55 schizophrenics used magnetic resonance imaging (MRI) of the brain at the time of diagnosis and about two and half years later. This study revealed that the internal spaces within the brain (the lateral ventricles), which are often found to be larger than normal in those with schizophrenia, had further increased in size only in those who had responded poorly to treatment or who had suffered a relapse. It is now abundantly clear that treatment should begin promptly and be continuous, and under no circumstances simply discontinued, if the best results are to be achieved.

Predicting the Outcome

Certain features assist in the prediction of outcome, which nevertheless remains difficult. A good or better outcome of treatment is likely when some or all of the following features are in evidence: sudden onset of symptoms; a short duration of illness; no previous psychiatric problems; the presence of a flat emotional tone; a lack of obsessive-compulsive behavior; no physically threatening behavior; good to normal functioning prior to the onset of illness; being married; having normal sexual function; having normal psychological function prior to illness; no abnormalities seen on brain imaging; and having no family history of schizo-phrenia.

Powerful predictors of an unhappy outcome include a history of behavior problems in childhood, including truancy and tantrums; a history of past psychi-atric problems; social isolation; a delay between the onset of symptoms and the beginning of treatment; and being single. Studies in different countries have shown that women treated for schizophrenia have a more favorable outcome than do men. Speculations about these observations include the fact that women were less likely than men to be socially isolated or institutionalized. The idea that there are fewer external demands and stresses may be the explanation for the observa-tion that individuals with schizophrenia in underdeveloped countries seem to do better than those in developed countries.

What Happens over Time?

The course and prognosis of schizophrenia in any individual will vary and depend upon many factors, including family history, age at onset, nature and severity of symptoms, promptness in obtaining treatment, whether or not iden-tifiable brain abnormality is evident, and various other factors. The onset of schizophrenia may either be barely perceptible and slow, or very abrupt. Social withdrawal with emotional detachment, remoteness, and aloofness are typical. There may also be disturbances of thought and impaired attention at this early stage, which may last from months to years. This initial period is usually followed by psychotic symptoms with characteristic hallucinations and delusions. This phase may be punctuated by acute and alarming psychotic symptoms alternating with remission or relapse, often preceded by mood disorder, sleep disturbances, anxiety, and delusions.

Despite a few outrageous and highly publicized crimes by schizophrenics (such as pushing a bystander onto the rails of an incoming train, and bashing in the head of a passerby with a rock, without any provocation), available data do not suggest an increased risk of serious crime by schizophrenics. There is, however, an increased risk of death due to suicide, accidents, side effects of medications, sub-standard care, and poor hygiene and safety. A 1994 study noted a suicide rate among schizophrenics some ten times greater than that for the general public. These researchers also estimated that more than half of all schizophrenics attempt suicide during their lifetime and that 10 percent actually succeed. In addition, it has been known for almost a quarter of a century that there is a markedly higher rate of violent deaths from suicide, accidents, and other causes among the blood relatives of schizophrenic adoptees, compared with the relatives of adoptees neither of whose parents had schizophrenia.

One review of 320 studies published between 1895 and 1992, of a total 51,800 schizophrenics followed for an average of six years, noted that 40 percent improved with treatment. However, variations in the definition of schizophrenia over time make interpretation of that analysis rather difficult. More recent, highly focused studies from Iowa of patients admitted between 1934 and 1944 and followed into the early 1970s showed that 20 percent were "psychiatrically well" at follow-up. However, 54 percent still had incapacitating symptoms; only 21 percent were married or widowed, whereas 67 percent had never married; and only 34 percent lived in their own home or with a relative, whereas 18 percent were in mental institutions. Some 35 percent were economically productive, but 58 percent had never worked. More than 10 percent had committed suicide. Clearly, schizophrenia is a devastating disorder for which early diagnosis and prompt treatment are crucial.

Costs

Beyond the overwhelming impact of schizophrenia on individuals and families is the enormous economic burden for society. A National Institute of Mental Health study estimated in 1991 that the total annual cost of schizophrenia nationwide was $65 billion dollars. Almost two-thirds of this amount was attributed to the costs of care, lost wages, and losses due to early death from suicide, and the balance was for treatment, public assistance, and other direct costs.

Extensive efforts are under way using new genetic technologies to recognize which patients would respond best to specific drugs. These so-called pharmacogenetic and pharmacogenomic studies aim to recognize how each drug is chemically handled by the body and which genes assist in predicting one or another advantageous or disadvantageous effect. In this way, and hopefully sooner rather than later, more effective drugs will become available for treating schizophrenia.

chapter **19**

Genes, Depression, and Other Mood Disorders

· ·

The task of unraveling the relative contributions of genes and environmental influences that jointly appear responsible for most diseases is proving long and arduous. Efforts at deciphering which genes are involved in mental illness or mood disorders have been especially difficult, given blurred distinctions between such disorders, their variable symptoms, and their changing definitions over decades. Notwithstanding these difficulties, two facts are abundantly clear. First, genes clearly play a major role in the causation of depressive illness and other mood disorders. Second, the frequency of depression, mania, suicidal impulses, and psychotic mood disorders has increased over the past 100 years. Moreover, not only have these disorders become more common but there is evidence that some are appearing at an earlier age.

Mood Disorders

The major mood disorders are depression (also called *unipolar depressive disorder*) and manic-depression (*bipolar disorder*). Some disorders that also have an effect on mood, personality, and/or behavior are also discussed briefly in this chapter.

Depression (Unipolar Disorder)

The main criteria for determining a major depressive episode are the symptoms of a depressed mood or loss of interest or pleasure, which must be present for at least two weeks, for most of the day, and nearly every day. The individual may feel sad or empty, or may be tearful. Children and adolescents may appear irritable. At least three or four of the following symptoms also need to be present to fulfill the diagnostic criteria: loss of more than 5 percent of body weight in a month, or loss of daily appetite; trouble sleeping, or excessive sleeping; agitation, or a feeling of doing everything slowly; excessive fatigue, or having no energy every day; experiencing feelings of worthlessness or inappropriate guilt every day; difficulties thinking and concentrating, and more pronounced indecisiveness every day; and recurring thoughts of dying with or without ideas of suicide.

The above criteria apply in the absence of drug abuse, other medical illness (such as an underactive thyroid gland), and normal bereavement after the death of a loved one. Depression after delivery of a baby (postpartum depression) occurs in about 1 in 10 mothers, especially among those with a history of a mood disorder, who did not plan their pregnancy, who are unemployed, and who are single.

About 4.7 percent of teenagers between the ages of 14 and 16 develop a major depressive disorder. Substance abuse and antisocial behavior are often present and complicate efforts at diagnosis. Behavior disorders and attention deficit disorders also complicate efforts at diagnosis.

Although unipolar depression might occur only once in a lifetime, recurrence is the rule rather than the exception. The risk of a second episode is about 50 percent. After a third episode, the risk of a fourth recurrence is about 90 percent. Recurring episodes tend to occur sooner and more abruptly and often involve new and more severe symptoms.

Manic-Depression (Bipolar Disorder)

The criteria for determining a manic episode depend first on a period of at least one week during which an abnormal and persistently elevated, expansive, or irritable mood is present. In addition, during this period of mood disturbance, three or more of the following symptoms are expected to be present:

1. Feelings of grandiosity (for example, writing checks for a million dollars) and of great self-esteem;
2. Marked decrease in the need for sleep (may only sleep a few hours);
3. Constant talking, or a sense of pressure to keep talking;
4. Racing thoughts and a flight of ideas on all manner of subjects;
5. Easy distractibility;
6. Intense goal-directed activity, socially, at work, or at school, with or without agitation;
7. Excessive pursuit of pleasurable activities with the potential of serious consequences (e.g., buying sprees, sexual indiscretions, or foolish business investments).

Again, determination of mania with these criteria excludes situations of drug abuse or other medical illness (such as an overactive thyroid gland).

In both unipolar and bipolar disorders the features may vary, and one should not expect all cases to be absolutely uniform. Moreover, different studies have noted various psychiatric disorders associated with bipolar or unipolar disorder among family members, suggesting that there is a varying spectrum that may be transmitted through some families. Within this spectrum are included mood swings (cyclothymia), melancholia (dysthymia), a schizophrenia-like disorder (schizoaffective disorder), alcoholism, eating disorders, and attention deficit/hyperactivity disorder. More recently, migraine also has been added to this list.

Compared with unipolar depression, striking aspects of bipolar depression include an earlier onset, more frequent episodes, a greater risk of suicide, a higher likelihood of psychotic symptoms, and a family history of mania and depression.

Bipolar depression afflicted many individuals with remarkable creative minds, including Charles Dickens, Lord Byron, Vincent Van Gogh, Robert Schumann, and Isaac Newton. It is in the manic phase that many artists produce their most creative work, and some prefer not to obtain treatment, for that very reason.

How Common a Problem?

The lifetime risk of a major depressive episode in the United States is about 6 percent, and that of a mood disorder, around 8 percent. Unipolar depression occurs about twice as often in women as in men. In addition, the incidence of major depression has been reported as being higher in separated or divorced individuals than in those who are married. Between 10 and 15 percent of people diagnosed with unipolar depression will eventually be diagnosed as having a bipolar disorder. Between 5 and 15 percent of adults with depression are eventually found to be bipolar. The risk of bipolar disorder is clearly higher among the young who have a history of depression, reaching at least 20 percent in adolescents and 32 percent in children of age 11 or younger.

Both culture and stress affect the prevalence of depression. A collaborative study among ten countries revealed a range of lifetime rates for major depression from a low of 1.5 percent in Taiwan (adults) to a high of 19.0 percent in Beirut. The number of adults with bipolar disorder usually ranges between 1.0 and 2.5 percent, although some studies report rates as high as 6.5 percent.

Suicide

The rate of suicide among young people (between the ages of 15 and 34) has increased dramatically in developed countries, and suicide now ranks second or third as the cause of death in this age group. Between 90 and 98 percent of adolescents who commit suicide had a psychiatric disorder before or at the time of their suicide. In the United States there are about 30,000 suicides annually. The given rate is estimated at 11 in 100,000 people. The lifetime risk of suicide in mood disorders is between 10 and 15 percent. Not unexpectedly, the risk of attempted suicide among those with unipolar or bipolar depression is extremely high (one study estimated an increased risk of 41 times the risk in the general population). Although more women than men attempt suicide, more men succeed. As so many television viewers have witnessed, patients with various states of bipolar depression, in which there is a combination of depression, rage, and grandiosity, may involve others in their suicide attempts. For example, there are individuals who taunt police with the hope and intention of being shot. It is estimated that about 4 percent of those who commit suicide murder someone else first.

Successful suicides and failed attempts occur about four times more often in the relatives of individuals who have tried to take or have taken their own lives. In many of these families there is serious bipolar disorder. Studies of identical twins also point to the involvement of genetic factors, with increased concordance (both twins committing suicide). Genes, depression, and family dynamics all play a role.

A remarkably large Dutch family was reported in which males with borderline mental retardation exhibited aggressive, violent, and impulsive behavior. Brunner's syndrome, named for its discoverer, is X-linked (see Chapter 7), affecting males only. The behavior reported included arson, exhibitionism, attempted rape, and suicide. A specific gene on the X chromosome was discovered that was believed responsible. Remarkably, in an experiment in which this gene was "knocked out" in a mouse strain, aggressive behavior followed.

Despite enormous efforts to detect signs that will assist in the prediction of suicide, results have largely been disappointing. Nevertheless, warning signs that should not be ignored include severe depression, expressions of hopelessness, psychosis with hallucinations (especially, hearing commanding voices), previous suicide attempts, a family history of suicide, and evidence of suicidal thoughts and planning.

ADHD and Bipolar Disorder

In childhood and adolescence, the distinction of attention deficit/hyperactivity disorder (ADHD) from bipolar disorder may not be easy. There are multiple, overlapping features, such as fluctuating moods, temper outbursts, disorganization, distractibility, excessive activity, excessive talkativeness, inattention, and impulsivity. Making both analysis and diagnosis more difficult is the finding that bipolar disorder and ADHD may aggregate in the same families.

Genes or Environment

It is all but certain that in the vast majority of people with depressive or manic-depressive illness there is no single, culprit gene. However, the earlier the onset of a mood disorder, the more likely and the more important is the genetic basis.

Family Studies

Mood disorders are familial. The risk of developing unipolar or bipolar disorder in first-degree relatives (parents, their siblings, and offspring) is 28 percent for either unipolar or bipolar disorder (see Table 19.1). If one parent has bipolar disorder, the risk of the parent's sibling or offspring developing bipolar disorder ranges between 4.1 percent and 14.6 percent. These figures are derived from major studies, and although they do not provide precision (most are given with a range), they do provide significant guidance. Certainly those with depression or anxiety have been noted to have more relatives who are similarly affected. Bipolar disorder is more often transmitted in families than are unipolar mood disorders. It is also well recognized that those with mood disorders have a higher family incidence of substance abuse.

Twins and Depressive Disorder.   Studies of twins have again supported the importance of genes in the causation of major depressive disorders. The range or average risk figures for twins are shown in Table 19.1. Most striking is the high risk of both twins being affected with bipolar disorder. Studies of identical twins reared apart from infancy yielded a concordance rate of 67 percent when bipolar mood disorder was considered, a rate in the same range as that found in identical twins reared together.

Individuals who have both a major depressive disorder and an anxiety state are found to have both of these mood disorders appearing with increased frequency

TABLE 19.1   Familial Risks of a Unipolar or Bipolar Disorder

If Affected	Unipolar or Bipolar	Risk (percent)	For Relative
One parent	bipolar	28.0	all offspring (uni- or bipolar)
One parent	bipolar	4.1–14.6	parent's sibling or offspring (bipolar)
One parent	bipolar	5.4–14.0	parent's sibling or offspring (unipolar)
Both parents	either	56.0–84.0	all offspring (uni- or bipolar)
One parent	unipolar	5.5–28.4	parent's sibling or offspring (unipolar)
One parent	unipolar	0.7–8.1	parent's sibling or offspring (bipolar)
One identical twin	bipolar	67.0–100.0	other twin
One nonidentical twin	bipolar	20.0	other twin
One identical twin	unipolar	50.0	other twin
One nonidentical twin	unipolar	50.0	other twin

in their families. In a study of 1,033 pairs of female twins, increased concordance has been found not only for major depression but also for generalized anxiety disorder, fear of public places (agoraphobia), and social phobia.

Adoptees and Mood Disorders.   Most studies have concluded that adoptees with mood disorders have a significantly greater likelihood of one of their biological parents being affected, especially when bipolar disorder is considered.

Predictably, the children of parents with major depressive disorders have high rates of various psychiatric problems. These have included major depression, ADHD, anxiety disorders, substance abuse, conduct disorders, and school problems. These disorders also make their appearance at an earlier age (12 to 13 years) in the children of parents with depression.

Patterns of Inheritance

Despite major studies aimed at determining one or more genes that cause or convey susceptibility to major depressive disorders, the effort has been largely unsuccessful. It is highly likely that multiple genes interacting with environmental factors (including stress) best explain the familial nature of depressive illness. DNA linkage analysis (see Chapter 6) has focused on one region of the X chromosome in a large collaborative study of bipolar disorder. In some families, red-green color blindness (a known X-linked trait) has been linked to bipolar mood disorders. In about one-third of cases this form of inheritance might apply with expected mother-to-son transmission. However, not all studies confirm this observation.

Beyond the issue of diagnostic definitions and distinctions between disorders within a spectrum is the likely reality that involvement of different genes, at one time or another, with variable environmental exposures, all eventually lead via different pathways to the same end—major depressive disorder.

Genes and Anxiety Disorders

Estimates by the U.S. National Institute of Mental Health suggested that about 1 in 8 adults have an anxiety disorder (neurosis). Common forms of neurosis include phobias, panic disorder, and obsessive-compulsive disorder. All of these mood disorders occur more frequently in the family members of an affected individual.

The brothers, sisters, and children of an individual with *panic disorder* have on average an almost 8 times greater likelihood of having the same problem. Actual rates have ranged between 2.6- and 20-fold more than in families without someone with a panic disorder. Research findings also support a genetic connection between panic disorder and mood disorders, especially bipolar illness. In several studies of panic disorder in twins, clear evidence of heritability emerged, mostly in the range of 30 to 40 percent. Once again, multiple genes and environmental influences likely govern susceptibility to panic disorder in families.

Family studies of *obsessive-compulsive disorder (OCD)* have not yielded consistent results. One key study focused on children with OCD. Fathers of affected children were noted to have OCD in 25 percent of cases, whereas 9 percent of mothers did. Genetic influences also have been implicated in studies of twins in the United States and in Japan. Identical and nonidentical twins were concordant for OCD in 87 and 47 percent of cases, respectively.

Twin studies of *generalized anxiety disorder* showed identical and nonidentical twin concordances of 28 percent and 17 percent, respectively.

Multiple studies have noted an increased risk of mood disorders among the first-degree relatives of individuals with *anorexia nervosa*. These individuals either starve themselves or vomit their food after eating (the latter disorder is known as *bulimia*) in order to remain thin (among other possible reasons). Considerably increased risks of major depression, bipolar disorder, or schizoaffective illness in first-degree relatives of individuals with anorexia nervosa have been reported as well. In families where one individual has been diagnosed with bulimia, high rates of mood disorders and eating disorders in other family members have been noted. Moreover, eating disorders have been linked to substance abuse in some of these families. Twin studies of anorexia and bulimia have found a markedly increased concordance (56 percent) for identical anorexic twins, compared with 5 percent for nonidentical twins.

Alcoholism

Psychiatrists classify alcoholism among the mood disorders. In a number of studies of the disorder, about 1 in 10 males and about 1 in 50 females have been considered alcoholics. Like other mood disorders, alcoholism has often been noted to be highly familial, and the question of family drinking habits or culture versus genetic susceptibility is not new. The risk of alcoholism occurring in first-degree relatives of an alcoholic is about 7 times greater than the risk in families of nonalcoholics. In a study of severe alcoholism, risks of 16 percent in the fathers and 7 percent in the siblings of severe alcoholics were found, compared with risks of 1.6 percent in the fathers and 0.5 percent in the siblings of relatives of nonalcoholics.

Various adoption studies also have pointed to genetic factors in alcoholism. Male adoptees (but not female ones) had higher rates of alcoholism when a biological parent was alcoholic. Studies of twins have produced variable results: The more recent ones have noted concordance rates for identical twin males of 59 percent compared to 36 percent for nonidentical twin males, and 25 percent versus 5 percent for twin females. The susceptibility genes for alcoholism, however, have not yet been identified. The search is made more complicated because alcohol abuse often coexists with antisocial personality, major depressive disorder, and other environmental or family problems.

A huge amount of research has examined genetic and environmental influences on human behavior. As noted repeatedly in this chapter and that on schizophrenia, twin and adoption studies have solidified the indisputable involvement of genes in the evolution of normal and abnormal behavior as well as mental illness. In particular, twin studies have emphasized the important and variable genetic influence on childhood disorders that include attention deficit/hyperactivity disorder, Tourette's syndrome (a disorder characterized by twitches and sudden vocalizations of sounds or even curses), autism, reading or cognitive disorders, and even juvenile delinquency. Appreciable genetic effects on personality development have been recognized for extroversion, agreeableness, conscientiousness, and neuroticism. Twin and adoption studies in both males and females also support a substantial genetic influence on the development of homosexuality.

Although we await the discovery of specific genes that contribute to our IQ and personality, and that when defective lead to mental illness, twin and adoption studies clearly indicate that most behavioral characteristics have an important genetic component.

chapter 20

Genes and Alzheimer's Disease

··

Almost a century ago in Germany, a professor of psychiatry, Alois Alzheimer, described the case of a 51-year-old woman with dementia. Alzheimer—who also could be regarded as an early neuropathologist, given his interest in and expertise on brain tissue—had extensively researched the disorder now named after him, even before his 1907 case report. The effects of this disorder on the life of the woman and her family were as devastating then as they are now in families afflicted with Alzheimer's, although the disorder then was far less widespread.

Alzheimer's Disease (AD)

Typically, loss of memory, uncharacteristic forgetfulness, misplacement of items, and inability to remember conversations or newly learned information all signal the onset of dementia. Memory may not, however, be impaired for complex material from the past, which may be recalled with great clarity.

Telltale Features

The telltale feature of AD is memory loss. This may be recognized by an individual repeatedly asking the same question, even though an answer already has been given, over and over again. Memory about where an item was left is often lost, as is recall of recent conversations. Difficulty in acquiring and remembering new information is typical, and forgetfulness of names and even the use of objects is characteristic. AD usually begins insidiously, and because forgetfulness can have many other causes, in many cases of AD it is difficult to identify the precise date of onset.

The abilities to plan, organize, or arrange activities and to solve problems are impaired. Affected individuals frequently develop a lack of insight, and many have difficulty in completing complex or new tasks. Driving, operating home appliances, and controlling even simple financial transactions become progressively impaired. Disturbances in language include difficulty finding words, and speech hesitancy, while difficulties writing steadily make their appearance. There may also be changes in personality and a developing disinterest about pursuits that were previously of direct interest. Social withdrawal may become obvious in some; but many with AD continue to function well enough to escape recognition for some years.

Various mood disorders may further complicate the course of AD, including anxiety, slowed thinking, apathy, sadness, talk of suicide, lack of energy, and feelings of guilt and depression.

Within 4 to 7 years after the onset of AD, clearly increased dependence on others becomes evident. The features described earlier become much more pronounced, and confusion and identification of even close family members living in the same house reflect neurological disturbance. Those afflicted can get lost even in familiar territory, and their language abilities, memory, and comprehension all progressively deteriorate. A wide range of uncharacteristic behaviors also may also occur in AD, being more frequent and problematic in those with more severe dementia. Other common behavioral symptoms include restlessness and overactivity; disruptive, bothersome, or repetitive behavior; verbal aggression; sleep disturbances; resisting help; wandering around the house at night; inappropriate or withdrawn behavior; loss of appetite; and physical aggression.

In some with AD, psychotic behavior, including paranoia and hallucinations, may provoke crises within families, leading to the hospitalization of the afflicted family member. Hallucinations and delusions are unfortunately fairly common, with some 40 percent of those with AD experiencing delusions at some point during their illness. About 25 percent of those with AD eventually experience hallucinations (see, hear, smell, or feel things that are not present). At this stage, self-care, which includes dressing, bathing, and toileting, may be neglected, and social graces are typically lost.

On occasion, AD may occur with atypical features, including loss of vision, inability to name objects, inability to write, an inability to recognize or comprehend written or printed words, or a loss of speech. These atypical cases are mostly explained by the typical pathological changes of AD (see below) occurring in less commonly affected regions of the brain.

In advanced stages of AD, individuals become completely dependent. Disintegration of memory and failure to identify spouse or children occur commonly, and language and personality traits steadily disappear. Difficulties in walking and incontinence of urine and feces may occur, and seizures may be an added complication. In the final stage, the individual with AD is usually bedridden, has a virtually total loss of comprehension, and frequently has difficulty in swallowing, which results in weight loss and in death within 7 to 10 years.

How Common a Problem?

AD is clearly the most common form of dementia in the United States and currently is thought to affect at least 4 million individuals. An estimated 360,000 new cases are recognized annually in the United States. The frequency of AD increases dramatically with age, and the number of cases in the United States doubles every five years. AD affects approximately 11 percent of the population older than 65. Among those of both sexes who are 85 years or older, some 30 to 50 percent have AD. About 20 million people worldwide are affected by the disorder.

The overall annual direct and indirect costs of care for AD in the United States are staggering—close to $100 billion.

Is It Aging or Alzheimer's?

Aging is often associated with loss of recent or short-term memory and with variable loss of intellectual functions. Difficulties in distinguishing the normal effects of gradually advancing age from the development of AD are due to wide variations in individual intelligence, in the effects of personal habits on the normal aging process (exercise, nutrition, smoking), in educational levels, and in established functional abilities, and to other factors. The normal changes in memory and learning functions among the healthy aged should not confound diagnostic efforts. In fact, recent studies show that intellectual performance may not really decline appreciably until at least 90 years of age. In contrast, those in the earliest stages of AD evidence changes that interfere with their daily functioning. Typically, those affected by AD are also unaware of their memory deficits, while continuing to act quite normally.

Because of the potential difficulties in separating symptoms of potential aging from those of senility with AD, the U.S. National Institute of Neurological and Communicative Disorders and Stroke and the Alzheimer's Disease and Related Disorders Association developed clinical diagnostic criteria for AD. This important step brought uniformity to the clinical diagnosis of AD, improved diagnostic precision, and, as a consequence, more accurate interpretation of new therapies. Three general categories were recognized (using the features described above): (1) probable AD; (2) possible AD; and (3) definite AD. The last category was reserved only for advanced cases in which brain tissue was tested after death tags, and body.

in order to confirm the diagnosis. Meanwhile, it is clear that AD is not inevitable with aging.

Trouble in the Brain Cells: Plaques and Tangles

Since AD only rarely occurs in those younger than 60, observations made on brain cells reflect both the effects of aging and those due to the disease process. Posthumous studies of brain tissues show that in the healthy aged person without any significant vascular disease, changes with age are relatively subtle. Changes in brain cells seen under the microscope in cases of AD, however, are far from subtle. Two features stand out—plaques and tangles. Under the microscope, plaques are small, dense accumulations of brain cell debris with or without a central core of a specific complex protein (amyloid). Tangles of fibrous filaments, similar to a ball of loose string, are caused by the accumulation of abnormal components of the walls and debris of dead brain cells. These changes occur predominantly in the upper, outer mantle of the brain—the cerebral cortex, or gray matter. Curiously, not all brain cells in the gray matter are involved; nor are plaques and tangles confined to the cortex alone. The distribution and density of these tangles and plaques are used to establish a tissue diagnosis of AD, either from a biopsy of live brain tissue or during autopsy.

Most, but not all, researchers have concluded that there is a significant correlation between the number of plaques and tangles in several areas of the cortex and the severity of dementia. Various other features are also seen under the microscope but may not be diagnostic of AD alone. Network connections within the cortex also undergo substantial degeneration in AD, contributing seriously to a decline in intellectual function. Variations in the density and distribution of plaques and tangles in the cortex and adjacent areas of the brain assist in differentiating elderly nondemented individuals from those with AD. Unfortunately, however, these diagnoses are not made without available brain tissue for study. Complicating interpretation under the microscope is the fact that typical dementia of the Alzheimer's type may occur even in the total absence of tangles. Loss of brain cell network connections in the cortex (synapse loss) is critically important in AD, and is thought to reflect the consequences of a disease process in somewhat more distant, though connected brain cells. Synapse loss does appear to be the immediate cause of dementia, which in turn is the primary result of brain cell disintegration in the connected, adjacent network.

Is Alzheimer's Disease Inherited?

The first description of AD occurring in more than one generation in a family, with full documentation, appeared in a 1932 report. Many families with multiple affected members have since been described. Nevertheless, detailed genetic analysis indicates that only about 10 percent of all cases of AD are directly inherited/transmitted via a single gene. In these cases, a dominant mode of inheritance due to a defect in a single gene is the cause. An affected person therefore has a 50 percent risk of transmitting the defective gene to each of his or her offspring. Precisely when an individual who has inherited an AD gene will develop the disorder varies from family to family. One recent, comprehensive study pointed to a 38 percent risk by age 85.

Culprit Genes

Thus far, four different gene defects have been identified that cause or that endow susceptibility to AD, and another gene has been localized (i.e., this gene's location is known, but it has not yet been cloned). Even more susceptibility genes are expected to be found. The first gene identified (the so-called amyloid precursor protein [APP] gene) was found on chromosome 21. This location was an obvious place to search, given that individuals with Down syndrome (trisomy 21) (see Chapter 3) almost invariably develop clinical signs of AD by the age of 40. Examination of their brains shows the features of AD. Mutations in this gene also carry a propensity to brain hemorrhage, for which no definite reason has yet been determined.

Since it was clear that the APP gene did not account for all directly inherited cases of AD, intense searches for the existence of another gene continued. The reward came with the discovery of the apolipoprotein E (apoE) gene on chromosome 19. Coincidentally, the apoE gene, while at work in brain cells, also plays a critical role in the regulation of cholesterol, as discussed in Chapter 11 on heart disease. Within this gene, three common structural variations (called polymorphisms) were found and are termed E2, E3, and E4. E2 is found in about 8 percent of Caucasians, E3 in 78 percent, and E4 in about 14 percent. Subsequently it was realized that the frequency of the E4 polymorphism in patients with AD approximated 40 percent. Moreover, it was noted that if an individual received the E4 variation from both parents, the onset of AD was earlier than in those who received only a single copy, and the risk of developing

AD was about fifteen times greater. This increased risk translates into about a 95 percent chance of AD by 80 years of age. Exceptions to these findings have occurred in blacks and Hispanics with AD, where conflicting results have been found.

When one copy of the E2 variation is inherited from each parent, a later onset of AD occurs. An E2/E3 combination results in AD after 90 years of age. The mechanisms by which any of these polymorphisms within the apoE gene affect the time of onset of disease remains a mystery. Clearly, there is some connection between the apoE gene and the tangles described earlier. It is thought that this gene normally functions like a cleanup crew, clearing away possibly toxic, accumulating protein fragments cleaved by enzymes responsible for dismantling a cell that is programmed to die (see discussion on apoptosis/cell suicide in Chapter 6).

The association between the E4 polymorphism and AD is robust among Caucasians, but this polymorphism is not specific to AD alone. Remarkably, individuals with the E4 variation have other risks as well. When the E4 polymorphism has been found among individuals with head injury, brain hemorrhage, and those having cardiac bypass surgery, all had a relatively worse outcome. In addition, there is evidence that the combination of a history of head injury and the possession of the E4 variation increases the risk of future AD tenfold, compared with individuals without head injury who have this variation.

After the discovery of the APP and the apoE genes, researchers soon realized that these two genes did not account for all cases of familial AD. Subsequent studies led to the identification of a third gene (presenilin-1), on chromosome 14. Defects in this gene were quickly recognized to be associated with particularly aggressive and early-onset AD—in fact, as early as 25 years of age, although average age at onset is 45 years. Incredibly, the presenilin-1 gene has been highly conserved through evolution, and a gene corresponding in structure has also been found in the simple worm *(C. elegans)** (discussed in Chapter 27) and the fruit fly. Ongoing studies with *C. elegans* have focused on the signaling process between cells and the functioning network in and between cells.

---

*The simple worm *C. elegans* was chosen by one of my early teachers, the famous Sydney Brenner, as a model organism for the laboratory study of genes, for several reasons: The worm grows from embryo to its one-millimeter, adult size in just three days! In addition, its body is transparent at all stages of the life cycle, allowing cell division, cell migration and tissue formation, and cell death to be closely observed. Due in part to the worm's suitability for microscopic study, the complete wiring pattern of its nervous system is now well known, and all of its more than 19,000 genes have been decoded.

Discovery of the fourth gene (presenilin-2) was facilitated after a gene structure similar to presenilin-1 was noted to be present on chromosome 1. Mutations in this gene appear to be rare. They are also associated with earlier-onset AD, but with considerable variability in the ages at onset, ranging from 40 to 85 years. Needless to say, individuals who carry one polymorphism in one gene and one or more in another AD gene are at risk of an earlier onset of the disease.

The four genes described thus far account for about one-half of the inherited cases of AD. A fifth gene associated with onset after the age of 60 has been identified on chromosome 12. It is likely that other AD susceptibility genes will yet be discovered.

Risks

About 11 percent of the general population in the United States will develop AD after the age of 65. For an individual with one affected parent and an affected sibling, the risk approximates 40 percent for early-onset AD. In a family with only a single affected member, the risk to first-degree relatives probably exceeds 10 percent but is less than 50 percent in a lifetime. Individuals who have a single E4 polymorphism have an increased risk of developing AD in their lifetimes, that risk being about 20 percent. Those who have inherited two E4 polymorphisms have about a 30 percent risk.

While for whites having one or two apoE-4 polymorphisms increases their risks of developing AD, this is not the case for African or Hispanic Americans. In fact, even in the absence of the apoE-4 variation, African Americans up to the age of 90 are four times more likely than whites to develop AD. Similarly, Hispanics without the apoE-4 variation have double the risk. The clear inference is that other genes (or risk factors), yet to be identified, are involved in AD in African and Hispanic Americans.

An increased concordance rate for AD in identical twins has been recognized, but precise rates are uncertain because of the large differences in the age at onset observed among some twin pairs. It is, however, already clear that the rate of AD in identical twins is less than 100 percent—confirming that environmental factors play a role also in the development of AD.

Unusual but well-documented multigenerational studies of families with many AD-affected members have been described. Simple, dominant inheritance operates in these families, in which an affected person has a 50 percent risk of receiving or transmitting AD. In these families, males and females have been equally affected, male-to-male transmission has occurred, and some variation (decreased penetrance) of the effects has been noted in some family members.

A few studies have noted an increased occurrence of Down syndrome in the relatives of those with AD. Although the first identified AD gene was found on chromosome 21 (which appears in triplicate in all cells in individuals with Down syndrome), the precise mechanism leading to the apparent association between Down syndrome and AD is unknown. Advanced maternal age at conception does not appear to increase the risk for AD among these offspring.

Complicating the assessment of risks of AD is the involvement of other genes that influence or modulate the susceptibility to AD. One such recently identified gene (interleukin-1) has been found to be particularly active in the brains of those with AD. This gene produces a product (called a cytokine) that strongly encourages inflammatory responses by tissues. Two structurally distinct forms of interleukin-1 have been recognized (type 1A and 1B), each of which is encoded by separate genes located on chromosome 2. Two specific variations (polymorphisms), one in each form (1A and 1B) of the interleukin gene, are now known to be important in the development of AD. An individual who inherits the particular variation in the interleukin gene for both the 1A and 1B forms from both parents has an increased risk of developing AD—almost eleven times that borne by the population at large. Given the pro-inflammatory function of interleukin-1, it is intriguing and exciting to note reports that indicate that the use of anti-inflammatory medications may delay the onset or slow the progression of AD. Knowledge of the orchestration of the degenerative process in the brain by interleukin may well be utilized in the development of therapies against such powerful modulating gene forces.

Actual mutations (not polymorphisms) in any one of the three identified genes (presenilin-1, presenilin-2, and APP) are responsible for early-onset AD and are dominantly inherited. Affected individuals have a 50 percent risk of transmitting this disorder to each of their offspring. These three genes are responsible for only about 2 percent of all AD cases. ApoE-4 is a recognizable risk factor in more than half of all AD cases.

Two other genes, so far, are known to convey a genetic susceptibility to AD, with the involvement of one or more "environmental" or acquired factors. The first of these genetic susceptibility genes is the apoE gene. The second recognized susceptibility gene is the α-2- macroglobulin gene on chromosome 12. About 30 percent of the population carries a common polymorphism in this gene that encodes a protein called α-2-macroglobulin. This gene clearly makes an individual more susceptible to developing late-onset AD and has an effect either equal to or stronger than apoE-4. One normal role for the α-2-macroglobulin protein is to function as a cleanup crew for brain cells by sweeping up proteins that could otherwise be toxic to nerve cells. These deposited and potentially toxic proteins are thought to be noxious to the brain cells, and their removal may avert the development of AD. A mutation in the α-2-macroglobulin gene may disable the cleanup crew, allowing the toxic protein to accumulate and thereby causing nerve cell death.

Ethnicity and Familial Alzheimer's Disease

AD has occurred in all ethnic groups studied, including West Europeans, Japanese, Mexicans, Hispanics, and Ashkenazic Jews. An important original founder of familial AD was recognized to be of Volga German ancestry. An analysis of 40 studies comprising some 5,930 elders with AD compared with 8,607 mentally normal control subjects revealed that the apoE-4 gene appeared more frequently among AD cases within every ethnic group. The association was weaker among African Americans than among whites. The apoE-4 association among individuals with AD was also stronger among Japanese subjects than among whites. Among Hispanics the E3/E4 combination was associated with a significant risk for AD, but the E4/E4 combination did not appear to confer an increased risk. The numbers supporting the latter conclusion, however, are too small to be definitive.

Acquired or Environmental Factors

Disturbances in fetal and early childhood development have been linked to many adult chronic diseases including heart disease, stroke, hypertension, diabetes, and chronic obstructive lung disease (see discussions in Chapters 11, 12, 14, and 15). Factors such as malnutrition, even mild and transient, have the potential for interfering with brain maturation. Brain growth occurs mostly during fetal development and early childhood, maturation being completed during adolescence. The very areas of the brain that take the longest to mature are those that show the earliest signs of AD. Growing up in a low socioeconomic area and having multiple siblings increase the risk of subsequent development of AD. One study claimed that for each additional child in the family, the risk of AD increased by 8 percent. The epidemiological conclusion from such studies has been that early life, childhood, and adolescent environments have the potential to increase risks of AD. Whether early malnutrition or other factor(s) are responsible remains to be determined.

Another environmental factor that has received close attention is cigarette smoking. A Dutch study concluded that smokers had more than double the risk of developing AD compared to those who had never smoked. Among individuals who smoked as well as possessed the apoE-4 gene, a 4.5-fold greater risk of developing AD was noted. Since smoking is a risk factor for heart and blood vessel disease, some effect through a vascular mechanism seems likely.

An early study to determine whether or not the female hormone estrogen protects against the development of dementia concluded that women with the E4 gene had no protection, whereas those with E2 or E3 did. More studies, however, are necessary to confirm these findings. A French study of 2,881 individuals over a five-year period noted that those who never married were three times more likely to develop AD than those who did marry. This protective effect was noted even after the death of a partner. It is not known what factor(s) is operative among people who stay single that makes them more vulnerable to AD later in life.

Should You Have a Predictive Test or Not?

Predictive or Susceptibility Tests

Presymptomatic and predictive genetic testing is fully explored in Chapter 24. For the individual with symptoms or signs suggesting the diagnosis of AD, efforts at securing a precise answer may include mutation analysis of the APP gene and the presenilin-1 and -2 genes. However, decisions about DNA tests for individuals without signs or symptoms but who may have a positive family history of AD are much more complex. DNA analysis of susceptibility genes such as apoE are not useful for predicting the subsequent development of AD in individuals who are cognitively normal. Individuals with the E4 variant, even in double dose, may never develop AD in their lifetimes. Analysis of the apoE gene at best can provide a relative risk estimate of likelihood for developing AD. Learning that you have an eightfold increased risk in your lifetime for developing AD is neither useful nor valuable and is not recommended. This opinion reflects consensus statements by many authoritative working groups and committees, including the Task Force for Genetic Testing of the U.S. National Institutes of Health and the Department of Energy Working Group on Ethical, Legal and Social Implications of Human

Genome Research. These recommendations can be accessed through the World Wide Web at med.jhu.edu/tfgtlsi.

Analysis of the apoE susceptibility gene is available but recommended only in the presence of symptoms or signs of AD. This analysis has been shown to improve the specificity of the diagnosis of AD from 55 percent to 84 percent. Again, the emphasis is on diagnosis, not on predictive testing. Moreover, many people who do develop AD do not carry E4 genes.

Ethical Considerations

The conclusions that can be reached reflect published authoritative consensus guidelines concerning genetic testing for AD. Diagnostic mutation analysis or susceptibility testing for the *symptomatic* individual is appropriate. Notwithstanding the availability of this diagnostic test, it remains an option for discussion, and could reasonably be accepted or declined. Gene analysis for AD of any sort is not appropriate or recommended for children, fetuses, or embryos. There will be occasions, however, when an individual subjected to head trauma (for example, boxing) may be considered for testing, given the association between later-onset AD and head injury. Individuals having a limited capacity to consent to a diagnostic genetic test for AD may need a family surrogate to assist. Nevertheless, decisionmaking may be complicated, due to the possibility of new information that may have serious implications for relatives. In these situations, other family members should be invited into the decisionmaking process. Predictive testing for susceptibility genes such as apoE is not recommended. This position would change only if and when proven remedies become available that could delay the onset of AD or be used as a beneficial treatment. In this connection, a ray of hope is seen in the very recent development of a beta amyloid vaccine, so far used with promising results in mice only. The strategy has been to interdict the accumulation of the harmful amyloid protein. Last, all considerations about genetic testing for AD should prompt the need for recognized, skilled genetic counseling. The need to ensure informed consent, confidentiality, and privacy cannot be overemphasized.

chapter 21

Genes, Aging, and Longevity

··

Why do we age? Is there a program encoded in our genetic blueprints, secreted deep in the nucleus of each of our cells? Are we therefore predestined to grow old and die according to a genetically predetermined plan? Does our lifestyle impact any such programmed scheme? Absent the common culprits of cancer and cardiovascular disease, could we live to be 200? Why not? Efforts to answer these questions in detail would fill an entire book. Here the purpose will be to provide the quintessential insights (mainly genetic) that have the potential to seriously impact our life span. First, however, we need to look at other species.

Lessons from Other Species

Circumscribed life spans are characteristic throughout the animal kingdom, as they are in humans. The lowly worm *C. elegans,* which has proven so vital to the understanding of gene function (see Chapter 27), has a life span of only fifteen days; however, mutations induced in this primitive worm have succeeded in doubling its life span. The life span of the mouse rarely exceeds two years; that of rats, four years; of cats, thirty years; of elephants, sixty years; and of horses, forty years. Bats, with an average life span of about thirty years, age much more slowly than mice, although the two species have similar body weights and cell chemistry (metabolic rates). Humans and virtually all animals age. Exceptions are those species that continue to grow in size after reaching adulthood. For example, several fish and reptile species show no biological age changes. They are not, however, immortal: They inevitably succumb to disease, predators, or accidents.

Laboratory rats fed diets that were sufficient in all constituents except calories showed retarded growth during the period of calorie restriction. After their caloric intake was increased, the rats proceeded to grow to adult size. They eventually exceeded the normal expected life span for that strain of rats. They reached about twice the maximum age achieved by rats whose diets were not interdicted.

Of particular interest was the observation that rats that were initially on the calorie-restricted diet had an associated delay in the onset of various tumors and chronic diseases related to aging. Calorie restriction also was shown to extend the life span of laboratory mice by 50 percent or more; but these mice were less fertile than normal mice, and were about 30 percent smaller. Similar results have been noted in chickens, bees, silkworms, and other species. The prolonging of life was most pronounced when low-calorie diets were started soon after birth. Removal of the sex glands of salmon early in their development also has been noted to prolong their life span.

Even the surrounding temperature in which animal species live has been questioned in its relationship to age. Fish raised at low temperatures have better growth and live longer. In contrast, rats reared at low room temperatures have a considerable decrease in their life span from all causes of death—including, for some strange reason, cancer.

Mice raised in a germ-free environment have been shown to have a longer mean life span, as have rats that have had their spleens removed early in life. The implication of both of these experiments is that the body defenses that act against infection play a role in the aging process.

That females tend to live longer than males suggests an influence of the sex hormones. This characteristic is not confined to the human species: Female spiders, fish, water beetles, houseflies, chickens, and fruit flies all live longer than their male counterparts. Yet in some bird species, especially pigeons, males reputedly live longer than females. Cats that have been castrated have attained the highest recorded ages. No equivalent human data exist (fortunately!). Some female fruit flies that are virgins, born without ovaries, or sterilized live longer than their normal female contemporaries. Virgin mice live longer than spayed females; but the oldest of all are castrated mice.

Why and How Do We Age?

Aging is an extremely complex process. Common sense tells us that genes and genetic mechanisms within cells play a key role and that lifestyle and environment are equally important. Studies of several hundred sets of twins from Sweden suggest that aging is influenced by genes about one-third of the time, and by lifestyle, some two-thirds.

The evidence (and theories) about aging can be grouped for consideration into two main categories: first, our internal, programmed body clocks; and second, the effects of wear and tear over time.

Death Clocks

We all know that our bodies march to an internal biological drumbeat that rhythmically controls key functions—for example, the timing of puberty, the cyclical arrival of menstrual periods, and the beginning of menopause. Indeed, women often mimic their mothers in the timing of menopause. From a host of hormone and other chemical studies, we know that with aging, considerable rhythmicity is lost. Experiments show that serious disruption of body rhythms decreases the life span in animals. Whether or not desynchronization of our body clocks, due to changes in gene expression as we get older, causes the effects of aging on the body remains to be discovered. We do know that our body clock is in the hypothalamus, in the central region in the brain from which our body rhythms (also known as *circadian rhythms*) emanate and oscillate.

Body Clocks and Programmed Aging

Forty years ago it was appreciated that human connective tissue cells (called fibroblasts) grown in the laboratory could undergo only about 40 to 50 population doublings before petering out. When such cells were interrupted after a fixed number of doublings, frozen away, and then thawed in order to grow again, they "remembered" how many cell divisions they previously had. Moreover, cells derived from adults consistently achieved fewer cell doublings than those derived from fetal tissues. Fibroblasts derived from individuals with genetic aging syndromes (see below) displayed severely limited growth potential. From these observations came the realization that there was an encoded program operating in our genes—at least, in cell cultures.

Exciting research performed at the Scripps Research Institute in LaJolla, California, focused on an analysis of over 6,000 genes in cultivated human cells from young, middle-aged, and elderly individuals as well as from those with an accelerated aging syndrome (progeria; see "Premature Aging Syndromes," below). Researchers noted that sixty-one genes were clearly more involved than the others with aging. Furthermore, they found that a smaller number of those same genes have functions that control cell division, especially chromosome duplication and separation (called mitosis; see Chapter 3). These results are especially intriguing, since defects in the genes involved in mitosis may cause chromosome instability, which can contribute to cancer or aging. It has been known for almost forty years that the rate of chromosome abnormality increases with aging. The

California researchers posit that aging is predominantly a disease of mismanagement of certain key points during cell division.

Chromosome Tips

Another marker of cell aging is rooted in the architecture of chromosomes, made up of the building blocks of DNA. The tips of each chromosome end—that is, the telomeres—are composed of hundreds of repeated copies of a six-base combination, abbreviated TTAGGG (see Chapter 6 for details about these bases). With each cell doubling, about fifty to two hundred base pairs are shed, resulting in progressively shorter telomeres, until virtually no telomeres are left and cell division ceases. Shorter telomeres are observed in fibroblasts and in white blood cells (lymphocytes) from older individuals, further pointing to the involvement of this cellular phenomenon in the aging process. However, this process probably has nothing to do with aging in cells that do not undergo cell division—for example, those in the adult brain.

Cell Suicide

Programmed cell death, or apoptosis, was discussed fully in Chapter 6. This process is initiated within a cell following DNA or other damage. It appears to play a central role in controlling cell numbers during early development, and for the removal of damaged or abnormal cells during later life. The normal process of apoptosis aims to remove defective, damaged, or precancerous cells before they can impair tissue functions. The aging process, however, may interfere with apoptosis. For example, a decrease in apoptosis has been linked to the development of cancer cells that are resistant to chemotherapy. In contrast, increased apoptosis has been noted in association with cancer suppression. Programmed nerve cell death increases in the affected areas of the brain in Alzheimer's disease, Parkinson's disease, and Huntington's disease, among other conditions—supporting evidence that a failure of apoptosis allows toxic protein fragments to accumulate and cause brain cell damage. Failure of apoptosis to remove impaired lymphocytes is also thought to contribute to the increase in autoimmune diseases (such as rheumatoid arthritis) with age.

As noted earlier, caloric restriction is well known to reduce the incidence of cancer in mice and rats. This phenomenon is thought to be due to the elimination of precancerous cells by apoptosis. Indeed, apoptosis is increased in calorie-restricted animals. The same mechanism is thought to be involved in the increased life span associated with calorie restriction in these animals.

Immune System and Aging

At least nine different, major changes occur in the immune system with advancing age. Our immune system works to protect and defend us from infectious agents (bacteria, viruses, and other organisms) and foreign proteins (for example, a mismatched blood transfusion). This system has the incredible ability to produce complex molecules called antibodies to entrap or "neutralize" foreign invaders. In addition, special killer squads of cells from the thymus gland, the spleen, and the bone marrow can also be marshaled for the body's defense. These squads of cells also produce certain proteins that have protective or anti-inflammatory effects. All of these and other immune functions diminish with advancing age. Consequently, as our immune responsiveness declines, the system becomes less able to discriminate between the body's own proteins and external invaders. This progressive failure to distinguish self from nonself not only increases our vulnerability to infection as we age but also means that we increasingly make antibodies against our own organs. As a result of this malfunction, chronic so-called autoimmune diseases emerge, including diabetes, rheumatoid arthritis, thyroid disease, and others. These misled immune responses are basically directed by our genes (especially the HLA genes on chromosome 6; see Chapter 10 for more detail about these genes). We do not really understand why the body turns on itself in this way. Unfortunately, the immune system alone cannot explain aging: Many organisms that age lack an immune system.

Aging Due to Wear and Tear

Long-term damage to cells, proteins, DNA, and other molecules is thought to produce the many manifestations of aging. As we age, DNA undergoes continual damage. This may occur in the form of breaks in the DNA strand, or chromosome breaks that result in rearrangements (such as translocations; see Chapter 5 for more about these). Translocations may produce a fusion gene the product of which promotes the development of a cancer. Other extrinsic factors including X-rays and powerful chemicals can damage DNA, as can factors within a cell, such as reactive oxygen radicals. The resulting DNA damage—from whatever source—accumulates, leading to malfunctioning genes, proteins, and cells, which results in organ deterioration and aging. The difference in longevity among species may be due to variations in the rate at which DNA damage occurs or is repaired.

Within our cells are enzyme systems capable of detecting and repairing damaged DNA. Failure to repair may be associated with genetic disorders, including those characterized by cancer or premature aging. Decreased repair of DNA has been noted in older individuals.

Problems in the Power Plants

Tiny particles suspended in the liquid (cytoplasm) that bathes the cell nucleus are responsible for generating most of the cell's energy. These particles, the mitochondria, were discussed fully in Chapter 7. Mitochondria are self-duplicating and have their own DNA, which is circular and contains some 16,569 bases that encode genes whose protein products have key functions. Each human cell may contain hundreds or even thousands of mitochondria. The mitochondria are fueled by oxygen and release potentially tissue-damaging, highly reactive electrons—the so-called free radicals. Free radicals have the potential to cause cell damage by cutting DNA, causing DNA mutations, or causing aggregations of molecules through cross-linking reactions. Mutations in mitochondrial DNA accumulate with time—in particular, at a location that controls replication. Certainly the frequency of mitochondrial DNA deletions or mutations rises progressively with age, not only in human tissues but also in monkeys, rats, mice, the fruit fly, and primitive worms. The greatest number of such changes are found in tissues with high oxygen utilization, including heart muscle, brain, and muscles. Notwithstanding the clear demonstration that specific mutations occur in the replicating region of the mitochondria, often with high frequency, it has not yet been proven that these mutations play a key role in aging. Certainly the mitochondrial furnace—the fire of life—may be slowly extinguished as we age.

The Protein-Sugar Mix

As we age, proteins increasingly combine with glucose (sugar) to form a product that has been linked to stiffening of connective tissues, hardening of arteries, development of cataracts, loss of nerve function, and diminished kidney efficiency. These features, especially blood vessel, cataracts, and kidney problems, occur with high frequency in insulin-dependent diabetics, who typically have high serum glucose levels. These complications are likely to be directly related to the increased protein/sugar combination caused by high blood glucose. Protein/

sugar combination is thought to be a potentially significant factor in the deterioration of certain tissues with age.

The Molecular Defense Against Aging

Our ability to resist and repair cell and DNA damage is essential to our survival and our longevity. Many of the cell mechanisms that keep our cellular machinery in balance have been discussed above. Key defenses include antioxidant mechanisms signaling cascades by cells in response to DNA damage, heat, cold, or other toxic stresses that activate protective reactions. Cells contain a variety of compounds that react with and inactivate various harmful agents. These include vitamin C and vitamin E. In addition to the multiple biochemical defenses are a wide range of genetic responses providing protection to our cells. In response to stresses that include heat, cold, trauma, heavy metals, and cortisone, our cells produce certain proteins (known as heat shock proteins) that turn on various heat shock genes. High levels of heat shock proteins are designed to protect cells against further damage, assisting in the removal of damaged molecules and activating repair systems. This heat shock response is diminished as we age, resulting in an increased sensitivity to cell damage and an increased rate of aging. Our ability to maintain a balance in our cell chemistry is achieved through these protective mechanisms or through repair, in which resides the key to our longevity.

Aging and Cancer

The U.S. National Institute on Aging has emphasized the cancer burden with aging. Some 60 percent of all cancers and 70 percent of all deaths from malignant tumors occur in the age group of 65 and older. This realization can be expected to have a very serious impact on the health status of aging Americans in the coming years. The numerical consequences will call for dramatic changes in health policy economics: If we are to avoid needless pain and suffering, then we must initiate preemptive and preventive—although not necessarily inexpensive—interventions. The elderly segment of the population (65 years and over) has increased from 12.3 million in 1950 (8.1 percent of the population) to 34.0 million (12.7 percent of the population) at the turn of the century. By 2030, the U.S. population will have doubled, reaching some 350 million, 1 in 5 (about 20.0 percent) of whom will be 65 years old or older. There will be an estimated 70 million elderly.

This age group has a cancer risk some 11 times greater than those under 65 years of age. This risk applies to all major cancers. Cancers of the colon, rectum, stomach, pancreas, and bladder account for some two-thirds to three-fourths of cancers occurring in elderly men and women. More than 65 percent of lung cancers and over 50 percent of non-Hodgkin's lymphomas occur in the elderly. Not unexpectedly, at least 77 percent of men with prostate cancer are older than 65. Contrary to media-generated emphasis, about 48 percent of breast cancers and 46 percent of ovarian cancers occur in older women.

Acutely aware of the serious public health implications of these statistical data, the National Institute on Aging has inaugurated multiple programs aimed at cancer control in the elderly. Meanwhile, seniors need to pay attention to their family history and personal health, and pay annual visits to their physicians. No symptoms or signs should be ignored, and blood tests for cholesterol, blood sugar, and prostate-specific antigen, as well as other tests including mammography and colonoscopy, should be pursued on a regular cycle and without delay.

Aging Sperm and Eggs

Sperm are made in a continual process, which takes about 64 days to complete. The original "factory" cells remain part of the key structure of the testes throughout life. These cells are not only subject to advantageous and harmful inherited genes but also to the environmental exposures that accumulate over a lifetime. The latter include toxic chemicals, viruses, X-rays, traumas, interference with blood supply, medications, excess heat, and other potentially damaging exposures.

Perhaps due in part to accumulated damage from various environmental exposures, advanced paternal age is known to be associated with an increased frequency of offspring with certain genetic disorders due to single point mutations. Examples include Marfan syndrome and achondroplasia (see Chapter 7). Why these two and other conditions might increase with paternal age is not yet fully known. An increase in the frequency of sporadic cases of hemophilia A is recognized in association with the advanced age of maternal grandfathers, the implication being that a mutation occurred in the germline cells of the mother's older father. There are a number of other, rarer, dominantly inherited genetic disorders due, or probably due, to mutations related to paternal age. All in all, however, the overall risk for all disorders due, or probably due, to advanced paternal age is small—almost certainly well below 1 percent.

The association of advanced maternal age with specific numerical chromosome abnormalities such as Down syndrome is better known. This association is explained in Chapter 3.

Premature Aging Syndromes

Pointing indisputably to the role of genes in aging are very rare, inherited disorders. Two are regarded as being recessively inherited, with the harmful genes coming from both parents. The first, called *progeria,* is characterized by startling features of advanced age during childhood, often evident before the child has reached 10 years. Growth is severely limited, and other aging features occur, including total hair loss, osteoporosis (thin bones), blood vessel and heart disease, and heart attacks in early adolescence. Bypass surgery or angioplasty (to unblock coronary arteries) have reportedly been performed on children as young as 14 years. Skin fibroblasts grow poorly, surviving through only about fifteen duplications rather than the expected forty to fifty. Researchers have not yet identified the defective gene causing this syndrome.

Like progeria, *Werner's syndrome* is characterized by advanced features of premature aging; but these features do not appear nearly as early as in progeria. The disorder is usually first noticed around puberty, when the normal growth spurt fails to occur. Over time, all of the well-known effects of aging become manifest, as do frequent complications of aging, including diabetes, calcified heart valves, and skin ulcerations. The median age of death in those affected by the syndrome is about 47 years. The culprit gene is located on chromosome 8.

Several other aging disorders involve defects of connective tissue due to mutations in collagen genes. In some affected by these disorders, premature aging is noted mostly in the face, hands, and feet; others may have characteristic mottling and darkening of the skin.

Although the recognized premature aging syndromes have been well studied and much is known about them, thus far this knowledge has been of little help to us in unraveling the intricacies of the normal aging process.

Is Longevity Inherited?

Human life expectancy has changed dramatically through the millennia. Prehistoric human remains rarely indicate ages greater than about 50 years. In ancient Rome, the average life expectancy at birth was only 20. Over the past 100 years, in developed countries, life expectancy at birth has increased from about 48 to 76 years. It took the previous *1,900 years* to achieve the same improvement in life expectancy! Clearly these advances have been directly related to the successful treatment and prevention of infectious diseases and, more recently, to the more accurate and more timely diagnosis and more effective treatment of cancers and cardiovascular disease. Enormous human and financial investments in biomedical research have successfully increased life expectancy through innovations in early diagnosis and treatment of serious diseases. However, the development of anti-aging interventions has not been as impressive.

Postponing Aging

Theorists have long suggested that delays in reproduction could eventually and significantly postpone aging. Certainly fruit flies that have had reproduction delayed across 10 or more generations have been shown to live 1 to 3 times longer than normal fruit flies and remain healthy much longer as well. Not only are these flies very active but they display superior physical capabilities, are better able to resist lethal stresses (for flies, that is), and even show more athletic prowess!

Is aging inevitable? In species that reproduce, the answer is probably in the affirmative. However, in the classic example, the asexual sea anemones show no ill health with age. Moreover, an Italian study of specially bred mice with a specific gene mutation showed that they were more resistant to oxidants (molecules that wear down tissues by the same process that causes metal to rust). Not only did this mutation induce resistance to stress; it was also associated with a 30 percent increase in life span. Clearly, genes make a difference.

Endless concoctions and remedies have been used from time immemorial to delay or remedy aging. Virtually all, including modern drugs and hormones, have proven useless. Indisputably valuable interventions that promote health and longevity have, however, been recognized, and should by now be common knowledge: These interventions involve long-term lifestyle changes, such as the cessation of smoking, initiation and maintenance of regular exercise, and consumption of nutritious, low-fat foods aiding the maintenance of a healthy body weight. Contradictory evidence exists for the use of high doses of vitamin E in reducing the risk of some age-associated diseases, such as cardiovascular disease.

The promise of gene therapy may well come to fruition in this century, and it is not difficult to anticipate genetic enhancement of our immune defense mechanisms. In fact, genetic techniques have already been used successfully to increase the life span of the fruit fly. With the completion of the Human Genome Project and new knowledge of the full complement of human genes, the pace of learning about how genes function and how their dysfunction can be remedied will emerge in this golden era of genetics. Remarkable remedies can be expected from the knowledge gained through the identification of all human genes. Humankind will benefit enormously from this incredible milestone—which nevertheless marks only the end of the beginning.

AVOIDANCE AND PREVENTION OF GENETIC DISORDERS

chapter 22

Prenatal Diagnosis of Genetic Disorders

···

If, of all words of tongue and pen,
The saddest are, "It might have been,"
More sad are these we daily see:
"It is, but hadn't ought to be."
—Bret Harte, *"Mrs. Judge Jenkins"*

The anguish shared by parents following the birth of a child with a serious birth defect, genetic disorder, or mental retardation is profound and everlasting. The feelings engendered by this sadness are compounded by distress and anger at the realization that it might have been avoided or prevented, and that a lifetime of suffering for the child, as well as the resultant burden on the family, could have been averted.

To avoid or prevent birth defects or genetic disorders, pregnancies should be *planned*. Such a plan presupposes careful consideration and knowledge of the family history as well as consultation with an obstetrician and, if necessary, a clinical geneticist. (See Chapters 2 and 25.) Armed with the knowledge of risks and informed about all the available options, parents have the freedom to choose their own approach. This chapter focuses on the critically important option of diagnosing a defect in the fetus in early pregnancy.

Early Warning

We are all comforted by the knowledge that about 96 in every 100 babies are born without evidence of major birth defects. The 3 or 4 percent who are not born sound are the root cause of our anxieties during pregnancies. Virtually all of us take on these background risks in having children. However, when there are indications of birth defects or genetic disorders in the family, or when other risk indicators exist (see Chapters 2–5; 7–9; 23), genetic counseling should be sought (Chapter 25) and prenatal diagnosis may be an option.

Preconception Planning

A careful, judicious couple will seek counseling prior to conceiving a child. Much grief can be avoided by couples who follow key recommendations, including folic acid supplements (see Chapter 9), adjustments of medications, and control of conditions such as diabetes and epilepsy (see Chapters 9 and 23). Clearly, once parents make a decision to have a pregnancy and elect prenatal diagnosis, the possibility of abortion of a defective fetus may need to be addressed. Those opposed to terminating a pregnancy should explore all of the options prior to conceiving so as to avoid even having to address the issue of abortion later. Preconception planning is the most important step would-be parents can take to avoid having a child with a serious birth defect or genetic disorder. This critical step is the very one that most couples omit, and may spend a lifetime regretting.

Screening in Early Pregnancy

Among the most important advances following the introduction of routine prenatal diagnosis in the mid-1960s was the development of screening pregnant women's blood samples for two specific classes of fetal defects—neural tube and chromosome defects. These are the two most common kinds of serious birth defects in the Western world. Neural tube defects include various disorders in the fetus that result from the failure of the tube forming the brain and spine to close. A failure of closure at the head results in exposed brain (called anencephaly), and an opening along the spine exposing the spinal cord is known as spina bifida. Anencephaly is not associated with survival much beyond birth (with rare exceptions), whereas spina bifida will almost always cause a lifetime of problems. In a series of 117 individuals with spina bifida, all had surgery to close the defect within 48 hours of birth. By 25 years of age, 48 percent had died. Expanded spaces within the brain (hydrocephalus) occurred in 85 percent. Visual problems occurred in 44 percent, including two children who were blind. Mental retardation was noted in 30 percent, who had IQs of less than 80. Leaking urine was a problem in 74 percent, and fecal incontinence occurred in 8 percent. Sixty-seven percent were wheelchair-dependent, and 54 percent required continuous, lifelong care. Chronic pressure sores (31 percent), epilepsy (23 percent), hypertension (15 percent), depression (7 percent), and obesity (26 percent) constituted the balance of the problems experienced at a mean age of 25 years. Early diagnosis and surgical repair while the fetus is in the womb has not led to improved function of the legs but may have reduced the frequency of surgical decompression (shunting) of the head with hydrocephalus, otherwise needed in many affected children. The various chromosome defects, including Down syndrome (see Chapter 3), result in mental retardation and multiple birth defects as their most serious potential consequences.

Before the advent of screening, about 95 percent of all babies born with neural tube defects and close to 80 percent born with Down syndrome were delivered without any advance warning of the fetal defect or recognition of any increased risk. The vast majority of other chromosome abnormalities also occurred without prior warning. Screening facilitating the recognition of pregnancies with increased risks for these defects represented an enormous advance. But even though maternal serum screening for neural tube defects has been possible since the early 1970s, it did not become standard practice in the United States until 1985. Even now, some Western countries do not provide such screening for all pregnancies. The screening of pregnancies to detect those at increased risk of chromosome defects also has been available for over a decade.

The prime purpose of screening is to identify pregnancies not otherwise known to be at risk for birth defects and then to provide precise diagnostic studies to secure the safest delivery, or the option to discontinue the pregnancy.

Blood samples are collected around the sixteenth week of pregnancy, and measurements are made of three blood constituents—AFP, hCG, and uE_3 (alpha-fetoprotein, human chorionic gonadotropin, and unconjugated estriol). A fourth constituent (inhibin) has been added to the chromosome screen. Measuring AFP enables the detection of virtually all pregnancies in which the fetus is later found to have anencephaly. Close to 90 percent of pregnancies are detected in which the fetus is subsequently found to have spina bifida. These neural tube defects are suspected when high levels of the protein leak from the open fetal defect into the surrounding amniotic fluid and thence into the mother's blood circulation, where elevated levels are found. Other defects that leak may also raise the maternal blood levels of AFP.

Since the level of AFP in the fetal blood is 150 to 200 times higher than it is in mother's blood, even a small leak directly into the mother's blood may raise the AFP level. Because the measurement of this protein is very accurate, finding an elevation of AFP in mother's blood in the absence of a demonstrable fetal defect usually means a small breach in the integrity of the placental interface between fetus and mother. Unexplained elevated AFP levels (see below) usually imply a breach in the placenta, which could prove problematic later in pregnancy. Raised AFP levels signal pregnancies at increased risk of premature labor, low birth weight, toxemia, and other potential complications, highlighting the need for increased surveillance during the last three months of pregnancy and throughout labor and delivery.

A high-resolution ultrasound study of the fetus is recommended if an elevated maternal serum AFP is detected. Targeted (high-resolution) ultrasound will reveal the diagnosis of anencephaly and close to 100 percent of spina bifida. The possible exception is the tiny defect near the end of the spine, which may be missed by the ultrasound imaging. Other defects that can cause elevated AFP also might be detected at this time. Twins, triplets, and other multiple pregnancies, with each fetus functioning as a "factory" making AFP, will obviously raise maternal blood levels and will also be detected by the ultrasound study.

An abnormal maternal serum screen showing an elevated AFP level should automatically lead to the recommendation of a high-resolution ultrasound study. Even if the subsequent ultrasound report is normal, I would also encourage an amniocentesis. The aim would be to measure AFP in the amniotic fluid as well as an enzyme (acetylcholinesterase) that is secreted from brain or nerve cells. The presence of this enzyme in a clear (nonbloodstained) amniotic fluid almost invariably points to a leak from the nervous system, such as that due to spina bifida. There are a host of different disorders that may cause elevations in amniotic fluid AFP. Hence, again, even in the presence of a normal ultrasound report, an amniocentesis would be recommended and should include chromosome studies of the cultivated cells derived from the fluid.

It should be noted that in about 5 percent of cases with spina bifida, the spinal lesion is covered by skin, does not leak, and will not be reflected by elevated levels of AFP in either the amniotic fluid or the mother's blood. In such cases, high-resolution ultrasound would not ordinarily be done except in cases where there is a family history of a neural tube defect such as spina bifida or anencephaly, or because the mother has taken medication that may cause spina bifida (such as the anticonvulsant valproic acid).

In contrast, low levels of AFP and uE_3 in conjunction with temporarily higher levels of hCG and inhibin together indicate increased risks of Down syndrome or another chromosome defect. Risks of Down syndrome are figured from complex calculations that take into account the mother's age, weight, race, the week of pregnancy, presence or absence of diabetes, and the level of the various measured blood constituents. When such a risk exceeds 1 in 270, I would recommend that an amniocentesis (see below) be done. By this method, over 80 percent of pregnancies in which the fetus has Down syndrome will be detected, as well as more than half of other chromosome abnormalities. A host of other, rarer fetal defects also may be detected through screening.

Screening findings that lead to an amniocentesis for Down syndrome or chromosome defects and conclude with normal chromosome and ultrasound results cannot be ignored: In close to two-thirds of such "unexplained" cases, late pregnancy obstetrical complications may occur; so in these cases, too, careful surveillance should be instituted. Since the hormones being measured are of placental origin, abnormal levels imply a dysfunction in this organ or a defect in the fetus. Hence, abnormal screening results signal a disturbance in normal function as well as the possible presence of a fetal defect. That is precisely their value.

Fortunately, the vast majority of abnormal serum screening results are not associated with a fetal defect or an adverse outcome to pregnancy. Nonetheless, considerable anxiety may be engendered when an abnormal result is obtained. Remember, however, that without screening, up to 80 percent of pregnancies with fetal Down syndrome and up to 90 percent with spina bifida would go undetected. There are also advantages to detecting pregnancies at previously unknown obstetrical risk, which can then be given more careful surveillance than they otherwise would have. Avoiding anxiety by not being screened does not limit a mother's risks, but it does limit her choices. Any anxiety generated "unnecessarily" during pregnancy is dwarfed by the anguish following the birth of a child with one of these defects. The ensuing emotional chaos and economic family burden, coupled with the suffering of the child, weigh heavily in favor of maternal screening.

Prenatal Diagnosis

All couples wish to have a healthy child. Couples who have an increased risk of bearing a child with a particular defect will most often seek ways to avoid or prevent such an outcome. Methods to avoid the development of defects that are not directly due to a chromosome or gene abnormality were discussed earlier (Chapter 9). Once pregnancy has begun, the only way to verify that a child has no chromosome or gene defect is through prenatal diagnosis (with the option of selective abortion). Fortunately, on average, over 95 percent of prenatal genetic studies yield normal results. Indeed, more children are born *because* of the availability of prenatal diagnosis compared to the number of pregnancies otherwise terminated due to defects. Prenatal diagnosis is therefore a life-giving approach, since many more couples choose to have children in the presence of high risks, given the assurance provided by prenatal diagnosis.

How Are the Studies Done?

The fetus floats within a cushion of fluid (amniotic fluid). Cells from fetal skin, intestinal and urinary tracts, lungs, and from the enveloping sac containing the fetus are shed into the fluid. Most of these are dead cells, but a few are viable. A needle introduced through the abdomen and through the wall of the uterus into the fluid (a procedure called amniocentesis) allows for the aspiration of about two tablespoonsful. Cells are removed from this fluid sample, placed in sterile plastic containers, fed with sterile broth (culture medium), and placed in warm, moist incubators. The modern, enriched broth stimulates the growth of cells that have adhered to the dish or glass cover slip. The cells slowly multiply, and about 7 to 10 days later, there are usually enough cells to analyze.

Amniocentesis is best (and safest) when done between the fifteenth and sixteenth menstrual weeks of pregnancy, when there is the largest number of viable cells, when the uterus is easily accessible by an abdominal approach, and when there is sufficient amniotic fluid to sample.

Procedural Risks

Introduction of a needle into the womb was first used in the early 1880s for the removal of excess fluid. Fetal evaluation by this method began in 1930, but modern prenatal studies as we know them today really began in the mid-1960s. Amniocenteses performed between the twelfth and fourteenth weeks are more hazardous to the fetus than the procedure described above, performed between the fifteenth and sixteenth weeks. The main risk is fetal loss. In highly experienced hands, the rate of loss approximates 1 in 1,000, with average rates ranging from 1 in 200 to 1 in 100. Clearly, the risk of the condition to be studied should exceed the potential rate of loss. Ultrasound guidance of the needle during amniocentesis is recommended, as it lessens the risks.

Amniocentesis for multiple pregnancy (twins, triplets, and so on) is best done by an experienced obstetrician. In good hands, the risks of fetal loss in twin pregnancies are about the same as for singletons. Amniocentesis in cases of multiple pregnancy is not unusual, given that the frequency of nonidentical twins increases with advancing maternal age. The rate of nonidentical twinning is between 7 and 11 per 1,000 births, whereas that for identical twins is 3 to 4 per 1,000 births. The widespread use of hormones to stimulate the production of ova (eggs) has resulted in a significant increase in multiple pregnancy, ranging as high as 30 percent. Whereas identical twins are almost invariably affected by the same genetic disorder, with nonidentical twins one might be affected and the other not. It is therefore not rare, for example, to find that one nonidentical twin has Down syndrome. The options in such circumstances are either to continue with the pregnancy as is or to have a selective reduction performed, eliminating the affected twin and allowing the other twin to continue developing until delivery.

Life-threatening maternal risks are extraordinarily rare. There have been two or three deaths in the world after millions of amniocenteses, with infection being the likely cause. Minor maternal problems occur in 2 to 3 percent of women and include transient vaginal bleeding or leakage of amniotic fluid, and uterine cramps. Potential fetal risks other than loss include needle puncture and injury, infection, or placental separation.

Women who have the blood group Rh negative and who may be carrying an Rh positive fetus should receive an immunoglobulin injection at the time of amniocentesis to prevent their becoming sensitized, leading to risks for a fetus in future pregnancies.

It is important for women to have a full discussion with their physicians about the indicators prompting prenatal studies and the associated risks, before undergoing any procedure. Written consent is usually required.

Diagnostic Accuracy

Prenatal genetic studies are among the most accurate of laboratory tests. Biochemical analyses are done on the cell-free amniotic fluid, and the cultivated cells are used for the study of chromosomes and genes. Accuracy rates are almost invariably in the 99 to 100 percent range. Nevertheless, care should be taken in selecting laboratories, with special reference to their experience and their ability to interpret results. Laboratory errors have certainly occurred, but fortunately they are infrequent. One potential pitfall is the inadvertent analysis of maternal cells in the sample of amniotic fluid. On those occasions where this possibility is recognized, a second amniocentesis study would be recommended.

Ultrasound Imaging

The use of sound waves (sonar) to visualize both the external and internal features of the fetus has brought about enormous advances in fetal medicine. Routine ultrasound studies are not aimed at identifying fine fetal anatomical details. Rather, the focus is on the number of fetuses present, measurements to assess fetal age and growth, and determination of the placental location, with the aim of finding a safe spot to introduce the needle for amniocentesis. Visualization of fine anatomical details such as the inner spaces of the brain (ventricles) or the four chambers of the heart requires high-resolution or targeted ultrasound, and should be performed by experienced obstetricians or radiologists. Despite the remarkable advances in fetal imaging, considerable limitations remain. Although there may be high degrees of certainty excluding a particular defect, most ultrasonographers will stop short of providing absolute guarantees. Reservations are common when reference is made to spina bifida, cardiac defects, and brain abnormalities.

Very Early Pregnancy Prenatal Diagnosis

Every couple with a significantly increased risk of having a child with a genetic defect, if they are willing to have prenatal diagnostic tests performed, would opt for doing the studies at the earliest possible stage of pregnancy. Indeed, diagnoses of chromosomal and single gene disorders, although first described by the Chinese in the 1970s, did not enter the clinical arena until the early 1980s. Genetic diagnosis began to be routinely achievable between the tenth and twelfth weeks of pregnancy, thereby reducing stress, maintaining privacy, and facilitating termination before pregnancy became obvious. Maternal bonding with the fetus is much less likely to be a problem in the face of necessary pregnancy termination in the first three months of pregnancy than between the sixteenth and twentieth weeks.

Procedures

Very early prenatal diagnosis is based on a tiny tissue sample from the placenta, which is used for chromosome or gene analysis. The procedure is like amniocentesis, with a needle being introduced into the uterus under ultrasound guidance. A tiny piece of the branch-shaped tissue (chorionic villi) of the placenta in the area of attachment to the uterus is removed (hence the procedure is called chorion villus sampling, or CVS). Given good ultrasound visualization, CVS for twins is also almost invariably successful.

The CVS tissue can be cultivated and chromosomes and genes thereafter analyzed. Alternatively, direct extraction of genes from the tissue can lead to even quicker gene analysis. Direct examination for chromosomes, although achievable, is less reliable than that performed on cultivated cells. Since the tenth week (when CVS is usually performed) is too early to obtain amniotic fluid, ultrasound studies to rule out spina bifida and similar defects should be done, where indicated, around the sixteenth week of pregnancy.

There are very few instances in which CVS tissue cannot be obtained. Perhaps the commonest situation is a malposition of the uterus. In such cases—and sometimes even for routine cases—some obstetricians choose to do the CVS procedure via the vagina and cervix. The presence of infection in the vagina, however, would add risk and would necessitate prior treatment.

Accuracy

Analysis of the tissue obtained at CVS for either chromosomes or genes is equivalent to that of cells cultivated from the amniotic fluid. Admixture of maternal cells is somewhat more likely to occur with CVS, increasing the possibility of a diagnostic pitfall. Moreover, since confined placental mosaicism (see discussion in Chapter 4) may occur in 1 to 2 percent of placentas, the chance sampling of chromosomally abnormal cells may lead to an erroneous diagnosis.

Risks

Most comparative studies of CVS performed between the ninth and twelfth weeks of pregnancy and amniocentesis performed between the fourteenth and twentieth weeks of pregnancy have not revealed statistically significant differences in fetal loss. Some studies, however, have pointed to an increased fetal loss rate from CVS over amniocentesis, with rates in CVS being around 2 percent, compared to 0.5 to 1.0 percent for amniocentesis. The complications that could occur are similar to those noted for amniocentesis.

In the early 1990s, concern arose about a reported cluster of four children born with limb and lower jaw abnormalities following CVS. Subsequently, major studies of more than 200,000 cases revealed no definitive association of such defects with the CVS procedure. The one recommendation that did emerge was that CVS not be performed too early. The optimal timing suggested was between the tenth and eleventh weeks.

Preimplantation Diagnosis

Some couples face a high risk (25 to 50 percent) that a future child will be born with a serious, possibly fatal genetic disorder. An ideal solution would be to check the fertilized egg before it implants itself in the lining of the uterus. This procedure has been accomplished in many hundreds of cases worldwide, and is called *preimplantation diagnosis.*

Procedure

The process involves in vitro fertilization, in which a harvested egg produced after hormonal stimulation of the mother is fertilized in a dish by contributed sperm. After fertilization, when the embryo is between four and eight cells in size, it is possible through micromanipulation to extract one cell for analysis of a single gene or specific chromosome abnormality. Since each of the eight cells possesses the entire genetic program for the future individual, the removal of one or even two cells does not harm the later development of the embryo and fetus. This process of embryo biopsy of individual cells (called blastomeres) has been followed by successful implantation and delivery of a healthy child in many pregnancies around the world.

Preimplantation gene analysis proceeds only after the precise mutation has been determined in both parents (recessive disorders) or in one parent (dominant or X-linked disease). The discovery that chromosomal mosaicism (see Chapter 4) may already be present in an embryo of between eight and sixteen cells in size adds a little more concern. If an abnormality is found, the fertilized egg is discarded. For those with an antipathy to abortion, preimplantation genetic diagnosis may be an important alternative.

Limitations

Preimplantation genetic analysis has inherent limitations, even beyond the considerable costs involved and the limited number of medical centers performing the procedure. Since extremely minute quantities of DNA are available from a single cell, chance contamination by extraneous DNA can cause and has caused diagnostic problems. The process of analyzing DNA from a single cell requires that the minute amount present be amplified. However, failure to amplify may occur in between 5 and 15 percent of cases. In addition, the diagnosis of cystic fibrosis has been missed, and the gender of the embryo has been mistaken. Nonetheless, many accurate diagnoses have been made on the basis of this type of testing, including diagnoses of Tay-Sachs disease, muscular dystrophy, and cystic fibrosis.

Pregnancies that occur after the rigorous efforts of preimplantation genetic diagnosis and in vitro fertilization are precious indeed. Nevertheless, careful consideration should be given to confirming the normal result by amniocentesis at sixrteen weeks of pregnancy. It is easy to comprehend that the majority of parents

TABLE 22.1   Reasons for Prenatal Diagnosis of a Chromosome Defect

1. Advanced maternal age

2. An abnormal maternal serum screening result

3. A previous child with a chromosome defect

4. One parent is a carrier of a structural chromosome rearrangement

5. A fetal abnormality detected on ultrasound study

6. Previous removal of an ovary (or sections of one or both ovaries)

in such hard-won pregnancies might resist taking any further risk of fetal loss by having a confirmatory amniocentesis.

Indications for Prenatal Genetic Studies

Prenatal genetic diagnosis is performed on a fetus to detect suspected chromosome abnormalities, single gene mutations, biochemical genetic disorders (due to single gene defects), and anatomical defects.

Chromosome Defects

A couple might consider having prenatal chromosome studies done in order to detect chromosome abnormalities for any one of the reasons outlined in Table 22.1. A brief look at these indications may be helpful. With advancing maternal age, there is an increased risk of having a child with one of five chromosome disorders, each characterized by the presence of an extra chromosome (see Chapter 3). The most common is Down syndrome (trisomy 21), the others being trisomy 18, trisomy 13, Klinefelter syndrome (XXY), and triple X (XXX chromosomes). Precisely why advancing maternal age is associated with an escalating risk of these disorders remains uncertain. Although the mechanism of sticky chromosomes (also called *nondisjunction*) is understood (see Chapter 3), the initiating mechanisms remain obscure. The escalating risk with age may be tied to the body's natural protective mechanisms against harboring chromosomally defective offspring. For example, the body is remarkably efficient at naturally discontinuing (by miscarriage) the vast majority of chromosomally abnormal embryos and fetuses. One likely reason for the increasing rate of certain numerical chromosome defects among older women may be a progressive breakdown in the body's surveillance (possibly a flaw in the immune mechanism), allowing a greater number of chromosomally defective embryos to survive.

Prenatal genetic studies are recommended routinely for women who become pregnant at 35 years or older. There is, however, a steadily, though slowly, rising risk of chromosome abnormality beginning in the late twenties. The arbitrary choice of age 35 is used since that is the time that approximately balances the risk of fetal loss due to amniocentesis and the likelihood of a chromosome defect.

Maternal serum screening, discussed earlier in this chapter, facilitates the recognition of pregnancies at risk for fetal chromosome defects. These screening studies serve as a very common indication for amniocentesis.

Parents who have had a child with the common trisomy 21 form of Down syndrome have a 1 to 2 percent risk of recurrence in a subsequent pregnancy when the mother is under 35 years of age. If this has occurred after 35 years of age, the risk of recurrence or of another chromosome defect depends on the maternal age at the time of the future conception. At 36 years of age, the risk of some type of chromosome abnormality approximates 1 percent, at 40 it is about 2 to 3 percent, and at 45 the risk ranges between 7 and 10 percent.

Women who have had a child with a chromosome defect, or a miscarriage in which a chromosome defect was detected, may have increased risks of recurrence in a subsequent pregnancy. After two miscarriages, the risk that one partner carries a chromosome abnormality is between 1 and 3 percent. In couples that go through three or more miscarriages, there is a 3 to 8 percent likelihood that one or the other carries a chromosomal abnormality. Blood chromosome analysis of both partners is recommended in these circumstances, to determine whether one or the other has or carries a chromosome abnormality.

The most recently recognized added risk of having offspring with Down syndrome is for women who have had an ovary (or part of one or both ovaries) surgically removed, or who were born without one of the ovaries. A report from the Centers for Disease Control and Prevention noted a 9.6-fold increased risk of Down syndrome among women with a "reduced ovarian complement." Although the mechanism is uncertain, the physiological state of the ovary is the focus of attention. The reported risk is clear enough to justify offering prenatal genetic studies to such women.

Single Gene Disorders

The chromosomal disorders just discussed are associated with risks of recurrence below 15 percent—mostly in the region of 1 percent. In contrast, risks of occurrence or recurrence due to disorders caused by single gene defects range (almost always) between 25 and 50 percent. In this group are many couples who in the past—prior to the availability of prenatal diagnosis—decided not to have a child (or another child), knowing that they faced high risks. Prenatal diagnosis now provides the reassurance that allows such couples to have as large a family as they wish. Specifically, prenatal diagnosis allows couples the opportunity to have unaffected children selectively, sparing the parents the agony of losing a child, and more importantly, sparing an affected child the pain and suffering of early death or serious disease or deformity.

I discussed disorders caused by defects in single genes (dominant, recessive, X-linked, and mitochondrial) fully in Chapter 7, including their risks of recurrence. For prenatal diagnosis in any of these four categories of single gene disorders, the first requirement is to identify precisely the gene defect or defects in one or both parents or in a previously affected child. Only with fulfillment of that requirement can prenatal diagnosis proceed to a specific, definitive conclusion. Prenatal studies for single gene disorders are recommended for couples who have had a previously affected child, or in which one partner is a known carrier of a recognized mutation (in dominant or sex-linked diseases), or both partners are known carriers (in recessive diseases). If a gene defect has yet to be described or cannot be determined (as may be the case with particularly large genes), DNA linkage analysis (see Chapter 6) could provide prenatal diagnosis with certainties exceeding 99 percent. DNA studies, however, require the cooperation of family members in the form of blood samples (for example, from parents, siblings, grandparents, great-grandparents, nieces, nephews, and cousins). Obviously, not everyone has to participate, but the more key members do participate, the greater the certainty of diagnosis, as long as there is clear evidence of who is and who is not affected among those family members. Given the variability of expression among certain single gene diseases (especially those that are dominant), DNA linkage analysis is sometimes difficult to accomplish. Because prenatal diagnosis of single gene disorders poses complex challenges, couples with a family history of such a disorder are advised to seek a consultation with a clinical geneticist prior to pregnancy.

Single gene disorders that manifest themselves through disturbed biochemical functions of cells are individually rare. Collectively, however, they occur about once in every 100 children born. Many of these cause severe mental retardation, seizures, stunted growth, and early death. Prenatal diagnosis is achieved by cultivating amniotic fluid cells, which are then subjected to analysis of the potentially defective or absent enzyme. The first prenatal diagnosis of a biochemical disorder was made in the late 1960s. That disorder was Tay-Sachs disease (see Chapter 8).

A similarly rare occurrence necessitating prenatal studies is when carrier detection in one of the parents proves inconclusive or impossible. In very specific cases, it may be possible to assess accurately the activity of a specific enzyme, or the lack thereof, in the cultivated amniotic fluid cells.

Prenatal Studies for Physical Defects in the Fetus

About 2 percent of all newborns have a significant physical defect. Included in this group are cleft lip and palate, spina bifida, heart defects, hand or limb defects, and brain abnormalities. Prenatal detection of these defects mostly requires imaging by high-resolution ultrasound around the sixteenth week of pregnancy. Such studies are not normally done unless there is a history in the family, a previous affected child, a report of an abnormal maternal serum screen, or a suspicion of fetal growth restriction or other abnormality on a routine ultrasound study.

Pitfalls and Problems

Diagnostic Errors

Despite an accuracy rate exceeding 99 percent for prenatal genetic studies, errors have occurred and do occur (see legal aspects, Chapter 26). Fetal sex determination has been incorrect a number of times, mainly because of admixed maternal cells in the amniotic fluid. Errors in biochemical diagnoses also have occurred in laboratories that have not performed their biochemical analyses with the required care and rigorous standards. Errors in chromosome analysis, although very infrequent, probably account for the majority of mistakes. Missed chromosomal mosaicism (see Chapter 4) and failure to detect small structural rearrangements are the two most common errors. Given the complexity of the process and the high technical skill required, the high degree of accuracy generally achieved is remarkable.

Aberrant, but Normal

Chromosome analysis of cultivated amniotic fluid cells sometimes turns up one or more cells, derived from a single dish, that have a specific chromosome defect. If no other such cells are found in any other dishes, and sufficient numbers of cells are analyzed, this finding is referred to as *pseudomosaicism*, to which no clinical importance should be attached. It is likely that the origin of these abnormal cells is placental or that they arose spontaneously upon cell division in the culture dish.

An Extra Fragment

About once in every 2,000 amniocentesis studies, an unexplained, tiny, extra chromosome fragment—technically called a *supernumerary marker*—is found in all or some of the fetal cells. This chromosome fragment may have been directly inherited from one of the parents. If so, and if the parent is outwardly normal, then there is implicit assurance that the offspring too will be normal. If neither parent possesses the fragment in question, there is about a 14 percent risk that the child will have birth defects or mental retardation. New DNA technology using FISH (see Chapter 5) now allows the precise identification of the origin of these fragments. Curiously, fragments derived from chromosome 15 occur most often. The clinical implications depend upon the chromosome origin of the fragment, its size, and the specific genes that constitute the fragment.

Disputed Paternity

From time to time—much more frequently than many would believe—clinical geneticists are confronted with the question of paternity. Using DNA analysis techniques and blood samples (or cheek scrapings for cells) from the mother and child and from two (or more) men, it is possible to discover who is and who is not the father, with degrees of certainty usually of either 99.9 percent or 100 percent. The reader can imagine the emotional pressures that place the fetus at risk through amniocentesis in order to determine disputed paternity. The question is best resolved after birth.

Multiple Pregnancy

As noted earlier, twins or multiple pregnancies occur in which one fetus is found to have a serious genetic disorder. In cases of nonidentical twins or multiples, selective reduction is possible by causing the cardiac arrest of the one affected. The dead fetus is slowly absorbed by the body, leaving only a tiny sac of tissue by the time delivery occurs. In the vast majority of such cases, the healthy twin survives to bring joy to the parents.

Culture Failures

The growth of amniotic fluid cells in plastic containers using highly enriched culture medium rarely results in failure of cell growth. On the very unusual occasions when this does occur, a few common reasons are recognized. The sample may have been too small (because of difficulty in aspirating sufficient fluid) or may be very bloody. Moreover, amniotic fluid sent over vast distances is often subject to extreme heat or cold, and the few viable cells present may have died. Further, careless laboratory techniques or lack of proper sterility when the sample was drawn may result in infectious contamination of the sample, ruining any chance for cell growth. Samples in culture must be kept in warm, moist incubators. Electrical failure or some other mechanical failure may cause a serious drop or elevation in temperature within the incubator, in which case the cells will almost certainly die. Larger laboratories, like ours, are wired directly to an emergency electrical generator to take care of just such eventualities. Some years ago, we and others discovered that the syringes and/or tubes into which amniotic fluid was placed contained chemical substances toxic to cells. Fortunately, such occurrences have been few and far between, and today they are rare.

Diagnoses of the Unexpected

Prenatal studies performed for advanced maternal age or by way of follow-up on an abnormal maternal serum screening result are the two most common reasons for amniocentesis. Although the search for Down syndrome and neural tube defects occupies central attention, an entirely unexpected defect may be discovered. I vividly recall one case in which we discovered that the fetus was an XYY male, with all the implications (including involvement with the police, as discussed in Chapter 4). The amniocentesis had been obtained for advanced maternal age (38). Ironically, both parents were FBI agents. Although they initially had different opinions about whether the pregnancy should be allowed to continue, they eventually opted for termination.

The Future Is Now

The analysis of genes permits us to detect mutations for disorders that otherwise might not have become apparent until many decades after birth. Examples include breast cancer, colon cancer, and degenerative brain conditions such as Huntington's disease. These issues are fully discussed in Chapter 24. Faced with a family history of one of these disorders and the ability to detect the mutation in your fetus or that of your partner, how and what would you decide? This will become an increasingly common challenge. Meanwhile, we are fortunate to have had the enormous technological developments in prenatal diagnosis that now assure couples at high risk of passing along serious genetic disorders to their offspring that they can selectively have unaffected children.

chapter 23

Genetic Disorders and Pregnancy

··

What You Don't Know Can Hurt You

To undertake pregnancy in the face of a genetic disorder of either partner is to invite potential personal risk as well as possible health hazards to the fetus or child. Not bothering to determine all of the risks and options prior to conceiving a child who may be doomed to a lifetime of pain and suffering could be construed as a form of child abuse. There is little solace in the realization that up to 50 percent of pregnancies are unplanned. Sexually active women should realize that pregnancy is always a possibility, even if they are using contraception; and if they become pregnant they should seek genetic counseling immediately if they have a genetic disorder or known increased genetic risks.

Preconception Genetic Counseling

The reasons for genetic counseling prior to pregnancy were outlined in the previous chapter. There are genetic disorders that interfere with, limit, complicate, or prevent fertility or conception. Much anguish, useless effort, and unnecessary expenditure could be spared in such cases by obtaining timely genetic counseling, thereby avoiding years of fruitless efforts to become pregnant, including in vitro fertilization and other assisted reproductive techniques. There are also literally thousands of genetic disorders in which pregnancy can occur—but in many cases, only with attendant complications. Since it is beyond the scope of this book to discuss or describe all of these disorders, the thematic message to those with a genetic disorder who are contemplating pregnancy is to obtain preconception counseling (see Chapter 9). By this route, any threat to personal health can be anticipated and avoided or treated. In addition, besides the recognition of unwanted risks, certain attendant options can be addressed (see Table 23.1).

TABLE 23.1   Options to Consider When Faced with a Serious Genetic Disorder

1. Consultation with a clinical geneticist
2. Decision not to have children
3. Consideration of vasectomy or tubal ligation
4. Adoption
5. In vitro fertilization
6. Egg transfer into the fallopian tube
7. Embryo transfer
8. Artificial insemination by donor
9. Receiving a donated egg
10. Considering a surrogate to carry your genetic offspring
11. Needle aspiration of sperm from the testis for in vitro fertilization
12. Inserting a single sperm into a harvested egg
13. Genetic disease carrier detection tests (for a partner)
14. Prenatal diagnosis
15. Preimplantation genetic diagnosis
16. Fetal treatment for selected disorders
17. Folic acid supplementation while planning pregnancy
18. Selective abortion

The period *before* pregnancy is also the best time to lay aside habits that may be harmful to a future fetus/child. Smoking, illicit drugs, certain medications (e.g., anticonvulsants), excess alcohol, and occupational hazards top the list. I and my research colleagues discovered a tobacco-specific carcinogen (cancer-producing agent) in the amniotic fluid surrounding the fetus as early as the fifteenth week of pregnancy, in mothers who smoked during pregnancy. We are now studying the potential lifelong implications, to determine whether exposed offspring are at increased risk of cancer.

Genetic Disorders That Pregnancy May Aggravate

Dramatic advances in medicine have resulted in more women affected by genetic disorders surviving to childbearing age and becoming pregnant. A number of these genetic disorders may become worse during pregnancy. Awareness of these disorders and concomitant hazards allows anticipatory care and guidance as well as treatment during pregnancy. There are some rare disorders of the body's chemistry that if untreated could, for example, result in a period of mental confusion or coma during or after pregnancy and/or delivery. Venous and/or arterial thrombosis could also be a complication in these disorders, especially following cesarean section. Women with a personal history of thrombosis in their legs or a family history of so-called deep vein thrombosis, with or without clots shooting off into the lungs (pulmonary embolism), should obtain blood tests to determine if they have a defective Factor V Leiden gene or prothrombin gene (see Chapter 10). Since 3 to 5 percent of whites have a mutation in the Factor V Leiden gene, and about 2 percent have a prothrombin gene mutation predisposing the individual to thrombosis, careful attention to past history and family history could be lifesaving. Either of these mutations may also increase the likelihood of miscarriage. Anticoagulants are used for prophylaxis and treatment.

Another group of disorders includes those that interfere with the integrity of connective tissue. Sophisticated care and counseling are necessary in such cases. Marfan syndrome and the Ehlers-Danlos syndrome are the most important disorders in this group. (See Chapter 12 for more detail on these disorders.)

Marfan syndrome—the cardinal characteristics of which include tall stature; long fingers; long, thin chest and arms; a dented or prominent breastbone; heart valve defects; aortic aneurysm (dilatation/ballooning of the aorta); dislocated lenses of the eyes; and hyperextensible joints—demands urgent attention if pregnancy is planned or already in progress (see Chapter 12). Marfan syndrome is transmitted as a dominant disorder (with a 50 percent risk for having affected offspring). If affected women plan to have children at all, they are encouraged to do so in their early twenties, because of the risk of rupture of the aorta. Pregnancy should not be undertaken if there already is significant ballooning of the aorta (precise measurements are usually made). Monthly echocardiograms are recommended in pregnancy, and vaginal deliveries are permitted only after precise measurements have been made of the aorta. Cesarean section may be the recommendation if some ballooning of the aorta has already begun. Associated high blood pressure has to be treated aggressively, and certain drugs (beta-blockers) are used to slow the rate of aortic dilatation. There also may be a need for prophylactic antibiotics, if heart valves are involved.

Similar care, advice, and treatment may be necessary for women with the Ehlers-Danlos syndrome, in which stretchable connective tissue is the problem. (See Chapter 12.) Close, early attention to these matters could be lifesaving.

Pregnancy may aggravate the difficulty with breathing in cystic fibrosis, necessitating hospitalization. Women with neurofibromatosis may find that the size and number of visible lumps under their skin increase, resulting in cosmetic changes. If internal lumps also grow, additional complications could occur. Hypertension could become a problem in pregnancy with this disorder as well as others (e.g., polycystic kidney disease). Mother, fetus, and child could be at significant risk for all manner of complications when the mother has myotonic muscular dystrophy. Moreover, this condition itself may worsen during a pregnancy.

Women who have a family history of hemophilia (classical type A or type B) should determine whether they are carriers of one of these genes. Carriers are more subject to hemorrhage following delivery; and if the fetus is male and affected, hemorrhage could also be a risk for the child during or soon after birth.

These few examples illustrate the critical importance of genetic consultation (especially preconception), particularly for women with genetic disorders. Male partners may also be affected with certain disorders that can lead to grave complications in the child at delivery. I well recall a consultation with a couple that had suffered three early miscarriages. In the course of our discussion, I inquired about the reason for the bandage on the man's hand. His casual response referred to a slight fracture, one of a number he had had in his life. His words had barely been uttered when I asked him to take off his spectacles. The whites of his eyes had a strikingly blue tint. The family history and subsequent studies confirmed the clinical diagnosis of brittle bone disease (osteogenesis imperfecta), and the 50 percent risk that he could transmit this disorder to each of his future offspring. This disorder, however, was not the cause of the miscarriages.

Maternal Genetic Disorders That May Threaten Fetal Health and Survival

Maternal diabetes, sickle cell disease, epilepsy, and lupus erythematosus are particularly common causes of miscarriage, fetal death, stillbirth, and malformations. The risk of malformations is especially significant in the presence of maternal diabetes or epilepsy (see Chapter 9). Lupus is in some cases associated with fetal heart rhythm disturbances (heart block) detected on the electronic fetal monitor in the last one-third of pregnancy and during labor. This detection is important, since treatment options exist.

Mothers with myotonic muscular dystrophy who have a 50 percent risk of having an affected child must anticipate the potential of serious complications. These include failure to breathe at birth, oxygen lack causing brain damage, joint contractures, and later, mental retardation. One study showed that 12 percent of the offspring of affected women are stillborn or die in the newborn period; 9 percent survive, although severely affected; and signs are present but less obvious in another 29 percent. Being fully aware of all of these pregnancy-related risks facilitates optimal pregnancy care, and everything should be discussed even before pregnancy has begun.

Increasingly, women with the biochemical genetic disorder called phenylketonuria are reaching childbearing age and having children. Untreated, they have a higher than 90 percent likelihood of damaging the fetus they are carrying, resulting in mental retardation and in heart and other defects. It is critically important for such women to follow a very strict, special low-protein diet, initiated *before* pregnancy begins and continued throughout. Very tight adherence to this diet is highly likely to secure the health and survival, without mental retardation, of their offspring.

Genetic Disorders That May Complicate Pregnancy

Once again, women with genetic disorders or conditions that require long-term medication (such as epilepsy and depression) are advised to seek a genetic consultation concerning the effect of their disorder and its treatment on pregnancy and the developing fetus. Women who are severely affected by cystic fibrosis may jeopardize their health and survival by becoming pregnant, and should seek expert advice prior to doing so. Moreover, should they decide to proceed, their partners should be tested to determine whether they are also carriers of a cystic fibrosis gene mutation (see Chapter 7). Special care and attention may be necessary throughout pregnancy, and hospitalization is common during the last three months. Women with severe sickle cell disease may also become sicker during pregnancy. In some women with epilepsy, pregnancy may bring on an increased frequency of seizures, threatening the life of both mother and child. Moreover, the anticonvulsant medication taken almost invariably has a teratogenic (malformation-causing) effect on the fetus. Consultations should take place prior to pregnancy, aimed at selecting a medication that poses the least risk to the fetus. Insulin-dependent diabetic women must have their diabetes under tight control before initiating pregnancy. The worse the control, the greater the frequency and severity of birth defects that can be expected (see Chapter 9). The better the control, the lower the risks of having a child with congenital defects. In some neuromuscular conditions (e.g., limb-girdle muscular dystrophy), muscle weakness may increase during pregnancy. Because of muscle weakness, cesarean section may be necessary. However, anesthetic risks may be increased, and spinal anesthesia may be recommended as the best option.

Pregnancy is not the time to first inquire about your family medical history, nor is it the most appropriate time to initiate or change a treatment schedule. Women who wish to exercise some control over their reproductive destiny and to guarantee the best possible outcome will plan their pregnancies and seek preconception care and counseling. After the birth of a child with a fatal or lifelong defect, no one wants to hear the refrain "It is, but hadn't ought to be!"

PERSONAL MATTERS, PREDICTIVE TESTS, GENETIC COUNSELING, AND TREATMENT ···

chapter 24

Presymptomatic and
Predictive Genetic Diagnosis

....

If you knew you were at high risk of developing a certain cancer, would you opt for careful surveillance and even preemptive surgical treatment? If your children were at the same risk, would you secure their health and lives by taking the same preemptive actions? Virtually all of us would respond affirmatively to these questions. Advances in biotechnology over the past decade have made it increasingly possible to know our health risks and take preventive action. DNA testing in the presence of high risk (e.g., 50 percent), to determine whether or not you will develop a specific genetic disorder sometime in the future, is termed *predictive testing*. Once a predictive diagnosis has been made, if no symptoms or signs of disease are evident, then a regular surveillance test (or tests) is recommended. Such *presymptomatic* tests aim at the earliest possible detection of overt disease and early preemptive treatment. Loss of the chance to be forewarned of a serious cancer risk can have devastating (and medicolegal) consequences, as the case on page 310 illustrates.

This case emphasizes the importance of knowing precisely the nature of the diagnosis in first-degree relatives. This information should also be communicated to physicians even if they fail to ask. DNA diagnosis is now feasible and precise identification of an inherited disorder is achievable. An individual known to carry the defective gene for familial adenomatous polyposis coli would be placed on a routine schedule of colonoscopy, beginning at the age of 11 if not sooner (see Chapter 17 for details). Since such an individual has a virtual 100 percent certainty of developing cancer, the awesome decision of electively removing the colon has to be addressed in the late teens and early twenties. Of course, each of J.C.'s children now has a 50 percent risk themselves of having inherited this serious form of cancer. Predictive diagnosis or presymptomatic diagnosis is the best option available to them.

J.C. was 40 years old when she began experiencing lower abdominal aches and pains. Visits to her physician yielded recommendations first for antacid treatment, then for bowel relaxants, and finally stool softeners. No family history was taken. Only after many months of these symptoms, accompanied by significant weight loss and a negative physical examination, did attention focus on the possibility of a serious intestinal disorder. This focus emerged because of J.C.'s observation of blood in her bowel movement. Examination of her bowel with a fiber-optic instrument (colonoscopy) revealed hundreds of polyps (mushroom-like growths) carpeting the inner lining of the lower bowel. At one site, however, a cancerous mass invading the wall of the bowel was found. These observations enabled the formal diagnosis of familial adenomatous polyposis of the colon.

Surgical treatment included total removal of J.C.'s colon, radiation, and chemotherapy. Treatment was to no avail, and the cancer spread uncontrollably. A medical malpractice suit was filed against J.C.'s physician for not having taken a family history, which would have enabled her to be placed on surveillance, given the 50 percent risk she had because of her father's identical diagnosis. Her father had died of the same disorder, which had been evident in his own mother and one of his siblings. All J.C. knew was that there had been bowel cancer in her family; but her physician could easily have determined the diagnosis that had been made, at least in her father's case.

Presymptomatic Testing

Familial Cancers

The history just described illustrates the critical importance of the family history, attention to which would have enabled predictive and presymptomatic testing in J.C.'s case. Gene studies by direct mutation analysis, DNA linkage analysis, or gene sequencing can precisely determine an individual's genetic destiny as far as certain cancers and other single gene disorders are concerned (see Chapters 6, 16, 17). For the important cancers tabulated in Table 24.1, even after a specific mutation is detected, no forecast is possible about when such a cancer might occur. However, these cancers have extremely high degrees of penetrance (e.g., up to 100 percent) in a lifetime. Hence, extremely careful attention is necessary after a positive DNA result is reported. The recommended course of action would include close attention to your personal health, with prompt visits to a physician for any symptoms or signs that raise concern; blood tests; imaging studies (CT scans,

TABLE 24.1   Examples of Familial Cancer Syndromes with 50 Percent Risks of
Inheritance or Transmission

Cancer Type	Features
Familial adenomatous polyposis	Hundreds to thousands of polyps in the colon and intestine
Hereditary nonpolyposis colon cancer	Multiple polyps in the colon
Bilateral breast cancers	Cancers beginning independently in each breast
Multiple endocrine neoplasia	Tumors in endocrine glands (thyroid, parathyroid, pancreas, adrenal, pituitary)
von Hippel-Lindau disease	Tumors especially in brain, eyes, kidneys
Retinoblastoma	Tumors in the eyes

MRIs, and mammograms); and visits to appropriate specialists (see Chapter 17). Experience has shown that the vast majority of individuals at risk have indeed chosen presymptomatic testing for themselves and their children in these high-risk cancer syndromes. These steps will facilitate a prompt and early diagnosis enabling individuals to have preemptive treatment and thereby secure their health and save their lives.

An Australian study of 461 women, 80 percent of whom had a high risk (1 in 2 or 1 in 4) of breast cancer, and 20 percent of whom had a 1 in 4 or 1 in 8 risk, focused on attitudes toward testing for the two common breast cancer genes—BRCA-1 and BRCA-2 (see Chapter 17). The researchers noted that 92 percent of women at risk indicated their interest in being tested—a figure consistent with earlier reports. Factors that were important in helping women decide to be tested included:

1. help in understanding how to reduce personal risk of cancer;
2. opportunity to learn about their children's risk;
3. obtaining an accurate risk figure;
4. planning for the future;
5. making childbearing decisions; and
6. assisting research.

The high-risk women in this Australian study concluded "that the benefits of genetic testing outweighed its risks."

Testing for Adult Polycystic Kidney Disease

Less well known but also very important is the availability of DNA testing for the adult form of polycystic kidney disease. This disorder usually manifests itself between the third and fifth decades of life, the key symptoms and signs including blood in the urine, high blood pressure, abdominal pain, and kidney failure. Cysts may also appear in the liver, ovary, pancreas, spleen, and brain. In addition, abnormalities of heart valves, including mitral valve prolapse, may occur; and more ominously, small balloon-like bulges (called aneurysms) may occur in the blood vessels of the brain in about 8 percent of cases. The danger is that these aneurysms may rupture and cause a potentially fatal hemorrhage in the brain. Preemptive surgery could save a life, if a DNA analysis were to indicate inheritance of this gene. If one of your parents has or had this disorder, even if you have no symptoms or signs, an immediate ultrasound or CT scan study of your kidneys is recommended. At least 11 percent of cases occurring before age 30 are missed by ultrasound study alone. DNA analysis will provide degrees of certainty that are likely to exceed 99 percent in most families (for the most common gene), but will require linkage analysis (see Chapter 6) employing blood samples from multiple, related family members. Mutation analysis of the gene itself, which is still too expensive, will undoubtedly become feasible with further technical advances.

Adult polycystic kidney disease is not rare, with a frequency of about 1 in 1,000. Despite its name, this disorder may appear as early as the first year of life. Complicating plans for DNA linkage analysis has been the discovery that three different genes can cause the disorder, all dominantly inherited, with 50 percent risks for each of the children of an affected person. Most cases arise as a consequence of a mutation in the gene located on chromosome 16.

Given three possible causal genes, care is necessary in the analysis of DNA linkage results. DNA observations must especially be reconciled with those who are known to be affected. In one study of 141 affected individuals, 11 percent decided against bearing children on the basis of risk.

The importance of accurate presymptomatic tests for potential at-risk kidney donors cannot be emphasized enough. Already on record are instances of individuals donating a kidney to a sibling, only to find some time later that they themselves had the disorder.

Given the well-known study from Johns Hopkins School of Medicine showing that about one-third of DNA reports are not correctly interpreted by physicians, individuals are advised to seek consultation with a clinical geneticist.

Predictive Testing

DNA-based predictive testing became a potential reality after the discovery of the gene location for the neurodegenerative Huntington's disease in 1983. Subsequently, the awesome implications of being able to predict, decades in advance, a disorder characterized by steady loss of all mental faculties with eventual vegetative state and death became a major concern. Guidelines for predictive testing were promulgated by the World Federation of Neurology Research Group on Huntington's Chorea and the International Huntington Association. Because of their importance, these rigorous and necessary guidelines are reprinted at the end of this chapter in full, and are self-explanatory. Moreover, these guidelines not only apply to all predictive tests for degenerative disorders of the brain, nervous system, and muscles but also to predictive tests for other potentially fatal or seriously disabling genetic disorders.

Individuals who have a 50 percent risk of having inherited the dominant Huntington's disease gene (or another neurodegenerative or fatal disorder) face enormously challenging personal questions. Should they marry or not? (I have seen couples planning marriage in whose case one partner leaves the relationship simply on the basis of the known risk, even without DNA analysis.) Should they have children, recognizing the 50 percent likelihood of transmission? Should they have prenatal diagnosis, with the option of selective abortion, even if their child's disorder might not become apparent for decades? What career decisions would they make if DNA results came back positive? What would be the point of obtaining such testing, if no meaningful treatment or cure was available or even on the horizon?

These and many other questions inevitably arise in consultations with individuals and couples during which this painful subject is discussed. Decisions for or against testing remain extremely personal. A 1999 report revealed that only 15 percent of those with a parent affected by Huntington's disease chose to learn their own fate. The majority of clinical geneticists would not recommend a "no hope" test when no meaningful treatment is available. This cautious approach is additionally vested in the recognition of increased suicide rates. A Vancouver group study of 107 centers in 20 countries assessed data on 5,781 individuals at risk for Huntington's disease. Suicide was attempted by 21, 5 individuals having succeeded. This was not, however, a long-range study. Moreover, the remarkable advances in

---

*Knock-out mice are specially bred after a specific gene has been targeted and "knocked out." This method enables experimentation and trials of treatments of mice with different genes removed, leading to specific disease studies. There are many different knock-out mice, including those with breast cancer, Huntington's disease, and similar genetic diseases.

genetics, including the engineering of a knock-out mouse* (see Chapter 27) with Huntington's disease, raise real hope for the first time that some form of intervention or gene therapy might eventuate in the foreseeable future.

Not surprisingly, few have as yet taken advantage of DNA predictive tests for Huntington's disease. In one study of those who did elect to be tested, some 40 percent required psychotherapy. Fortunately, the vast majority of those found to be carrying the gene returned to their baseline psychological functioning level within a year after receiving the news. Although good news was provided to about half of those who were tested, of these fortunate people some had significant psychological problems requiring therapy. These difficulties were largely in the province of survivor guilt, with noncarriers experiencing a lack of relief, numbed emotions, and problems in developing a life-plan, whereas no such problems had previously existed.

Rare but remarkably difficult cases have emerged in which one of an identical twin pair wanted to be tested and the other did not. Clearly, in such a poignant situation, even without communicating the actual test result, the actions and reactions of the tested twin would be revealing. In the Western world, it is the responsibility of the individual at risk to make decisions about personal testing. In contrast, in Japan, informed consent is provided not by the individual but by the family. This cultural difference subverts an individual's wishes to the "best interests" of the family. More often than expected, geneticists have encountered both affected and unaffected individuals reluctant to contact their relatives about the potential need for their being tested. I have personally seen patients who, despite having discovered that they are affected, have failed to inform relatives of their risk and of the availability of a new DNA test—only to learn later of the birth of an affected child with an ultimately fatal disorder. What a sad commentary this is on the nature of human relations.

Although all predictive tests in the United States must be voluntary, experience has revealed the existence of family pressures. Insistence about testing has come from a partner in some cases and from children in other instances. Some professionals have exerted pressure for presymptomatic testing in cases where a susceptibility to a particular cancer has been established. These examples flout the principles of autonomy and informed consent.

One especially difficult issue in predictive testing for breast cancer among younger women, particularly with a family history or a proven mutation, relates to the timing of testing and their reproductive plans. Women who have directly and personally witnessed and suffered along with an affected mother are often intent on not subjecting a future child of their own to the same pain and suffering. These women, once their own mutation has been precisely diagnosed, may want to select prenatal diagnosis with the option of elective abortion prior to their having elective bilateral mastectomy and removal of both ovaries (see Chapter 17). Similarly difficult, though less awesome and self-injurious, are decisions for prenatal diagnosis and selective abortion when either prospective parent is a proven carrier of a colon cancer gene mutation. Since Ashkenazic Jews with a family history of colon cancer face about a 28 percent risk of having the common mutation, and a 50 percent risk of transmitting it if they do, the need for decision-making is not uncommon in that ethnic group.

It is certain that with the identification of mutations for more and more serious, or fatal, single gene disorders, prospective parents may well elect prenatal diagnosis even if the disorder is not likely to become overt until decades after birth. Among the many serious disorders of this type are various disorders of the heart, blood vessels, connective tissue, and kidneys, and disorders that impair brain, nerve, or muscle function. Genetic disorders with variable penetrance, or those characterized by some uncertainty about whether the disorder in question will appear at all, are much less optimal candidates for predictive testing or prenatal diagnosis.

Guidelines for the Molecular Genetics Predictive Test in Huntington's Disease*

Recommendations

1. All persons who may wish to take the test should be given up to date, relevant information in order to make an informed voluntary decision.
2. The decision to take the test is the sole choice of the person concerned. No requests from third parties, be they family or otherwise, shall be considered.
    2.1. The test is only available to persons who have reached the age of majority (according to the laws of the respective countries).
    2.2. Each participant should be able to take the test independently of his or her financial situation.
    2.3. Persons should not be discriminated against in any way as a result of genetic testing for Huntington's disease.

*Promulgated by the World Federation of Neurology Research Group on Huntington's Chorea and the International Huntington Association.

2.4. Extreme care should be exercised when testing would provide information about another person who has not requested the test.

2.5. For applicants with evidence of a serious psychiatric condition, it may be advisable that testing should be delayed and support services put into place.

2.6. Testing for HD should not form part of a routine blood investigation without the specific permission of the subject.

2.7. Ownership of the test results remains with the person who requested the test. Legal ownership of the stored DNA remains with the person from whom the blood was taken.

2.8. All laboratories are expected to meet rigorous standards of accuracy. They must work with genetic counselors and other professionals providing the test service.

2.9. The counselors should be specifically trained in counseling methods and form part of a multidisciplinary team.

3. The participant should be encouraged to select a companion to accompany him or her throughout all the different stages: the pre-test, the taking of the test, the delivery of the results, and the post-test stage.

3.1. The counseling unit should plan with the participant a follow-up protocol that provides for support during the pre- and post-test stages, whether or not a person chooses a companion.

4. Testing and counseling should be given within specialized genetic counseling units knowledgeable about molecular genetic issues in Huntington's disease, preferably within a university department. These centers should work in close collaboration with the lay organization of the country.

4.1. The laboratory performing the test should not communicate the final results to the counseling team until very close to the time the results are given to the participant.

4.2. Under no circumstances shall any member of the counseling team or the technical staff communicate any information concerning the test and its results to third parties without the written permission of the applicant.

4.3. Neither the counseling center nor the test laboratory should establish direct contact with a relative whose DNA may be needed for the purpose of the test without permission of the applicant and of the relative. All precautions should be taken when approaching such a relative.

5. Essential information
  5.1. General information
    5.1.1. On Huntington's disease, including the wide range of its clinical manifestations, the social and psychological implications, the genetic aspects, options for procreation, availability of treatment, etc.
    5.1.2. On the implications of non-paternity (and non-maternity).
    5.1.3. On lay organizations, including their documentation on HD, their addresses for help and social contacts, etc.
    5.1.4. Psychosocial support and counseling must be available before the test procedure commences.
  5.2. Information pertaining to the test
    5.2.1. How the test is done.
    5.2.2. Possible need for DNA from one other affected family member and the possible problems arising from this.
    5.2.3. The limitations of the test (error rate, the possibilities of an uninformative test, etc.).
    5.2.4. Although the gene defect has been found, the counselor must explain that at the present time no useful information can be given about age at onset, on the kind of symptoms, their severity, or the rate of progression.
    5.2.5. The predictive test indicates whether someone has or has not inherited the gene defect, but it does not make a current clinical diagnosis of HD if the gene is present.
  5.3. Information on consequences
    5.3.1. For the person, him or herself.
    5.3.2. For the spouse or partner, and children.
    5.3.3. For the affected parent and his or her spouse.
    5.3.4. For the other members of the participant's family.
    5.3.5. Socioeconomic consequences, including employment, insurance, social security, data security, and other problems that may occur as a consequence of the test result.
  5.4. Information on alternatives the applicant can adopt.
    5.4.1. Not to take the test for the time being.
    5.4.2. To deposit DNA for research.
    5.4.3. To deposit DNA for possible future use by family and self.
    5.4.4. DNA deposited under 5.4.2 above would be made available to the donor's family members at their request after

the death of the donor if it is essential to obtain an informative result.

5.4.5. In the case of DNA deposited under 5.4.2 and/or 5.4.3
above, the unit collecting the DNA must provide a written
declaration that samples will not be used for purposes
other than specified in the said declaration with the exception of the provisions of 5.4.4.

6. Important preliminary investigations

6.1. It is important to verify that the diagnosis of HD in the person's
family is correct.

6.2. Neurological examinations and psychological appraisal are considered important to establish a baseline evaluation of each person. Any other specialized tests are always noncompulsory;
refusal may not affect participation in the test.

7. Prenatal diagnosis

7.1. It is essential that prenatal testing for the HD mutation should
only be performed if the parent has already been tested. For a
possible exception see 7.3.

7.2. The couple requesting prenatal testing must be clearly informed
that if they intend to complete the pregnancy if the fetus is a carrier of the gene defect, there is no valid reason for performing
the test. Furthermore, this situation is contrary to recommendation 2.1.

7.3. Test centers may still perform an exclusion test for a future pregnancy if a 50-percent at-risk person specifically requests it. For
this test the person at risk, partner, parents, and fetus are tested
only with adjoining DNA probes.

8. The Test and Delivery of Results

8.1. Excluding exceptional circumstances there should be a minimum interval of one month between the giving of the pretest
information and the decision whether or not to take the test. The
counselor should ascertain that the pre-test information has been
properly understood and should take the initiative to be assured of
this. However, contact will only be maintained at the applicant's
request.

8.2. The result of the predictive test should be delivered as soon as reasonably possible after completion of the test, on a date agreed upon
in advance between the center, the counselor, and the person.

8.3. The manner in which results will be delivered should be discussed between the counseling team and the person.

8.4. The participant has the right to decide, before the date fixed for the delivery of the results that these results shall not be given to him or her.

8.5. The results of the test should be given personally by the counselor to the person and his or her companion. No result should ever be given by telephone or by mail. The counselor must have sufficient time to discuss any questions with the person.

8.6. All post-test provisions (see section 9) must be available from the moment the test results are given.

9. Post-Test Counseling

9.1. The frequency and the form of the post-test counseling should be discussed between the team and the participant before the performance of the test, but the participant has the right to modify the planned program. Although the intensity and frequency will vary from person to person, post-test counseling must at all times be available.

9.2. The counselor should have contact with the person within the first week after delivery of the results regardless of the test result.

9.3 If there has been no further contact within one month of the delivery of the test result, the counselor should initiate the follow-up.

9.4 It is essential that post-test counseling is made available regardless of the person's financial situation.

9.5. The lay organization has an important role to play in the post-test period. The information and support that it can provide should always be offered to the participant, whether of not he or she belongs to that organization.

Comments

1. The highest standards of counseling should be available in each country. It is recommended that informed consent for the test be documented with the signature of the person to be tested and the professional responsible for the counseling as a standard medical practice.

2. The person must choose freely to be tested and not be coerced by family, friends, partners, physicians, insurance companies, employers, and governments.

    2.1 A prenatal test may be an exception to this rule. Testing for the purpose of adoption should not be permitted, since the child to be adopted cannot make a personal decision. It seems appropriate and even essential, however, that the child when reaching the age of reason should be informed about his or her at-risk status.

    2.2. Each national lay organization should use its influence with government departments, and public and private health insurers to reach this goal.

    2.4. This will arise when a child at 25-percent risk requests testing with full knowledge that his or her parents does not want to know the status. Every effort should be made by the counselors and the persons concerned to come to a satisfactory solution of this conflict. A considerable majority of representatives from the lay organizations feel that if no consensus can be reached the right of the adult child to know should have priority over the right of the parent not to know.

    2.6. Such a specific permission should in principle also be required for symptomatic persons.

    2.7. The consent form should address this issue. Local legal opinions may be helpful.

    2.8. The lay organizations can provide an inestimable service in inquiring about the rigorous standards of the laboratory and can assist persons who want to be or have been tested with their inquiries and concerns.

    2.9. Such a multidisciplinary team should consist, for example, of a geneticist, neurologist, social worker, psychiatrist, and somebody trained in medical ethical questions.

3. This companion may be the spouse or partner, a friend, a social worker, or any person who has the confidence of the participant. It may not be appropriate that the companion should be another at-risk person

    3.1. Support should be available close to the person's community.

4. Often the test will be conducted at a site different from the counseling center. If no lay organization exists in the country, the center should contact the IHA.

4.1. The aim is to protect the participant from the possibility of counseling bias at any time (see also comment 5.2.5).

4.2. Only in the most exceptional circumstances, for example, prolonged coma or death, may the information about the test result, if so requested, be provided to family members.

5. "Essential information" means information that is absolutely vital to the whole test procedure.

5.1. This information should be both written and oral and be provided by the team responsible for the test service.

5.1.1. It must be pointed out that at this time neither prevention nor cure is possible.

5.1.2. Genetic testing may show that the putative parent is not the biological parent; this aspect should be drawn to the attention of the applicant and discussed. With the presently available techniques of in vitro fertilization even occasional nonmaternity may occur.

5.1.3. If no lay organization exists in the country, contacts can be made with the IHA or lay organization of a neighboring country.

5.1.4. Lay organizations should be mentioned as an additional source of support and information.

5.2.2. Asking an affected person, who may be unaware of or unwilling to acknowledge his or her symptoms, to contribute a blood sample may be an invasion of privacy.

5.2.4. Much more information will be needed about implications of the number of repeats.

5.2.5 Particular care should be taken with participants who are believed to be showing early symptoms of HD; however, persons with established, unacknowledged symptoms should not automatically be excluded from the test and should receive additional counseling.

5.3. All consequences have to be discussed, the presence or absence of the gene defect as well as not taking the test.

5.3.2. If the companion of the participant is not his or her spouse or partner, special consideration should be given to such spouse or partner.

5.3.3. The feelings of the affected parent, who may well become aware of the results, must be taken into account.

5.3.4. Whatever information is obtained, it will influence the feelings of and the relationship with other relatives.

6.2. Refusal to undergo these and other additional examinations will not justify the withholding of the test from applicants.

7.1. It is highly desirable that both parents should agree to a prenatal test. If there is a conflict, every effort should be made by the counselors and the couple to reach an agreement. Exceptional circumstances (for example, rape or incest) may justify deviating from this recommendation.

7.2. Testing a fetus carries with it a small additional risk of miscarriage and, possibly, of congenital abnormality.

7.3. The purpose of the exclusion test, which was frequently performed before the gene defect itself had been found, was to permit a 50-percent at-risk person to exclude the possibility of affected children without changing his or her 50-percent at-risk status. This does include the termination of pregnancies of a 50-percent at-risk fetus and continuation of pregnancies of a low-risk fetus only.

8.1. Prenatal testing may be such an exception. Such an interval is necessary to give the person sufficient time to assimilate the pre-test information in order to make an informed decision. During this interval specialists from the test center must be available for further consultation.

chapter 25

Genetic Counseling

································

Dismay and distress pervaded the office. I had just finished explaining the genetic basis of the cancers that had wracked the family. The consultation had been arranged by the granddaughter, then 28, who had asked her maternal grandfather to join her when we all met for genetic counseling.

Her maternal grandmother died because of breast cancer. She had lost her mother to ovarian cancer. One maternal uncle died within six weeks of diagnosis from pancreatic cancer. Two other maternal uncles were living but were battling prostate cancer. Another uncle and two aunts on that side of the family were healthy. The daughter of the uncle with pancreatic cancer developed breast cancer in her early thirties.

My patient had two children—a boy and a girl. Urged by her grandfather, who had watched a television program on the subject, she decided to come in for genetic counseling, concerned about her personal risk and that of her children. This was our third session. Our first meeting was devoted to exploring the family history and explaining the high likelihood that in this Ashkenazic Jewish family a gene mutation in one of the breast cancer genes transmitted from the maternal grandmother may have caused the ovarian cancer as well as the pancreatic and the prostate cancer in her family. She herself then had a 50 percent risk of possibly having inherited this breast cancer gene mutation from her mother. If she had this gene defect, each of her children had a 50 percent risk of inheriting it from her. A significant portion of time in the first meeting was devoted to discussion about whether or not she should have the direct gene mutation analysis.

The second meeting was entirely devoted to the question of testing for the gene mutation. (See Chapter 17 for details.) We further explored the potential consequences if she indeed was found to harbor this culprit gene. She faced up to an 85 percent lifetime risk of developing breast and/or ovarian cancer. Her options were clear. The conservative approach was to initiate detailed and frequent personal and physician surveillance, including mammography, as well as intermittent imaging of her ovaries. Added to this was a blood-screening test (see Chapter

17) that was useful but not very effective. The second approach she faced was the decision to have elective bilateral mastectomy as well as removal of both ovaries. Data from a major Mayo Clinic study showed that women who opted to have their breasts removed averted cancer in at least 90 percent of cases. That study, however, did not have the benefit of breast cancer gene mutation analysis for the subjects studied.

Our third session was devoted to all of these issues, including an assessment of the psychological strength that she would need in order to withstand the consequences of surgery. Beyond the medical and genetic issues, we discussed her feelings, concerns about self-image, femininity, the effect on her husband and their intimate relationships, breast implants, and of course the question ultimately of testing her children.

Some time later she let me know that her decision was to avoid surgery, establish surveillance, and return to see me on an annual basis.

What Is Genetic Counseling?

The consultation just described is a typical example of genetic counseling. This is a communication process that focuses on the occurrence or risk of recurrence of a genetic disorder, at the same time addressing related personal concerns. The counselor's aim is to provide the fullest understanding of the disorder in question and its broader implications, as well as the range of options available. The overall purpose is to help families through their problems, their decisionmaking, their possible anguish, and their adjustments. The goal is *not* to make decisions for them. The essential goal of genetic counseling is to secure a good understanding of a particular genetic disorder so that the person at potential risk is able to make rational decisions. Although every geneticist hopes to prevent or avoid both the occurrence and recurrence of serious genetic disorders or mental retardation, the consensus in the Western world is not to direct the decisions of those who seek information about genetic diseases.

Who Needs Genetic Counseling?

There are thousands of genetic disorders or traits for which genetic counseling may be sought. In general, these reasons break down into defined categories, shown in Table 25.1. Clearly, from a review of this table, it is clear that anyone with a genetic disorder or with a close, affected relative is likely to benefit from genetic counseling. Although it might seem obvious that genetic counseling is

TABLE 25.1   Reasons for Seeking Genetic Counseling

1. You or your partner has a genetic disorder.

2. A previous child, a parent or close relative has a genetic disorder.

3. You know or suspect that you are or might be a carrier of a specific genetic disorder, or require a test to determine whether you are a carrier.

4. You have a genetic disorder about which you need more information relating to features, prognosis, and treatment options.

5. You or your partner has an ongoing pregnancy with known genetic risks and options for prenatal diagnosis.

6. You are planning pregnancy and wish to explore options including artificial insemination by donor and other methods of assisted reproduction, prenatal diagnosis, adoption, or selective abortion.

7. You need assistance in the care and treatment of an already affected child or close relative.

8. A prenatal screening or diagnostic test reveals a fetal abnormality.

best sought prior to marriage, prior to conception, and certainly prior to having a child with an avoidable genetic defect, the overwhelming number of people in the Western world neither seek nor receive genetic counseling. This may be because they do not realize there is a genetic disorder in the family or because their physician has failed to recognize it or neglected to refer them for an appropriate consultation. Doctors the world over have until recently been inadequately trained in medical genetics. This basic inadequacy has been compounded by the phenomenal explosion of knowledge in human genetics, which has even geneticists scrambling to keep up.

If there is a disorder affecting at least two family members, that alone might be sufficient reason to seek genetic counseling. You should consult with your physician if you are uncertain, or telephone a medical geneticist to determine whether or not it makes sense to come for a genetic evaluation and counseling. You will usually obtain answers, and more often than not, you will be reassured.

What Should You Expect from Genetic Counseling?

People may have different expectations of genetic counseling, and many don't know what to expect. An individual or couple should come away from genetic counseling with a clear understanding of the disorder in question, its features, the mode of inheritance involved, and all of the options available. A full inventory of all clinical features might be unnecessary, but certainly the major manifestations that directly impair health and affect prognosis must be understood. Information and recommendations about specific tests, including their efficacy and limitations, should be relayed together with an understanding of who in the family might need testing. Where appropriate, related resources and support groups should be discussed. The counseling process is also aimed at helping families cope with their problems and at assisting and supporting them in their decisionmaking.

What to Expect from a Genetic Counselor

You should feel confident about the factual knowledge that the counselor has about the disorder under discussion. This knowledge is expected to encompass diagnostic or confirmatory methods, risks of occurrence and recurrence, the mode of inheritance, the availability of carrier tests, the variability of key features of the disorder, and all matters relating to prognosis and quality of life. Where relevant, knowledge about treatment and its efficacy is also important. Knowledge of all of the relevant options and available resources should be evident. This would include cognizance of and about genetic testing in adoption, and guidelines about banking DNA (both of which are discussed in Chapter 26).

Key ingredients in the counseling process include the transfer of information in nontechnical language (with the assistance of an interpreter, if necessary), and an adequate allotment of time. Faced with serious implications of a genetic disorder, people frequently suffer anxiety-block. This means that assimilation of the factual content of the consultation may be largely lost. For this reason, medical geneticists invariably follow up a consultation with a letter both to the referring doctor and the patient. Errors have occurred in genetic counseling, some leading to litigation, as discussed in Chapter 26.

We would all prefer a knowledgeable counselor who is caring, sensitive to our needs, and empathetic to our problems. A warm, caring, sympathetic, and understanding individual with insight into the human condition is the choice of a counselor we would all make—someone who is able to anticipate and respond to our unspoken fears and questions and make the counseling experience as beneficial as possible.

Miscarriage and Stillbirth

Women who have suffered a miscarriage look to their doctor and later their genetic counselor or clinical geneticist (when referred) for empathy and understanding. Historically, physicians have managed the aftermath of miscarriage poorly. They have frequently been inattentive and insensitive, have not provided opportunities for discussion of the woman's feelings, and have been slow to recognize her grief. This grief is complex, being an admixture of a loss of developing bonding with the fetus, a loss of self-esteem, dashed dreams, guilt, and, not infrequently, considerable anger at her physicians. Women should seek physicians, clinical geneticists, or genetic counselors who recognize their grief, who can explain their fluctuating moods and anxieties after miscarriage (due to hormonal flux), address potential causes, and, one would hope, assuage feelings of guilt and concerns about future pregnancies.

Women who experience stillbirth may experience even greater levels of distress than women who suffer miscarriage. Any physician or counselor who fails to recognize and address a woman's feelings of sheer emptiness after a stillbirth, and on leaving the hospital empty-handed, deserves to be abandoned by the patient. The depth of grief following a stillbirth may last a lifetime, more especially if improperly managed.

Infertility

In recent years, with the enormous growth in assisted reproductive techniques aimed at helping infertile couples, counselors have recognized the considerable stress before, during, and after infertility treatments. Repetitive failures to achieve pregnancy may be associated with grief and depression, and women should seek counselors who understand the need to psychologically prepare for the emotional and stressful periods that can easily be anticipated during infertility treatments.

Communicating Bad News

We all have expectations of a caring, empathetic physician or genetic counselor vested with the responsibility of communicating bad news. After four decades of experience I can personally relate how difficult it is to tell a patient face-to-face that he or she has a fatal, nontreatable, genetic disorder. One such example would be a neurodegenerative disease like Huntington's, in which slow but steady disintegration of mental and neurological faculties proceeds to dementia and eventually death. In these awfully difficult situations, patients and their loved ones should expect clear and direct explanations in simple language that they can understand, with total avoidance of technical jargon. A quiet, private office (not a corridor), with everyone seated and without outside interruptions, are the minimum prerequisites.

People should not expect (or demand) communications of this sort over the telephone. Those being counseled need an approach that is gentle, compassionate, and caring. There should be no doubt about the certainty of the diagnosis and the expertise of the physician communicating the information. It is vitally important for the individuals affected to have support persons accompany them, although that sometimes turns out to be impossible. Certainly parents should be together when serious communications are necessary concerning their child. The realization that no one can assimilate all the bad news and the factual information at such a time should be vocalized, and a suggestion made for a return visit to take up the same issues and information again as well as to answer more questions.

Along with these prerequisites, at least two other considerations should be kept in mind. No genetic counseling can really begin without an accurate diagnosis. Those seeking genetic counseling should know that even in the presence of information about the family medical history there may still be a need for the physician/counselor to see photographs, autopsy reports, laboratory reports, medical records, and possibly radiology reports. Second, as mentioned earlier, genetic counselors do not tell their patients what to do. This important principle is tested when the counselor is confronted by a couple planning to have their second, third, or even fourth child with a serious genetic disorder. The guiding principle in providing impartial and objective counseling is not to have the physician visit upon the patient his or her religious, eugenic, racial, or other dictates of conscience. Ultimately, it is the individual who will have to live with his or her decisions, and not the physician or counselor.

Who Provides Genetic Counseling?

Physicians and scientists with doctoral degrees can, after special training, become certified for genetic counseling by examination through the American Board of Medical Genetics and the Canadian College of Medical Genetics. Graduates of master's degree programs in genetic counseling can become certified through the American Board of Genetic Counseling. Clinical geneticists, all being physicians, frequently work with genetic counselors as a team. The majority of such teams are located in major medical centers. To determine the nearest location of a clinical geneticist, contact the American College of Medical Genetics, the Canadian College of Medical Genetics, or the National Society of Genetic Counselors (see Appendix C).

Risks and Odds

We all have different perceptions as far as risks are concerned. These perceptions are influenced by education, culture, comprehension of mathematical probability, intelligence, basic personality types, and so forth. I have been staggered, for example, by patients who express relief upon hearing from me that the risk of having a child with a serious disorder is 25 percent. Upon inquiry, I have learned that they came to the consultation expecting that the risk that all of their children would be affected was 100 percent. In contrast, communication about a 1 or 2 percent risk for a specific abnormality has been met by great consternation even after a prospective parent (usually a father) learns that on average all couples have a 3 to 4 percent risk of bearing a child with a major birth defect, mental retardation, or genetic disorder. Matters are complicated further when it is recognized that the appreciation of risk and the interpretation of odds vary among different personality types. A pessimistic parent may reach very different conclusions from those of an optimistic parent with the same odds in mind. There is also a basic difference in attitudes toward risk among risk-takers and more cautious persons. Further complicating matters is the fact that attitudes toward risk can change with changing moods.

In addition, people might easily be confused if they are told, for example, that their risk is 10 percent greater than normal risks. If the explanation is not more explicit than that, they might not realize that the normal risk is only 0.1 percent and that 10 times that risk puts them at a 1 percent risk level—meaning that they have a 99 percent likelihood of having no problem whatsoever!

Chance Has No Memory

An explanation indicating that a recessive hereditary disease has a recurrence risk of 25 percent might be totally misinterpreted. It means that for *each and every* pregnancy there is a 25 percent risk of the particular disease occurring again— and this risk does *not* change with the number of children already born, or whether they are affected or unaffected. Some parents make the mistake of thinking that since their first child already has the disease, they will be safe in their next three pregnancies. Clinical geneticists and genetic counselors frequently encounter individuals whose understanding of their personal risks, even after counseling, is totally erroneous. If you or your loved ones are at risk, be sure that you understand the precise risks and options and that you obtain a letter confirming this information.

Contact and Recontact

Contact with a clinical geneticist by any person affected with a genetic disorder (or any couple with an affected child) is very important, given the remarkable and rapid advances being made in human genetics. From that initial consultation, guidance will emerge about the panoply of issues discussed in this book, including the potential need to remain in touch. Millions of people have had genetic counseling, but my impression is that relatively few have returned years later to find out about new discoveries about their genetic disorder. Pertinent new discoveries may lead to a revision of the diagnosis, a more precise diagnosis, an opportunity for presymptomatic or predictive diagnosis, an opportunity to determine carriers of certain gene defects, and recommendations for future anticipatory care, prenatal diagnosis, or treatment.

For many years I and my colleagues have advised those who come to us for genetic counseling to stay in touch, anticipating that further advances will be made that might affect their lives and health, especially as regards possible future childbearing. We believe that our patients' family physicians should share the responsibility for reminding them of the need to recontact a clinical geneticist when appropriate. This is especially important because so many millions change their addresses every year—and their doctors. Virtually all extended families have disorders with at least some genetic basis (e.g., cardiovascular disease or cancer). Whether or not genetic counseling is necessary can be determined by contacting the family physician, a clinical geneticist, or a genetic counselor.

chapter 26

Genes, Ethics, Law, and Public Policy

··

The deciphering of the human genetic code has brought with it untold opportunities for the good of humankind. This technical achievement ranks with splitting the atom and reaching the moon. For more than two decades, steady advances have made major inroads in our ability to achieve precise genetic diagnoses. However, just as for all significant opportunities, significant risks abound. These risks are not about the creation of monsters but rather about our ability to marshal this new knowledge for the betterment of all. The focus in this chapter is on the quintessential practical issues that occupy the frontier of our concerns about the use of personal genetic information. Most of the other medical, legal, moral, ethical, religious, and public policy issues have been dealt with in three other academic volumes (*Genetics and the Law*, eds. A. Milunsky and G. Annas, 3 vols. [New York: Plenum, 1975, 1979, and 1985]).

At the root of our concerns lie the potential misuses of the new genetic information that could cause personal grief, invasion of privacy, breaches of confidentiality, psychological distress, threats to autonomous decisionmaking, stigmatization, and discrimination. None of these consequences are necessary, and all are avoidable. To achieve what should not be a utopian goal but rather a normal expectation of every citizen will only be possible by orchestrated actions taken in our joint interests by individuals, physicians, legislators, lawyers, ethicists, theologians, and others dedicated to human health and welfare. This chapter aims to alert you to potential personal issues and to inform you about guidelines, expectations, and ongoing legal and legislative actions in this area.

Goals and Ethical Principles

This book has been dedicated to the primary goal of medical genetics—that is, to help individuals and families achieve a healthy life with healthy children by being well informed about the subject and its new developments. Emphasized throughout is the abiding need for nondirective genetic counseling. In this way, physi-

331

cians are expected not to tell individuals what to do but rather to identify all the issues needed for an individual to make decisions.

The fundamental principles and tenets that form the ethical foundation of medical genetics are well recognized. The *autonomy* of an individual calls for recognition of the need for self-determination and protection—especially for protection of those with diminished autonomy. This principle emphasizes the freedom of choice and in particular highlights a woman as the most important decisionmaker in matters concerning reproduction. Implicit in this principle are the assumption that genetic testing should always be voluntary and the fact that there should be no compulsory testing of individuals or populations. Hence, it would be unacceptable for services or testing to be provided under coercion by physicians, public health authorities, or government.

The principle of *beneficence,* or of doing good, is the prime goal and obligation of physicians and public health authorities, acting in the interests of the welfare of individuals and families.

The Hippocratic oath taken by all physicians to "do no harm" encompasses the principle of *nonmaleficence.* In the context of medical genetics, this tenet calls for respect for the dignity and diversity of all people, including those whose views are in the minority. This principle is incompatible with discrimination of any kind and in any arena, including employment, insurance, and education.

The ideal of *justice* requires that everyone be treated equitably and that public resources be allocated fairly, with guaranteed equal access for all.

Embellishments of these basic principles as applied to genetic counseling are explored fully in Chapter 25.

Informed Consent for Genetic Testing

The World Health Organization Human Genetics Program has proposed seven fundamental elements that underwrite the process of informed consent for voluntary genetic testing.

1. The individual to be tested must fully understand its purpose. It is a common experience to encounter individuals who have no idea of what test they had, let alone its purpose.
2. The likelihood that a test will provide a correct answer should be clarified before samples are obtained. There are many reasons why a test might not yield the expected information, reasons that have been identified and discussed in many chapters. Critical examples include

the unrecognized fact that more than one gene may account for the same disorder; that the initial clinical diagnosis is incorrect; that there is an unexpected mode of inheritance; that the wrong test was ordered; that there is an unexplained and nontested mutation; and that a lab error occurred.

3. The implications of the test results for the individual and family must be clarified beforehand. The unexpected discovery that a family member is affected who did *not* wish to know can have devastating consequences. Detection of a carrier, although benign in and of itself, may have psychological ramifications. Predictive tests for the individual could produce devastating news of an unexpected, fatal disorder. The reservations and limitations concerning predictive testing are outlined in Chapter 24.

4. Before testing, an individual needs to recognize his or her options and alternatives. For example, a person with high risk might not wish to have a test relative to the issue of reproduction but might instead prefer to research the possibility of adoption. The discovery that a couple are both carriers of the same harmful mutation may result in the decision not to marry, if they do not consider prenatal diagnosis and selective abortion, or assisted reproduction, viable alternatives.

5. Anyone tested should be informed about the potential benefits and risks, including issues related to marriage and reproduction, and the potential for psychological distress.

6. Available information about potential discrimination by health and life insurers and even employers (although illegal) requires discussion.

7. Emphasis is necessary in providing informed consent that regardless of the decision an individual or family makes, their care will not be jeopardized.

Confidentiality

This is a serious issue in medical genetics. The utmost care is usually taken to protect confidentiality, especially with reference to third parties (e.g., insurers and close family members). Difficulties arise, however, when genetic test results imply likely important risks to close relatives. Most often, and sometimes with encouragement, the individual tested will communicate the important information and recommendations to relatives who are at risk. On occasion, because of long-term family squabbles and alienation, an individual refuses to communicate with relatives. These situations, though uncommon, present considerable ethical dilemmas, not the least of which is breaching the confidentiality (at least by inference) of the tested person. The World Health Organization recommends that contact may be made with relatives at risk in these difficult cases, especially when effective remedies or preventive measures are available. Such contact aims to keep the identity of the tested relatives secure, but recipients of the information can draw their own conclusions. Other serious dilemmas may also arise when testing for a genetic disorder, such as when nonpaternity is unexpectedly discovered. The principle of doing no harm is fully stretched in these not-so-infrequent situations, discussed in Chapter 7.

DNA Banking

Over the past two decades, during the frenetic search to identify human genes, blood and tissue samples have been taken from many families for DNA extraction, testing, and banking. Most of the banking has taken place in academic institutions, but it increasingly occurs in commercial operations. The collection of samples for research purposes requires prior, written consent from the donor. Straightforward diagnostic samples other than those used for predictive tests (see Chapter 24) were often studied without such consent requirements. Acquisition of DNA samples for research purposes often included multiple family members.

The DNA analyses yielded information that became part of the DNA data bank in the particular institution that performed the studies, constituting an important repository of genetic information. The banked DNA samples also became important sources for further research. For the most part, informed consent was given for the use of samples in research studies, and shared DNA data or samples have mostly not created problems, as long as the individual's identity was deleted from records (such samples are described as *anonymized*).

Ethical Guidelines

The World Health Organization Human Genetics Program has proposed ethical guidelines for access to banked DNA. In addition to the requirement of informed consent prior to sample use in any project, the WHO recommends several other measures aimed at securing an individual's rights and an institution's responsibilities. Unexpectedly, it recommended that blood relatives be given access to stored DNA in order to learn their own genetic status, but without the ability to learn the donor's status. This access is considered ethically acceptable, regardless of whether the relatives have contributed financially to the banking of the DNA. Donors are encouraged to keep their addresses current with the DNA bank, but the WHO proposes that the burden of contacting families at regular intervals about new developments in testing and treatment be borne by the DNA banking institutions. Access to DNA banks by spouses without donor consent is not supported by the WHO. However, the organization finds that the donor has a moral obligation to provide the spouse with any information relevant to childbearing. Although the donor's consent is regarded as primary, exceptions are recognized for forensic reasons (for example, investigation for murder) or in situations that directly relate to public safety. A compelling statement excludes access by all third parties (such as insurance companies, employers, schools, and government agencies) that may be able to coerce consent—even if an individual consents to such access. Last, the guidelines support storage of DNA for as long as it might benefit living or future relatives, allowing for its destruction only after all relatives have died or all attempts to contact survivors have failed. Research on banked samples is encouraged as long as identifying characteristics are removed.

Ethical and Legal Issues

A surprising number of issues have been encountered over the past two decades necessitating the promulgation of the guidelines just discussed. In a number of research studies where many family members' samples were needed, individuals were pressured to provide a blood sample, thereby negating the requirement of voluntariness and informed consent. It later became clear that they had been reluctant to participate because they did not wish to learn (or have others learn) about their genetic status or to divulge evidence of nonpaternity, incest, or undisclosed adoption.

DNA samples obtained for a specific disease study may later be used to analyze an entirely different disorder. Although permission might have been given for the initial study, no additional consent might have been sought for the later study, and indeed such a process may have been considered extremely burdensome. In this situation, the discovery of a completely unrelated, serious genetic condition may place researchers in an ethical bind: Are they obligated to contact the individual to whom this information is directly relevant, or are they obligated not to divulge the information to the individual, who has not requested it and may not wish to know?

My view about researchers' duty in this situation is clear, at least on one point. If such information has personal health consequences for the individual for which there are treatments or preventive remedies, then the researcher should have a duty to give the donor the appropriate information and assistance. These events are not uncommon. For example, in our own laboratories in the Center for Human Genetics at Boston University School of Medicine, we provide DNA diagnostic analyses for many disorders, including a common genetic disorder predisposing those affected to thrombosis (Factor V Leiden deficiency; see Chapter 10). When it had become technically feasible to analyze a second gene that also predisposes the affected to thrombosis, as part of a new research study we reanalyzed all of the samples that had been submitted for the Factor V gene analysis, this time for the prothrombin gene mutation. Predictably, analyses of a number of samples showed prothrombin mutations in the second gene. Since anticoagulation (blood thinning) could be lifesaving, we did not hesitate to contact these individuals, who were already under evaluation for a predisposition to thrombosis, with the recommendations for prophylactic treatment. In the situation where samples have been anonymized and a fatal condition is subsequently uncovered, no contact with the original donor would be possible.

In the extremely commercial era in which we live, a person's DNA may also have great economic potential. The unsettling, precedent-setting case of *Moore v. University of California* provided an early example of the problems that can ensue: Mr. Moore developed an uncommon form of leukemia that resulted in considerable enlargement of his spleen. As part of his ultimately successful treatment, physicians removed his spleen. Later it was used as a source for the development of permanent cell lines for research studies. After considerable effort, the researchers discovered that Mr. Moore's cells produced unique proteins that clearly had commercial potential.

Mr. Moore, during a routine follow-up visit to the physicians who had saved his life, discovered by accident how valuable his cell lines had become. Indeed, their value was assessed at tens of millions of dollars! Mr. Moore consulted his attorney, whose position in court was that Mr. Moore had a property right to his own cells. The Supreme Court of California, however, demurred, holding that he could not bring such a claim but that he could claim breach of fiduciary duty or lack of informed consent. The court's reasoning was that a physician should disclose personal interests, including economic interests, which could affect his or her professional judgment.

Issues about DNA ownership arise in other contexts as well. Although many of these quandaries are covered by the guidelines noted above, some are not. If there are storage fees for the DNA sample, what are the consequences of nonpayment? Are there guarantees for a donor's anonymity in other research studies? In this era of bankruptcies, mergers, and gene patenting, who owns the DNA sample if institutional ownership changes?

DNA banks were established initially for the purpose of uncovering genes that cause serious genetic disorders. Subsequent use has included diagnosis, disease prediction, and treatment, as well as the identification of criminals, soldiers missing in action, and other missing persons. DNA banks and associated databases are of enormous value, but they need strict and enforceable guidelines in order to guarantee the safety of donors and of the public.

Forensic DNA Data Banks

DNA fingerprinting—that is, identification of individuals through the unique pattern of their DNA—has proved of incredible value to law enforcement agencies. The vast majority of U.S. states have enacted statutes establishing DNA data banks of samples taken from various categories of criminals. Samples of blood, cheek cells, or saliva have been systematically collected from violent sex offenders, pedophiles, and other violent felons, from which DNA has been extracted and used to create a profile for each offender. These profiles have been digitized and entered into linked computers, providing a powerful tool for the subsequent identification of an individual. Micro-modification of these analyses by means of new technology has allowed recognition of these DNA profiles from tiny specks of semen or blood obtained at a crime scene. The DNA Identification Act, a federal law passed in 1994, has led to the establishment of state laboratories for DNA analysis. That law also formally authorized the Federal Bureau of Investigation to develop the national computer network CODIS (Combined DNA Identification System), aimed at facilitating the exchange of DNA profiles among databases in different states.

Many criminals have been identified through these new DNA tools. Moreover, repeat offenders have been apprehended through DNA profiling. For example, Minnesota's DNA data bank successfully identified an individual responsible for eighteen separate sexual assaults. Of enormous importance, DNA profiling has enabled absolute exclusion of wrongly convicted and imprisoned individuals, including some on death row. Many of these individuals had been incarcerated for up to twenty years due to a miscarriage of justice. In England, DNA fingerprinting of virtually all criminals has led to the apprehension of thousands of offenders. Notorious rape and murder cases occasionally have resulted in police collecting DNA samples from all males in the towns where the crimes occurred, as first documented in Joseph Wambaugh's *The Blooding*.

DNA Testing of Children

All medical interventions in behalf of children have as their basis the best interests of the child. Genetic testing is no different. The matter may be complicated, however, by the misguided wishes of parents, who might demand inappropriate predictive testing. Of course, the appropriateness of any DNA or other test aimed at diagnosing a disorder for which the child is already showing symptoms and signs is not in question. Moreover, predictive testing in a child at high risk for developing a disorder that can be successfully treated is not only acceptable but is imperative. DNA tests to determine the carrier status of a child where no immediate medical benefit will ensue are not recommended. Predictive tests that can determine a genetic disease of later, adult onset and for which no useful treatment is available are also contraindicated in children. Some protocols allow for but do not encourage such testing for individuals who are 18 years or older.

Circumstances may arise, however, in which decisionmaking about the genetic testing of minors is less straightforward. For example, in the case of an only child who is at significant risk of having inherited an ultimately lethal disorder, the parents may seek testing to help them determine whether or not to have another child in order to obtain a surviving heir. It may prove difficult to identify and weigh the potential benefits and risks that could result from testing in such circumstances; here, the advice of a clinical geneticist could be of considerable help to parents. Other examples might also try the limits between parental authority and the best-interests-of-the-child standard. Parents might also seek to determine a child's status for a specific genetic disorder with the primary aim of either discouraging sexual activity or encouraging sexual responsibility. However, a positive test result in a teenager could lead to a devastating alteration in self-image as well as to stigmatization by peers if the disorder became common knowledge.

Parents might also request that the results of a test not be disclosed to the child. Depending upon the age and maturity of the child and the particular test requested, secrecy may not be justified. When the child is eventually informed, his or her realization of the initial nondisclosure might result in distrust of the medical establishment as well as of the parents.

The "new genetics" will soon provide opportunities for susceptibility gene testing. Such testing in very early childhood can be expected to become commonplace, founded on the goal of identifying predispositions that allow therapeutic interventions for the good of the child. Such interventions might include dietary advice, medications, strict avoidance of obesity, avoidance of contact sports, anticipatory desensitization for serious allergic reactions that could cause death, and other lifestyle changes. Such testing would, of course, be in the best interests of the child.

Genetic Tests in Adoption

Who among us would not wish for a perfectly healthy child? Who among prospective adoptive parents would not wish the same? These uniform echoes in the affirmative suggest that genetic testing of the child to be adopted should be a simple and straightforward matter. The truth is far from it. The potentially conflicting interests in the process include those of the child, the prospective adoptive parents, the adoption agency, geneticists/pediatricians, and the state. All maintain that the "best interests of the child" is the determining standard. How each party to the process attempts to achieve this goal feeds the potential conflict.

The Child

The best interests of the child include physical and mental health, personal privacy, and social development. Genetic testing may affect all of these interests. Understandably, prospective adoptive parents may pressure their pediatrician to arrange for such genetic tests, although they cannot guarantee the child's future health. Genetic testing of children for treatable or preventable disorders is completely appropriate; testing for disorders that may become manifest later in life is not. Equally unacceptable is genetic testing that may be of use in behavioral assessment or in determining a child's carrier status for disorders that might affect reproductive decisions that child makes later in life, as an adult. In the event of a genetic test revealing a high risk for a specific genetic disorder, undesirable consequences for the child may follow. Adoptive parents may decline adoption. If adopted, the child may be parented in a way that prejudices development. Ultimately, disclosure about an adoptive child's genetic disorder could seriously limit the child's future opportunities for insurance or employment.

Adoptive Parents

Even as the best interests of the child are the key concern of adoption laws, the laws are also designed to protect the interests of adoptive parents, guaranteeing them the same right that biological parents have, with broad discretion for making medical decisions for their children. Adoptive parents typically request full disclosure of a child's medical background, and they could argue that in order to make a fully informed adoption decision, they will need to have certain genetic tests performed. The results of such tests, if they reveal a genetic disorder, could provide legal grounds for an annulment of adoption before it goes through, sparing the adoptive parents from finding themselves unable to cope later on and from the anguish of a postadoption annulment. The prospective parents may argue that their finances limit their ability to care for a child with a serious disorder.

Birth Parents

Birth parents invariably hope that the child is placed in a good home with loving parents. Frequently, however, their primary concern is their personal anonymity and privacy. They may not wish to divulge aspects of their family's medical history. The opening up of adoption records that were previously sealed, such as in Oregon, will undoubtedly cause great consternation and grief for some birth parents and bring unexpected pleasure to others.

Adoption Agency

A good placement is the adoption agency's main goal. The agency makes every effort to ensure the privacy of the child, the birth parents, and the adoptive parents. Since adoption agencies have potential liability, they too are interested in genetic tests of the child and sometimes the birth parents. More than ten U.S. states have recognized a cause of action for wrongful adoption in cases where claims against agencies have alleged negligence or fraud in placing children whose health status and genetic background were not fully disclosed. Many states have imposed an affirmative duty on adoption agencies to disclose medical information that they possess about the child to be adopted or the biological parents.

The State

Even as it seeks to uphold the legal standard of the best interests of the child, the State also aims to encourage adoptions. The goal is the successful placement of as many children as possible, preferably from among eligible children within the U.S. social service system.

Professional Guidelines

The American Society of Human Genetics and the American College of Medical Genetics, the premier professional organizations in human genetics, have jointly published three specific recommendations concerning genetic testing in adoption:

1. All genetic testing of newborns and children in the adoption process should be consistent with the tests performed on all children of a similar age for the purposes of diagnosis or of identifying appropriate prevention strategies.
2. Because the primary justification for genetic testing of any child is a timely medical benefit to the child, genetic testing of newborns and children in the adoption process should be limited to testing for conditions that manifest themselves during childhood, or for which preventive measures or therapies may be undertaken during childhood.
3. In the adoption process, newborns and children should not be tested for the purpose of detecting genetic variations of, or predispositions to, physical, mental, or behavioral traits within the normal range.

Law and Ethics

The philosophical and practical issues concerning the "new genetics" encompass a discussion that could fill many volumes. In this section, I provide a succinct, highly focused view of four pertinent topics: the issues surrounding the patenting of human genes; the potential for genetic discrimination; the ongoing efforts at federal legislation; and the expanding territory of medical malpractice vis-à-vis medical genetics.

Patenting Human Genes

Law tends to lag significantly behind advances in science and medicine. Only after harm or damage has occurred do we see legal or legislative involvement. It would be refreshing to see anticipatory legislation passed on the biomedical sciences, establishing a legal and ethical framework dedicated to the public good and aimed at catalyzing continued positive developments. One wonders whether the paucity of scientists and physicians in legislatures, which are overpopulated with lawyers who are disinclined to pursue science, is the reason for the absence of such legislation. Whatever the reasons, since the identification of DNA in 1953, we as a society have had fifty years to plan and anticipate virtually all of the developments in genetics that now occupy center stage in the legal arena; yet we have failed to achieve timely, balanced legislation that safeguards the public good, avoids genetic discrimination, and secures the health of individuals and families.

General public awareness of commercial efforts to patent human genes only dawned in the mid-1990s. Many have questioned the ethical, legal, and moral grounds on which any private firm or individual might apply for a patent on a human gene. The resolution of the Human Genome Project and the anticipated identification of all disease-causing genes opened the floodgates for patent applications. Trouble erupted when commercial efforts surfaced that aimed at patenting tiny scraps of genes that had yet to be fully identified and the functions of which were still unknown. Many took affront to this approach, which clashed with precedents earlier established in gene patenting. Originally, the process depended upon steps that were the very opposite of those in use today. Because of technical limitations, the process of identifying genes began first with recognizing the specific protein product produced by the gene, such as insulin. Thereafter, working backward, gene identification was achieved, and only then was patenting of the gene and its product with known function deemed acceptable. This should still be the case today, in the opinion of most research scientists. If a few giant corporations were allowed to patent virtually the entire human genome, there would be a stultifying effect on scientific progress. No one would wish to work on gene identification and determination of gene function if a corporation already held the patent.

Patenting, which is a monopoly right, is granted for a specific period (usually, seventeen years) so as to allow society to benefit from the invention while safeguarding the economic interest of the patent holder in benefiting from the results of his or her labor. Some have argued that human genes should not be patented, since they occur in nature and have not been invented. However, patent offices have responded that isolation of a gene and determination of its product could be regarded as a "new" finding, which, if a specific use is delineated, could be patented. Meanwhile, many human genes, such as the breast cancer gene, have been patented; others are awaiting similar action by the Patent Office. Needless to say, the effect on individuals who need gene analysis will be to increase the personal cost of such analysis, given the license and royalty fees that must be paid to the patent holder. The ethical and moral issues surrounding the patenting of human genes are still hotly debated, and probably will be for some time.

Phenomenal health benefits can be anticipated as the study of human genes and their products continues; but the pursuit of knowledge for the public good is in danger of being swamped by the profit motive, with biotechnology companies operating in secret to protect their stockholders. Carefully planned legislation is needed to simultaneously protect intellectual property rights and encourage both scientific progress and business success.

Genetic Discrimination in Insurance

A handful of cases in the early to mid-1990s drew attention to reported instances of employers and of health, life, and disability insurers using genetic information to deny jobs and insurance coverage, raise rates, or limit the extent of coverage. A wave of legislation followed, aimed at restricting health insurers from using genetic information that was of a predictive (not diagnostic) type. More than half of the states have imposed such restrictions. The U.S. Congress was slow to enact similar, federal-level legislation. While a slew of bills lie stagnant in Congress, the 1996 Health Insurance Portability and Accountability Act was passed. This act is aimed at prohibiting group health insurers from excluding coverage for those with positive predictive tests who were not symptomatic. Most states nonetheless allow insurers to use family history information and to base their actuarial assessment on the clinical status of the applicant.

In February 2000, former president Clinton signed an executive order prohibiting U.S. federal agencies from using genetic information in hiring or in promoting employees. The president, in moving to protect personal genetic privacy, expressed the hope that his executive order would further support the proposed Genetic Non-Discrimination in Health Insurance and Employment Act of 1999. The executive order stated that it was the U.S. government's policy "to prohibit discrimination against employees based on protected genetic information." A major recent study shows that very few cases of genetic discrimination have occurred.

Litigation in Medical Genetics

The remarkable advances in human genetics and the speed with which they have occurred present an enormous challenge to the practicing physician. This is especially the case for physicians who graduated between ten and twenty years or more ago, when the foundations of the "new" genetics had barely been laid. Keeping abreast of only those advances that directly impact patient care is a major challenge. Consequently, those at risk may want to consider a consultation with a clinical geneticist, especially when the possibility exists of new DNA tests.

In the hope of keeping abreast of the latest developments, many individuals now surf the World Wide Web. Caution should be exercised in acting on information gleaned from the Internet. Time and again I have encountered patients completely misled by piecemeal or inaccurate information taken from the Internet. Even if the material is correct, its significance may be misinterpreted or its applicability to the individual misconstrued.

Given the enormous number of advances, it is not surprising that medical departures from an expected standard of practice are occurring with increasing frequency in medical genetics. Not only are there the well-known areas of malpractice; but the advent of newly available knowledge has opened up areas not previously frequented by malpractice litigators. Malpractice has occurred (and continues to occur) in four distinct areas:

1. negligence in genetic counseling;
2. negligence in communication;
3. negligence in diagnostic laboratories;
4. negligence in prenatal diagnosis.

Negligence in Genetic Counseling.   Negligence has occurred in genetic counseling, in all of the following circumstances:

- Failure to advise about increased risks of having a child with a chromosome defect.
- Failure to offer specific prenatal genetic studies that could help avoid the birth of a child with lifelong mental retardation or other serious defects.
- Failure to study the chromosomes of a parent with a family history of Down syndrome, resulting in the failure to identify a familial structural chromosomal rearrangement (e.g., translocation; see Chapter 5), with the consequence of a child born with Down syndrome.
- Failure to advise about the likely mode of inheritance of a certain genetic disorder (e.g., microcephaly, characterized by a very small head with mental retardation), with the consequence of the birth of a second affected child.

- Failure to advise about the risk of a particular medication's causing serious birth defects (e.g., anticonvulsants resulting in spina bifida).
- Failure to recognize a disorder as genetic (for example, polycystic kidney disease, congenital deafness, and the eye cancer known as retinoblastoma).
- Failure to counsel correctly about the implications of a carrier test for risks in a future pregnancy (e.g., sickle cell disease; Tay-Sachs disease; cystic fibrosis; thalassemia; fragile X syndrome).
- Failure to recognize significant genetic risks simply because of an individual's ethnic origin (e.g., Tay-Sachs disease; sickle cell disease; thalassemia; see Chapter 8).
- Failure to offer routine and timely maternal serum screening for birth defects (e.g., for spina bifida and Down syndrome).
- Failure to counsel a pregnant woman whose diabetes is out of control about her increased risk of having a child with serious birth defects.

Negligence in Communication. Instances of negligence in communication include:

- Failure to communicate a prenatal diagnosis result in a timely fashion allowing parents the opportunity to electively terminate a pregnancy where the fetus has a serious defect.
- Failure to correctly interpret and accurately communicate maternal serum screening results, the consequence being the birth of a child with serious congenital defects.
- Failure to communicate the recommendation for a second amniocentesis, depriving parents of the opportunity for a successful prenatal diagnosis and resulting in the birth of a child with a serious birth defect.

Negligence in Diagnostic Laboratories. Cases of negligence on the part of diagnostic laboratories have been found in:

- Failure to diagnose a fetal chromosome abnormality.
- Failure to detect that both parents are carriers of the gene for Tay-Sachs disease, cystic fibrosis, or another disorder.

- Failure to achieve an accurate prenatal diagnosis of a biochemical genetic disorder (e.g., Hurler syndrome; Canavan disease).
- Failure to detect a chromosome abnormality carried by a mother who subsequently gives birth to a chromosomally abnormal child.
- Labeling error resulting in the wrong sample being analyzed, an incorrect result provided, and a child born with Tay-Sachs disease.

Negligence in Prenatal Diagnosis.   Prenatal genetic studies are remarkably accurate. Yet even though accuracy rates exceed 99 percent, they do not necessarily reach 100 percent. Knowledge about the potential pitfalls should enable both those who are undergoing tests and analyses and those who are administering them to take preventive steps. The various errors that have occurred include:

- Failure to offer prenatal diagnosis. Prenatal genetic studies are usually offered/recommended when the risk of fetal abnormality is significantly greater than random risk. Usually this risk indicator is about 1 percent, since there is a need to balance the likely risk of a fetal defect against the potential loss of the pregnancy due to the prenatal diagnostic procedure (amniocentesis or chorion villus sampling; see Chapter 22). Failure to offer the appropriate prenatal diagnostic test in the face of risk exceeding or equaling 1 percent and associated with an adverse outcome to pregnancy invites a medical malpractice lawsuit.
- Failure to provide accurate information regarding risks of occurrence or recurrence of a defect. Legal case literature is replete with cases in which entirely incorrect information was provided regarding the future risks of having a second or even a third child with the same genetic defect. Physicians' statements suggesting that the event was a "fluke" or a "one-in-a-million" chance fly in the face of a 25 percent risk of recurrence. These cases also reflect a common failure of physicians to refer their patients for necessary genetic counseling.
- Failure to explain significantly abnormal results, with catastrophic consequences. On occasion, abnormal results may emerge from prenatal genetic studies, which are either misinterpreted or misjudged by the physician. Failing to fully explain their potential meaning or to refer the parent(s) for additional genetic counseling in these circumstances may deprive the parent(s) of the opportunity to discontinue a pregnancy with a serious fetal defect.

- Failure to provide timely results of prenatal diagnosis. Delayed tests, failure of cells to grow in the laboratory, or other reasons may result in communicating a prenatal diagnosis of a serious defect too late for elective abortion of the fetus.

- Failure to determine the correct fetal sex or genetic disorder because maternal cells have contaminated the amniotic fluid sample. (See Chapter 22.)

- Failure to diagnose a defect because of a mix-up in samples or slides. Although the risk of this happening is small, it is real. The risk is greater in high-volume laboratories where many samples are analyzed simultaneously or in rapid sequence.

- Failure to order an indicated test. For example, a mother who has had three miscarriages prior to giving birth to a child with a chromosome defect may subsequently learn that she or her partner has or carries a particular structural or numerical chromosome abnormality. Expected standards would have required both parents to have had chromosome analyses, at least after three pregnancy losses.

- Failure to analyze the fetal chromosomes correctly. There have been some egregious errors in straightforward diagnoses. Other cases have involved more subtle aberrations of the chromosomes.

- Failure to recognize significant chromosome mosaicism. Unnecessary speed or inattention in the laboratory may result in a few abnormal cells admixed among normal cells being missed. The implications for serious birth defects may, in many cases, be the same as if all cells were chromosomally abnormal.

- Incorrect performance or interpretation of a biochemical or DNA assay. Practicing physicians are rarely able to make their own interpretations of DNA results. Even with relatively straightforward tests, physician reinterpretation of the laboratory results has led to erroneous guidance and to the subsequent birth of an affected child.

- Failure to understand a laboratory report coupled with failure to clarify the result by contacting the laboratory. A simple failure to telephone a laboratory for clarification about a report instead of making a wrong assumption has led to erroneous communication and the subsequent birth of a child with serious defects.

- Failure to detect obvious fetal defects on ultrasound.

- Failure to recommend to women who are planning pregnancy that they take folic acid supplements through the first three months of pregnancy as a preventive measure against spina bifida and other neural tube defects. (See Chapter 9.)

- Failure to deliver a blood sample to the laboratory in a timely manner, with the subsequent birth of a child with a congenital defect.
- Delay/failure in making a timely diagnosis of a serious genetic disorder in an existing child, thereby depriving parents of risk data and of the options for prenatal diagnosis (among others) in a subsequent pregnancy, resulting in a second (or even a third) affected child.

Clearly, couples undertaking prenatal genetic studies because of any indications of significant risks of fetal defects would benefit by consultation with a clinical geneticist.

Future Litigation

Many different kinds of legal challenges will emerge in the coming years as a consequence of the new genetic discoveries. A bewildering array of lawsuits can be expected, including personal injury cases involving all aspects of genetic testing; discrimination cases based on invasion of privacy; gene therapy cases based on claims of irreparable harm; and litigation regarding genetic data banks, DNA banks, and various other DNA forensic technologies. The good news is that after a quarter century of advances in genetics, many judges are actively pursuing their own education in this subject.

chapter 27

Treatment of Genetic Disorders

Prevention is better than cure, especially where genetic disorders are concerned. Knowing your family history, seeking genetic counseling, and having appropriate genetic tests will provide you with opportunities to prevent or avoid serious, life-long, or fatal genetic disorders. Predictive genetic tests (see Chapter 24) help establish critical surveillance, early diagnosis, and preemptive treatment aimed at cure. Fortunately, for those with a genetic disorder, a remarkable array of therapeutic approaches are available, and further major progress can be anticipated once the Human Genome Project is completed and all disease-causing genes have been identified.

Since the most successful treatment for a genetic disorder will depend upon our understanding of its molecular basis, enormous progress can be expected in the development of new therapies as we learn about the normal and abnormal products of normal and defective genes in the coming years. Meanwhile, we are already benefiting from these advances. Many important new treatments are available that can save lives, extend life spans, dramatically improve the quality of life, and even result in cures. This chapter provides a review of current therapeutic modalities and an examination of gene therapy and other new approaches.

Modifications of the Diet

Dietary modifications for various genetic disorders can be achieved by restriction and substitution, exclusion and augmentation, and supplementation and replacement.

Restriction and Substitution

Remarkable success has been achieved by dietary modification using restriction and substitution. When an essential food is toxic to an individual because of a hereditary disease, it should be restricted or excluded from the diet. The classic example is the biochemical recessive disease called phenylketonuria (PKU). This recessive disorder, inherited from both parents equally, is characterized in its untreated state by severe mental retardation, seizures, eczema, and other complications. It is caused by deficiency of an enzyme whose function it is to chop up accumulating toxic products. Because of this enzyme deficiency, these toxic products accumulate, and their accumulation—in conjunction with the deficiency of necessary products not produced—permanently damages the brain. What must be restricted in such cases is the essential amino acid called phenylalanine, which is present in all high-grade proteins. The body normally needs this amino acid in order to manufacture its own proteins.

All newborns are routinely tested for this biochemical disorder during the first days of life. Since the initiation of newborn screening for PKU, over 150 million infants have been screened in the developed world and over 10,000 have been detected with PKU and have received dietary treatment. Early detection allows immediate initiation of a low-protein diet that limits the production and accumulation of the brain-damaging toxic products. Timely diagnosis and treatment avoids the otherwise certain development of severe mental retardation. The diet, however, contains a synthetic supplement (made up of an amino acid mixture, calories, minerals, and vitamins) that is offensive in odor and taste. Forcing children affected by PKU to adhere to this unpalatable diet, which in addition deprives children of many of their favorite foods, inevitably leads to serious emotional stress in families. But failure to adhere to the diet results in reduced cognitive ability, neurological abnormalities, and personality/behavioral disturbances; therefore, considerable effort is necessary to enforce its use through the early childhood years. In addition, the repeated visits to the clinic, repetitive blood sampling, and the special aura of caution that surrounds the child all take their toll on the parents and the child.

A generation of experience with this disorder has led to the realization that life-long dietary control is necessary. Discontinuation of the special diet in later childhood has resulted in cognitive and behavioral decline. As it is, even with dietary control, intellectual accomplishments tend to be below par.

Females with PKU who decide to have children need very special attention. Their situation is no longer uncommon, since newborn screening and immediate and continuing treatment enables normal development and procreation. When planning pregnancy, however, women with PKU need to ensure that they are on an extremely strict low-phenylalanine diet aimed at avoiding the accumulation of toxic products that could permanently damage the developing fetal brain, causing miscarriage or serious birth defects (especially in the heart and brain). Very strict attention to the special diet and constant monitoring of the blood for toxic products allows women with PKU to have healthy children.

Carrier detection and prenatal diagnosis through DNA analysis are now available for those at risk of having children with this disorder.

There is also new hope since the PKU knock-out mouse was created as a model for human PKU. The therapeutic use of an enzyme (called phenylalanine ammonia lyase) has been shown to reduce circulating toxic phenylalanine levels in the laboratory mice. The hope is that this substitute enzyme treatment will help control PKU, diminish the need for the rigorous diet, and improve the quality of life for those affected.

A similar but much rarer condition called galactosemia is also managed by dietary restriction. Untreated, this recessive disorder (which is also inherited equally from each parent) results in mental retardation, cirrhosis of the liver, cataracts, and early death. The cause is a deficient enzyme whose failure to chop up a milk sugar leads to the accumulation of one of its components (galactose), which proves toxic especially to the brain, liver, and eyes of the fetus and infant. Treatment is based on the exclusion of milk and milk products from the diet. Early newborn diagnosis is critical and lifesaving, and continued dietary treatment allows normal development. Unfortunately, despite adequate dietary restriction of milk and milk products, learning, speech, and behavioral problems are subsequently encountered in most of those affected. Women with galactosemia who are hoping to have a child must avoid milk and milk products when planning pregnancy and throughout pregnancy in order to have the best chance of the fetal brain not being damaged by the accumulation of galactose. Both biochemical and DNA methods are available to detect carriers, as well as for prenatal diagnosis.

Augmentation and Exclusion

Certain foods and liquids can be used to protect a patient against biochemical disorders. Some of the instructions are remarkably easy to follow. In sickle cell anemia, the blood cells become sickle-shaped when dehydration occurs. It is therefore very important for an affected person to keep well hydrated by drinking water. Water is also essential in the treatment of cystinuria, the genetic condition that produces kidney stones. Taking alkalis provides additional help in preventing stone formation in both kidney and bladder. A number of hereditary disorders involve low blood sugar, or hypoglycemia; simply eating fruit and other foods may protect an affected person against the complications of the disorder.

Certain sugars, such as glucose, sucrose, and fructose (sugar in fruit), may individually cause serious problems for persons with rare hereditary disorders involving the body's carbohydrate chemistry. In the condition called fructose intolerance, the body is unable to digest the natural sugars in fruit. Dietary therapy is facilitated by the fact that the affected person often develops an aversion to sweet-tasting foods.

Other rarer disorders involving amino acids (the body's protein-building blocks) may also require strict diets very low in protein. For some of these disorders, particularly those for which newborns are routinely screened, strict, very low-protein diets may be lifesaving. Cutting out saturated fats, drastically limiting fat intake, and using low- or no-cholesterol foods are important maneuvers in aiding the medical treatment of heart and blood vessel disease.

Obvious restrictions apply to individuals with a serious allergy that could cause an acute reaction, shock, collapse, and death (called anaphylaxis)—for example, peanuts, penicillin, chestnuts, and red currants. Other drugs (e.g., sulfa drugs) or foods (e.g., shellfish) that result in allergic reactions obviously also must be excluded from the diet. All of these allergic responses are inherited either via a single dominant gene or multiple genes acting in concert with the extrinsic allergic factor.

In the disorder called acute intermittent porphyria, commonly seen in white South Africans of Dutch descent, barbiturates may precipitate severe attacks of abdominal pain and may prove fatal. Many hospitals in South Africa routinely test every white patient for porphyria on admission, in order to avoid potential catastrophes. This dominantly inherited disorder carries a 50 percent risk of the affected individual transmitting the disease to each of his or her offspring. Similar symptoms may be precipitated not only by barbiturates but also by many other drugs, including sulfa drugs, anticonvulsants, hormones, and antifungal drugs.

The examples discussed of disorders in which restriction or substitution methods may be used are not all-inclusive, but provide a framework to understand this therapeutic approach.

Supplementation and Replacement

Vitamins added to the diet may be extremely important in preventing certain birth defects and treating some genetic disorders. The most prominent advance in this context is the routine use of folic acid (0.4 mg) as a daily supplement for women planning pregnancy and also taken through at least the first three months of pregnancy. This simple supplementation, the value of which has been proven by the experience of more than a decade, facilitates the avoidance of 70 percent of cases of spina bifida or anencephaly (see Chapter 9), among the most common birth defects seen in the Western world. It is a sad commentary indeed that most women still do not take such supplements in a timely manner or at all. The embryonic neural tube (forming the spinal cord and its coverings) closes on a schedule between 26 and 29 days following conception. Folic acid taken for prevention after that time might as well be rubbed on the belly button, for all the good it will do!

Supplementation of the fat-soluble vitamins A, D, E, and K is critically important in the management of cystic fibrosis. Absorption of these vitamins is impaired in this disorder and supplementation is necessary in order to avoid vitamin deficiencies. In certain rare and complex biochemical disorders, vitamin supplementation will stimulate compensatory chemical reactions in the body and help it overcome the basic derangement. Since vitamins are vital to certain body reactions, massive doses are sometimes needed. Note, however, that in routine pregnancy the idea that a small dose is good and a higher dose is better can result in grave consequences. Our research group has shown that megadoses of vitamin A taken in early pregnancy, for example, may cause birth defects.

For common diabetes mellitus, the supplement is insulin, necessary when the pancreatic cells fail to produce sufficient insulin or when the body builds a resistance to the available insulin.

Another type of supplementation, used in the treatment of diabetes insipidus, involves water plus a hormone. The genetic type of diabetes insipidus is a sex-linked disorder carried by females but affecting only males. It renders the kidney unable to concentrate urine because of a pituitary hormone deficiency. The affected male passes excessive quantities of urine, becomes dehydrated, and could even die if not treated. Water supplementation temporarily saves the day; however, the specific hormone also has to be provided. This hormone is now available in the form of a fine powder to be inhaled through the nose. The hormone is absorbed through the nasal membranes into the bloodstream and circulates to the kidney, where it helps concentrate the urine.

A host of diseases exist in which a specific enzyme is absent or deficient, resulting in the progressive and systematic accumulation of complex fats, proteins, or carbohydrates. These so-called storage diseases (e.g., Tay-Sachs disease, Gaucher's disease, Niemann-Pick disease, Fabry disease, Refsum disease, adrenoleukodystrophy, Hurler syndrome, Hunter syndrome, and many others) would be treatable if the deficient enzyme could be given with ease, and without causing complications. Thus far, enormous success has been accomplished in the treatment of Gaucher's disease by the administration of the deficient enzyme glucocerebrosidase. This enzyme replacement therapy has been extraordinarily expensive, but has been effective in restoring and maintaining the health of individuals with this disorder. Major efforts continue, and clinical trials are in progress for the treatment of other serious storage diseases.

Retinitis pigmentosum is a disorder that ends in blindness and that is caused by various gene defects, which may be recessive, dominant, or sex-linked (see Chapter 7). The initial symptom is visual difficulty at night or at twilight, followed by progressive loss of peripheral vision. Progressive disease affecting the retina of the eye results in "tunnel" vision, and eventually in total blindness. Most are legally blind by the age of 40. About 1.5 million people worldwide are affected by this disorder. The use of vitamin A supplements (15,000 IU daily) has been shown to slow the annual decline of visual loss by about 20 percent. For example, those taking this megadose of vitamin A from the age of 32 have been estimated to retain useful vision until age 70, compared with those not taking this supplement, who typically have lost useful vision by age 63. In individuals with liver disease, however, such high doses of vitamin A are not recommended. An additional caution is for those with retinitis pigmentosum to avoid taking high-dose vitamin E supplements—a practice that seems to hasten visual loss.

Hemophilia A is a typical example of the value of replacement therapy. This disorder, occurring in about 1 in 10,000 males, is due to mutations that occur in the gene responsible for the clotting factor VIII, which is located at the tip of the long arm of the X chromosome. This very large gene and the hundreds of different mutations that have been found within it lead to the sex-linked disorder almost exclusively in males. Very rare cases in females may be due to an entire missing X chromosome in a female, who would otherwise have been a carrier (the gene defect being on the remaining X chromosome), or may result when a male with hemophilia fathers a child with a partner who by chance is a carrier of the gene for hemophilia A. (See Chapter 7 for more detail on the process of inheritance of this disorder.)

A mutation in the hemophilia A gene renders the affected individual deficient in the critical clotting factor VIII. This factor can be derived from concentrating plasma from blood donors, purifying the needed fraction, and administering it intravenously as a so-called concentrate, or cryoprecipitate. In the 1980s, due to the general failure to inactivate common viral contaminants in blood plasma (including hepatitis B and C, cytomegalovirus, and the AIDS virus), many hemophiliacs became infected after receiving cryoprecipitates. Procedures for inactivation now allow virus-free administration of these factor VIII concentrates. Moreover, with the cloning of the factor VIII gene, DNA methods for the manufacture of factor VIII have produced a commercial product in use since the early 1990s. Unfortunately, the body may react adversely and produce antibodies to administered factor VIII, making management extremely difficult. About 10 to 20 percent of patients develop such antibodies.

About one-third of those born with hemophilia have a new mutation. In cases where there is an inversion within the gene (similar to inversions within a chromosome; see the discussion of structural rearrangements in Chapter 3), the mother is almost invariably a carrier, the inversion having originally occurred in the sperm cells of the mother's father.

Other examples in which replacement of proteins is critical to health and survival include disorders in which the body fails to manufacture gamma globulin, which protects against infection. Regular, lifelong injections of gamma globulin may be necessary in such cases.

Treatment with Medications

An arsenal of drugs exist for the treatment, albeit noncurative, of a huge number of genetic disorders (see Table 27.1). Most such drugs are dedicated to remedying troublesome or worrisome symptoms; but particularly valuable are drugs that interdict or prevent the most serious consequences or complications of a genetic disorder. Once a diagnosis has been definitively made, either using cardinal clinical signs (e.g., Marfan syndrome) or by DNA analysis (e.g., Factor V Leiden deficiency), preventive therapy can be initiated. As noted in the table, beta blockers are used to decrease the risk of aortic rupture in Marfan syndrome. For those who have had deep vein thrombosis in the legs (or elsewhere) and have Factor V Leiden or a prothrombin gene mutation, lifelong treatments with anticoagulants may be necessary. The aim would be to prevent recurrent thrombosis in which there is the potential of dislodging a clot that could cause sudden death.

More recently, new anti-inflammatory drugs (so-called cyclooxygenase inhibitors) have been found to reduce the number of polyps in the colon and rectum in individuals with the dominantly inherited familial adenomatous polyposis, in which there is a virtual 100 percent risk of colorectal cancer.

Hormone treatment for genetic deficiencies in the endocrine glands (thyroid, pituitary, pancreas, adrenal, parathyroid) are all remedied by administration of the necessary deficient hormone or provision of the deficient protein. Certain forms of gout are inherited. The use of specific drugs will block the formation of the accumulating uric acid, which causes excruciating attacks of pain.

356

TABLE 27.1   Examples of Medication Types Used to Treat Selected Genetic Disorders

Genetic Disorder	Type of Medication	Purpose
Biotinidase deficiency	Vitamin (Biotin)	Newborn diagnosis avoids mental retardation, seizures, vision loss, hair loss, balance problems, skin eruptions
Congenital adrenal hyperplasia	Steroids (e.g., cortisone)	Save life; remedy growth and other problems
Coronary heart disease	Anti-inflammatory (e.g., aspirin)	Reduces frequency of heart attacks
Cystic fibrosis	Pancreatic supplements	Remedy malabsorption of fats
Diabetes mellitus	Hormone (insulin)	Controls blood sugar
Diabetes insipidus	Hormone (antidiuretic)	Concentrates urine to stop water loss
Factor V Leiden gene mutation	Anticoagulant	Prevents thrombosis
Familial hypercholesterolemia	Statin drugs	Reduces blood cholesterol and frequency of heart attacks
Familial polyps of the colon	Anti-inflammatory (e.g., Cyclooxygenase inhibitor)	Reduces polyp formation in the colon
Growth hormone deficiency	Hormone	Remedies lack of growth hormone, restores, growth and other functions
Hypothyroidism	Protein (e.g., thyroxine)	Remedies underacting thyroid gland
Long QT syndrome	Beta-blockers	Reduce the frequency of heart rhythm abnormalities
Malignant hyperthermia	Muscle relaxant (e.g., Dantrolene)	Prevents death from very high fever during anesthesia
Marfan syndrome	Beta-blockers	Reduce risk of rupture of the aorta
Osteogenesis imperfecta	Biphosphonate pamidronate	Improves bone density
Prothrombin gene mutation	Anticoagulants	Prevent thrombosis
Retinitis pigmentosum	Vitamin A	Megadoses delay visual loss

(continues)

TABLE 27.1   (*continued*)

Genetic Disorder	Type of Medication	Purpose
Sickle cell disease	Increases fetal hemoglobin production e.g., hydroxyurea	Reduces painful crises and frequency of blocked arteries
	e.g., vaccine	Reduces frequency of bacterial infection (by pneumococcus)
Smith-Lemli-Opitz syndrome	Cholesterol	Improves growth and behavior
Spina bifida/anencephaly	Folic acid (vitamin B complex)	Prevents about 70 percent of cases
Tyrosinemia	Enzyme inhibitor	Improves liver and kidney function and prevents eye and skin complications
Wilson's disease	Chelating agent	Rids body of excess copper

In sickle cell disease (see Chapter 8), hydroxyurea raises the concentration of the fetal type of the oxygen-carrying protein hemoglobin and significantly ameliorates this disease. Its use has markedly reduced the frequency of painful crises and the need for repeated hospitalizations and blood transfusions. Since infection by a common bacterium (*Streptococcus pneumoniae*) is a leading cause of death among infants with sickle cell disease, a vaccine against this invader has been recommended for affected children at 2 years of age, with booster doses at age 5. Unfortunately, however, infection has occurred in some children even after vaccination. Until better vaccines are developed, routine use of prophylactic penicillin twice daily from the age of 2 or 3 months to 3 years of age, followed by 250 mg twice daily until 5 years of age, has been recommended.

With the identification of all human genes, the science of pharmacogenomics may make it feasible to activate or stabilize an abnormal enzyme or protein using a type of genetically based drug, or by actually making the normal protein and administering it as a replacement for the defective protein.

Depletion

Excessive accumulation of toxic elements may cause irreversible damage to certain organs in the body and eventually prove fatal. The most common genetic example is that of the recessively inherited hemochromatosis. (See the section on human leukocyte antigens [HLA] in Chapter 10.) About 1 in 10 whites carry this gene. Between 1 in 200 and 1 in 250 have this lifelong chronic disorder, which in many cases goes undiagnosed for decades. The key problem is an excessive accumulation of iron in the body and its deposition in key organs, which can be especially damaging to the liver, pancreas, and heart. The most valuable treatment is phlebotomy—the removal of a pint of blood once or twice a week initially, with the monitoring of specific markers of iron concentration and deposition, and then a schedule of lifelong phlebotomy at intervals of 2 to 6 months. Early treatment is extremely important to secure health and ultimately to save life. The important aim of early diagnosis is to avoid the development of irreversible liver damage (cirrhosis) or an increased risk of liver cancer, either of which could shorten life expectancy. Chelating drug therapy, the goal of which is to bind iron and assist in its excretion, may also be used in conjunction with phlebotomy.

Wilson's disease, due to a defective gene transmitted by each parent, results in the excessive storage of copper, particularly in the brain and liver. Signs of this disorder include tremors, incoordination, loss of mental faculties, and cirrhosis of the liver. All foods with a relatively high copper content, such as cherries, chocolate, or beef, must be excluded from the diet. Drugs are also necessary in order to rid the body of copper as well as to interfere with its absorption.

Avoiding Harmful Factors

We all make efforts to avoid illness; but those with a genetic disorder or a genetic predisposition need to exercise special care. For many of the thousands of genetic disorders and traits, advice about what to do and what not to do can be provided by clinical geneticists. This book deals with many examples and offers advice appropriate to each case. For example, individuals with albinism should avoid sun exposure because of the increased risk of skin cancer. Persons with alpha-1-antitripsin deficiency, a recessive disorder, absolutely must not smoke or work in polluted environments because of the danger of lung disease. Those with pseudocholinesterase deficiency need to avoid certain muscle relaxants given during general anesthesia, since they may fail to resume breathing on their own for two to three hours after the cessation of anesthesia. This is a recessive disorder that is particularly common in certain populations, such as Eskimos, many of whom have a family history of such problems following surgery and general anesthesia. In the Alaskan Eskimos, about 1 in 10 carry the gene for this disorder. About 1 in 30 whites are carriers.

Infants born with a specific mitochondrial gene mutation are especially susceptible to becoming deaf if they are treated with a particular class of antibiotics called aminoglycosides (such as streptomycin or gentamycin). Tiny, premature infants or sick newborns may well be treated with these powerful antibiotics, which would lead to deafness or hearing impairment in those with the genetic mutation. A family history of deafness (especially on the mother's side) dating back to the earliest weeks of life, or later, should serve as a potential alert for this uncommon disorder.

Repair and Reconstruction

Surgery is a common form of treatment of many birth defects or genetic disorders. Operations are performed on the head (for hydrocephalus—enlarged head, with excess fluid accumulation); the heart (ventricular septal defect or "hole-in-the-heart"); cleft lip or palate; exposed bowel (called omphalocele); genital abnormalities; limb defects; and so on. Surgical reconstruction—for example, of a missing ear due to certain genetic disorders—is at least cosmetically achievable.

Enormous energy has been expended in attempts to repair the open spine defect (spina bifida) while the fetus is in the womb. Unfortunately, with experience in about 100 such cases, no significant improvement has been gained as far as the use of the legs is concerned. However, there are early claims that surgery on the fetus decreases the incidence of shunt-dependent hydrocephalus but increases the incidence of premature delivery; and low-birth-weight infants have their own set of other potentially serious complications. More time and experience are necessary before the results of this type of fetal surgery can be definitively assessed.

Of course, surgery to remove cancers of genetic or other causes is self-evident. The indicated surgical therapies for specific genetic cancers are referred to in Chapter 17.

Transplantation

Although every genetic disorder is characterized by the defect being evident in the DNA of every cell in the body, quite often one or more organs are affected more than others. For example, the adult form of polycystic kidney disease (see Chapter 24), if not complicated by cysts or aneurysms elsewhere in the body, can be effectively treated by removal of the diseased kidneys and transplantation of healthy ones. Similarly, transplantation has been valuable for certain defects in the heart (cardiomyopathy; see Chapter 12), liver (cirrhosis or cancer), bone marrow (leukemias and lymphomas), and corneas (various eye diseases). More recent techniques involving the growth of connective tissue cells (fibroblasts) in sheets in the laboratory for use as grafts to cover burns or other skin defects have seen early success.

Bone marrow transplantation, used for the past thirty years in treating severe combined immunodeficiency disease, has been successful for hundreds of affected children. This otherwise fatal disorder caused by the body's inability to resist infection has been cured by transplantation of bone marrow cells obtained from closely matching siblings or parents (see the discussion on HLA in Chapter 10). One study recorded a 100 percent survival rate and a healthy life quality in those who as infants received HLA-identical marrow. Survival rates were 78 percent in those whose marrow was not exactly compatible. Frequently, however, even after transplantation, some may need continued injections of gamma globulin to assist their bodies in warding off infection.

More recently, physicians in France successfully used gene therapy to treat the X-linked form of severe combined immunodeficiency disease (SCID). This otherwise fatal disease, whose male victims typically succumb to infection, was remedied by using a virus to deliver the needed gene. In preparation for the procedure, key cells (lymphocytes) were extracted from two affected boys. The cells were infected with the viral vector; several days later, they were reinfused into their patients—without the necessity of ablating the bone marrow. The cells containing the healthy gene became engrafted and the gene began manufacturing the necessary protective product.

Bone marrow transplantation for various genetic storage diseases has been performed in the hope that newly received blood cells would carry the necessary enzymes into brain cells, relieving them of their burden of stored material. This method has so far proven unsuccessful. More promising have been techniques used in the treatment of Parkinson's disease, in which specific brain cells derived from embryonic pigs have been transplanted into the affected region of the patient's brain. Improvements in gait and in performance of other activities of daily living have been noted in early reports. These studies of so-called xeno-transplantation continue.

Stem Cell Transplantation

Our bodies are made up of organs and tissues the cells of which differ greatly from one another. What process allowed such different tissues to develop after fertilization, from a single cell? This original cell divides rapidly and repeatedly into a ball of cells, each having a full copy of the blueprint for the future human being. During the first fourteen days or so, these cells have not yet been designated to any specific body part. In fact, each possesses the ability to become any organ system. Called totipotent stem cells, these are the progenitors of the next set of stem cells (called pluripotent). Both totipotent and pluripotent stem cells are capable of forming different kinds of cells but not an entire body. A biological genetic event then occurs that directs the stem cells to activate the blueprint, which calls for different stem cells to assume different organ destinies. The totipotent stem cells are found in the very early embryo, whereas the pluripotent cells can be found in the slightly later embryo, eventually in the blood from the umbilical cord, and in the bone marrow and tissue of the fetus.

Our ability to isolate pluripotent stem cells in order to repair or replace organs or tissues that are impaired by illness (such as diabetes, heart disease, stroke, burns, or arthritis), injury (e.g., of the spinal cord or brain), degenerative disease (e.g., Huntington's, Parkinson's, or Alzheimer's), or even aging is still very limited. Nevertheless, proof of principle is reflected in the successful treatment of some cases of Parkinson's disease by fetal cell implantation into the brain. The fetal cells used were already "committed" nerve cells, harvested at the time of early abortion. Uncommitted stem cells would have even more potential benefits in research and treatment, particularly now that we know it is possible to grow embryonic stem cells in the laboratory. Hope recently was spawned among paraplegics by reports of a successful experiment in which paraplegic rats were able to walk again after embryonic brain stem cells were implanted in their spinal cords, below the level of the injury.

We all retain very active stem cells carrying self-generated instructions, making it possible, for example, for our skin to re-form and heal after injury, our intestinal lining to constantly replace itself, and our bone marrow to work continually and precisely in producing blood cells of the same lineage. Perhaps it will prove possible to coax these "old" cells to learn new tricks. Adult-derived stem cells could possibly be manipulated to differentiate into any desired cell type—such as nerve, muscle, liver, or blood. Early experiments in mice hold promise, but it is far too soon to assess the future of this approach.

Stem cells harvested from the umbilical cord blood of a matching sibling have successfully cured the leukemia of the donor's older brother. Some parents have even gone so far as to harvest stem cells from the cord blood of their newborn, aiming to store the stem cells in case of future need, due to the development of leukemia or cancer, which could then be remedied by the use of stem cells.

The disorders that undoubtedly have potential for successful treatment using stem cells include certain leukemias, lymphomas (such as Hodgkin's disease and non-Hodgkin's lymphoma), aplastic anemia due to bone marrow failure, failure of the immune system (e.g., severe combined immunodeficiency disease), thalassemia, and sickle cell disease. All stem cells collected for banking purposes and for use in treatment must of course be tested for infectious diseases and for inherited blood diseases.

Despite the incredible, lifesaving benefits that have already accrued in the cure of fatal genetic diseases, major controversy has arisen about the ethical use of stem cells. The controversy festers around the thorny question of the cells' derivation. To date, these cells have been obtained from embryos that couples have created during fertility treatments and that were thereafter discarded. Couples undergoing these fertility procedures create more embryos than they normally use in trying to produce a child. Consequently, an estimated 150,000 frozen embryos are thought to be stored in the United States at present. Clearly, the use of an embryo to derive stem cells makes it no longer viable. However, embryos used for this purpose were destined to be discarded anyway, without opportunities of saving many lives. Those opposed to abortion are against the use of stem cells derived from embryos, whether donated and frozen or following elective abortion.

A number of actions are currently pending in the U.S. Congress that could scuttle the use of this incredible resource for diminishing pain and suffering and saving the lives of many. Although current federal law allows support for research on human fetal tissues as well as on stem cells derived from dead fetuses, three restrictions apply:

1. There is a criminal prohibition against the sale of human fetal tissue for a price exceeding the expenses of procuring and delivering the tissue.
2. Informed consent from the donor is required if fetal tissue is to be used for therapeutic transplants.
3. The donation of fetal tissue to a designated recipient is prohibited.

Genetic Engineering

Gene Knock-Outs and Knock-Ins

Once a gene has been identified and its structure is known, it can be cloned and targeted for manipulation. Knock-out mice are those especially bred or "engineered" in whom the gene of interest has been deleted. Knock-in mice are bred with a specific gene inserted. The purpose of these genetically engineered mice is to understand gene function, including how genes are regulated or influenced by other genes. This newfound technology of gene targeting has provided incredible opportunities for creating new strains of mice with specific gene mutations. Hence, mice have been created with desired mutations that in humans cause specific genetic disorders. There are, for example, knock-out mice bred for phenylketonuria, Huntington's disease, cystic fibrosis, Tay-Sachs disease, obesity, hypertension, breast cancer, and many other disorders.

Using this method, research scientists first introduce the desired mutation into the cloned copy of the chosen gene. The mutation is then transferred into a pluripotent mouse embryo-derived stem cell. These stem cells containing the targeted mutation are then injected into other early embryos and placed in foster mothers. The offspring that are generated are capable of transmitting the targeted mutation. In addition to gaining a better understanding of the target gene's functioning, researchers also add to current knowledge of the interaction of neighboring regulatory genes. These methods allow one to study how genes are inactivated or how their function can otherwise be altered purposefully. Through this technology, specific and selected gene functions in the mouse can be used to learn about very complex biological processes, including normal development and learning. These genetic engineering techniques have much to offer humans regarding detailed knowledge about cancer, neurological diseases, and virtually all other human genetic diseases. Needless to say, however, gene manipulation that has been applied to mice is not applicable to humans. These targeted (or knock-out or knock-in) mice nonetheless make repetitive treatment trials possible by providing representative models of human disease that are available in a continual supply.

Cloning for Health

Worldwide attention several years ago was riveted on the announcement in Edinburgh, Scotland, of the cloning of Dolly the lamb. This breakthrough had been achieved by transferring the nucleus of an ordinary sheep's body cell (from the udder) into the egg of a sheep the nucleus of which had been removed beforehand. Later, using the same techniques, a bull was also cloned. Of particular interest is the use of skin cells (fibroblasts) taken from the bull, which were first cultivated and passaged through some thirty cell doublings. This period in cell culture did not compromise the capacity of these cells to generate healthy animal clones.

By combining gene cloning with gene targeting, animal models could be designed to produce extremely valuable gene products. These procedures will facilitate the production of proteins and enzymes with enormous commercial value, which could also be of great benefit to people with specific genetic disorders. For example, cloned and gene-targeted sheep, goats, and cows can be used to produce pharmaceuticals such as human insulin, or the enzymes needed to treat human biochemical genetic disorders such as Gaucher's disease. This same technology also could be used for the conservation of endangered species. Cells could be frozen for many years and used later to generate animal clones of that species.

Since genes can be modified in domestic animals, pigs could be engineered that possess a genetic makeup compatible with the human immune system. This might facilitate the transplantation of certain pig organs (e.g., kidneys) to humans, without risk of rejection. Of course, environmental systems excluding any possible animal viruses would also be necessary for specially bred, genetically engineered animals.

Ethics and Laws

No sooner had the brouhaha broken out about the cloning of Dolly the sheep than our airwaves echoed with questions and concerns about the cloning of humans. Lost in the hullabaloo was the realization that identical twins are clones. In fact, many animals, plants, and greenflies are clones, usually of their mothers. Imagine that such techniques were available to humans. It would be possible to have a "twin," but separated by any number of years. An adult, having a child cloned from him or herself, would be able to follow a personal copy through early growth and development. And that child growing into adulthood would be able to witness his or her future in the life of the parent! We can easily anticipate the serious potential risks of this approach, which would include greater frequency of diseases such as cancers and premature aging occurring, due to the accumulation of gene mutations with age that might be contributed to the cloned individual. Human cloning would, however, benefit couples who were unable to have genetically related children by any other means. Could such parents ethically be denied the right to have their own, genetically related children?

There is no doubt about the potential value of cloning technology for human health and fertility. In contrast to the poorly considered and impetuous bills introduced into the Senate and House following the cloning of Dolly, the Prohibition on Cloning of Human Beings Act of 1998 introduced by Senators Diane Feinstein and Edward Kennedy underscores the power of rational thought and insight. The bill would ban, for at least ten years dating from passage into law, the implantation of a specifically developed embryo into a human uterus for the purpose of creating a child; but it also would protect research on nuclear transfer technology.

Gene Therapy

A person with a disease due to dominant, recessive, or X-linked inheritance possesses a copy (or two) of the defective gene(s) in every cell of the body, or an admixture of normal and abnormal cells (called a mosaic). Any remedy to replace or repair the defective gene would constitute gene therapy. The major challenges have been and continue to be the safe transfer of a healthy gene into the cells of the diseased organ(s), its placement in a specific location, and its activation to normal function. Even more challenging are efforts to devise techniques to repair a gene mutation.

Gene Delivery

Transporting a healthy gene into the cells of a genetically defective organ (or a tumor) has proved an enormous challenge. A number of different viruses, including some that frequent the respiratory tract, have been used as carriers or vectors for this purpose. In this process, the gene to be transported is inserted or integrated into the viral DNA. Viruses and their DNA quickly spread through the body, are taken into cells, and are incorporated into their nuclear DNA. A few obvious difficulties have been encountered and continue to cause concern because they effectively delay progress.

The hope is that the required gene will be inserted or integrated into the sick cell's DNA. The optimal goal would be to have the healthy gene inserted precisely where the defective gene is located. This type of targeting is not easily accomplished. These viruses, modified as they might be, may themselves cause infection and illness, or, by combining with other viral DNA, develop a cancer-causing potential. Since genes work as an orchestra (see Chapter 6), the insertion of a transported, healthy gene into an inappropriate site on the wrong chromosome could result in a combination with a neighboring gene that would then produce a protein that has cancer-causing potential. In addition, viruses may elicit strong immune reactions by the body, limiting their value in gene delivery. Some of the viruses used experimentally have also proved toxic to cells. Some transport viruses (called vectors) will only be integrated into cells that are dividing, which again limits their value. For these and other reasons, nonviral gene delivery systems have been sought.

Tiny vesicles called liposomes (resembling microscopic fat droplets), which are chemically synthesized and are microscopic in size, have been used to deliver drugs into the body. They have also been used to package DNA, having the advantage over viral vectors of not eliciting the body's protective immune response and not being as toxic. In addition, they pose no cancer risk. Initial efforts were thwarted, however, since the body's immune system tends to gather up these packages, preventing them from reaching their targets.

A clever trick by researchers was then to use a "stealth" technology, coating these vesicles with the chemical that normally coats our red blood cells. This technique has been more successful in gene delivery and holds considerable promise for gene therapy. Various other vectors also have been developed, all with the ultimate purpose of delivering a healthy gene to an appropriate organ site, disease-bearing cell, and specific chromosome location.

Disease Candidates for Gene Therapy

Many different genetic diseases affecting specific organ systems such as the bone marrow, liver, brain, muscles, and heart could be candidates for gene therapy. Thus far, efforts to introduce healthy genes or replace defective genes by inserting normal gene segments have seen some success. The bleeding disorder hemophilia B (also called Christmas disease) is due to a deficiency in factor IX, which normally helps blood to clot. A virus containing factor IX, injected into affected individuals in a clinical trial, resulted in some improvement.

A more novel approach is aimed at triggering the inherent genetic ability of cells to commit suicide (apoptosis; see Chapters 6 and 16). Another approach involves techniques for suppressing genes that stimulate the growth of blood vessels that are vital for the growth of tumors. The use of genetic techniques to induce the body's immune system to destroy specific cancer cells may also eventually prove successful. The activation or introduction of genes that interfere with the products of other genes that stimulate cell growth opens up yet another avenue for fighting uncontrolled cancer growth.

Gene therapy aimed at altering the body's immune response to foreign proteins may well prove a boon in organ transplantation. Failure to obtain matching donor organs results in thousands of deaths each year. There is a real possibility that gene therapy techniques could ultimately be used to alter the fate of those who need organ replacement.

Gene-delivery vectors traditionally have been created by essentially gutting a virus of its genes, getting rid of those genes that might be harmful to the host and retaining genes that enable the virus to insert DNA into the chromosomes of the host. Thus, the virus carrying the healthy gene for delivery will be capable both of infecting cells and of splicing the corrective gene into the chromosomes. The correct insertion of the gene, however, does not solve the problem. The vector also needs a mechanism by which to switch on the healthy gene, as the process is not automatic. Usually, the on/off switch (or promoter) of a gene is located near the gene. Promoters themselves are stretches of DNA that may be extremely complex and even large, making their placement into the gutted virus quite difficult. To counter this problem, instead of using the corrective genes' own promoter, researchers have used promoters of the virus DNA. One vexing problem is that the newly inserted gene can only be weakly activated. Pharmaceutical techniques also have been used successfully to switch on the corrective gene's function.

Another approach has combined traditional chemotherapy with gene therapy for otherwise untreatable cancers of the head and neck. Researchers directly injected tumors with viruses engineered to kill cancer cells and to leave normal cells unscathed. Complete responses were achieved in 8 of the 30 treated, and 19 showed partial responses.

No sooner has one problem been anticipated and corrected than another has arisen. In the insertion of the corrective gene, no control so far has been possible over the number of copies of the gene that will be integrated into the chromosome. Moreover, the gene to be inserted may be integrated at any spot along the chromosome, disrupting or altering its function or disturbing the function of another normally functioning gene. Some years ago, in an effort to darken the purple flowers of petunias, additional copies of pigment genes were added. The result: The flowers of the altered plants were white! Now we know—at least for plants, worms, and flies (among other organisms)—that an extra dose of a gene can silence the expression of that gene.

Gene Therapy Before Birth

Many serious genetic disorders cause irreversible damage to the developing fetus long before birth. Gene therapy aimed at preventing irreversible change in the early stage of fetal growth would have to be safe and used early enough to avoid irrevocable damage to vital fetal organs. One excellent candidate disease is alpha-thalassemia, which causes a profound hemolytic anemia leading to fetal death as well as toxic symptoms in the mother. Various gene-engineered cell transfer approaches have been used in different animal models, in which the ability to deliver corrective genes has been demonstrated. However, many questions need to be answered before gene therapy in utero proceeds. Primary among these questions are the safety of the mother, the risk to the fetus, and the inadvertent change of genes in the germline (eggs or sperm) of the developing fetus. There is still much work to be done.

Germline Gene Therapy

Insertion of a gene(s) into ova (eggs) or the cells that manufacture sperm is known as germline gene therapy. Once it is inserted, the new gene will be transmitted through subsequent generations. The technique for insertion of corrective genes into the mouse germline has been perfected. In essence, immediately upon fertilization of the egg by the sperm, the single-celled mouse embryo is injected with the corrective gene. The embryo is then placed in the uterus of the recipient female mouse. Mice with specific genetic defects (e.g., beta thalassemia) have been dramatically cured of their diseases via this method.

Tinkering with the genetic future of mankind through germline gene therapy has been prohibited largely for moral and scientific reasons. Primary among these scientific concerns is that a potential effect of gene transfer may go unrecognized for many years and become apparent only a generation or more later. Examples already exist in mice of improperly regulated gene function resulting in offspring with a high incidence of cancer in adult life. Beyond the scientific concerns lie deep-rooted theological traditions, and cultural and legal barriers. Notwithstanding these barriers, extensive experimental work continues and is expected to reach a point within twenty years where the potential for curing genetic diseases may be such as to warrant a reexamination of the issues and the balance of benefits.

A Challenge for the Future

The promise of gene therapy was seriously set back in 1999, following the death of a young man with a rare biochemical genetic disorder (ornithine transcarbamylase deficiency). He had received the corrective gene via a modified virus, which in retrospect appeared to have activated his entire immune system, resulting in his death from respiratory distress syndrome. A few other deaths also have occurred during gene therapy for other disorders, but the outcomes in these cases were not as clearly due to the novel, experimental therapy. Despite these setbacks, there is room for optimism about the future of gene therapy, as reflected in ongoing successes.

Definitive cures of serious genetic disorders using gene therapy are still rare, but the remarkable advances in genetic biotechnology promise further successes in the foreseeable future. Meanwhile, know the details of your family history, and determine the need for a consultation with a clinical geneticist, with the help of your doctor or by a direct call to a geneticist or genetic counselor. For now, this is the surest way to secure your health and even save your life.

APPENDIX A

Some Frequently Asked Questions

Given the more than 8,500 genetic disorders and traits that have been catalogued, there may well be an infinite number of questions stemming from your personal and family experience. Because space limitations exist, I have selected questions commonly asked of me during consultations. Also, rather than attempt all-inclusiveness, I have given the quintessential information in response, emphasizing the type of inheritance (dominant = D; recessive = R; sex-linked = X-L; mitochondrial = M; multiple genes and environment = P), and where possible providing information that will assist you in securing the best genetic health guidance possible. Consultation with a clinical geneticist is recommended for more complete information pertinent to you and your family.

My father developed Parkinson's disease at 49 years of age. He is now 60 and seriously disabled. What are my risks of developing this disorder?

Parkinson's disease (D, rarely R) is a degenerative disorder of a specific area of the brain that influences movement. As such, it is among the most frequent of the degenerative disorders, occurring in 1 or 2 percent of those older than 65 years of age. A constant tremor, stiffness, and slowing of movement are characteristic. In about 90 percent of those who fall victim, the cause is not known. At least 4 susceptibility genes have been identified for the dominant form of Parkinson's disease. Early-onset (before 50) Parkinson's disease is almost always due to a single gene (D or R). Mutations in one particular gene (called *parkin*) account for about half of those with early-onset Parkinson's disease. The *parkin* gene causes the recessive form of Parkinson's disease and the very early-onset type of isolated juvenile type Parkinson's disease that is seen at or even before the age of 10 years.

In response to the question: The risk is likely to be very small if recessive inheritance applies (see Chapter 7). The affected father must have received a defective *parkin* gene from each of his parents.

The son, who is asking the question, will, however, be an obligatory carrier of the *parkin* gene mutation. Only if his wife also carries a *parkin* gene mutation would they have a 1 in 4 likelihood of bearing a child who would eventually develop Parkinson's disease.

Medications are valuable in controlling the movement disorder in this condition; and recently, some success has been seen following injection of fetal cells into the ailing portion of the brain in affected individuals.

In his late teens, my brother developed a constant twitch involving his face, head, neck, and shoulders. He also swears uncontrollably while twitching. No one else in the family seems to have this problem, although I do notice that my father has an occasional twitch. I am now over 25 years of age and am still worried about developing this disorder. Is this condition genetic?

This disorder, first described in France in 1825, is called the Gilles de la Tourette syndrome, having been named for the physician who described its various features. Primary among these are chronic, uncontrollable twitching involving the face, head, neck, and shoulders, which in severe cases may be accompanied by obscene, sudden, involuntary shouts. The vast majority of those who develop this troubling disorder do so between 2 and 15 years of age. It is therefore highly likely that the brother who is asking the question does not have the disorder.

Extensive studies to determine the precise mode of inheritance for this familial disorder have failed to identify any single gene. Various reasons are advanced to explain this failure, including the range of different features within a family, the lack of a precise definition in many studies, and the likely effects of various similar disorders, all possibly caused by different and interacting genes. Recently, multiple genes involving the chemical transmitters within the brain have been implicated. Individuals with Tourette syndrome or members of their close family may manifest any one or more of the following features: attention deficit disorder, learning disorders, conduct disorders, phobias (fears) and panic attacks, obsessive-compulsive disorder, depression, manic depression, and sleep problems.

My sister has multiple sclerosis (MS). Is this a hereditary disorder? What are my risks?

Multiple sclerosis (P), a chronic disease of the central nervous system, is regarded as a multifactorial disorder that results from the interaction of an environmental agent, such as a virus, with susceptibility genes including one or more in the HLA region (see Chapter 10). The near relatives of an affected person are at increased risk, and both the age of those related and the age of the affected person at onset of the disorder are factors that influence risk. The lifetime risk for developing MS in the general population approximates 1 in 500. A major Canadian study of families with at least one affected member provided guidance about risks to siblings.

The risk of MS to siblings who have one affected parent is higher than when there is no affected parent. Brothers with no affected parent have an age-unadjusted risk of 0.8 percent, compared with brothers with one affected parent, whose risk is 3.4 percent. Sisters without an affected parent have a risk of 2.4 percent, whereas those with one affected parent have a risk of 8.1 percent. Risks appear greater when a father is affected. For example, brothers with an affected father have a risk of 8.7 percent, in contrast to those with an affected mother, for whom the risk is basically 0 percent. Sisters with an affected father have a risk of 16.7 percent, compared with 5.4 percent for sisters with an affected mother.

Nieces, nephews, and first cousins of an affected individual on either side of the family have a risk of between 1 and 3 percent until ages 17 to 23, when the odds diminish. There is a high concordance rate in identical twins, and a higher rate than expected among nonidentical twins compared to other siblings supports the idea that exposure to an infectious agent might be triggering the disorder early in life.

We have one child with epilepsy of unknown cause. What is the likelihood that our next child will also have epilepsy?

Epilepsy is common and affects between 1 in 200 and 1 in 300 individuals. There are many known causes—both genetic and acquired. Genetic disorders associated with epilepsy include chromosomal abnormalities, single gene disorders, and specific genetic syndromes. Acquired injuries include infection, direct injury to the brain due to infection, hemorrhage, or lack of oxygen. Where no cause is discovered after full evaluation, the observed risk of recurrence, given healthy parents without epilepsy, is about 5 percent. If one parent has epilepsy of unknown cause and one child is already similarly affected, the risk of recurrence is thought to be about 10 percent.

My hobby is genealogy, and I have tracked six generations of my very large family. One great uncle had amyotrophic lateral sclerosis (ALS). Is this a genetic disorder?

ALS (also called Lou Gehrig disease, after the baseball great) is rare, affecting about 1 in 50,000 people. Weakness and wasting away of the muscles of the hands and feet, and later of the tongue, throat, and diaphragm, are the most common signs. About 90 percent of cases are sporadic, and the remaining 10 percent are familial. Most of these inherited cases are due to a single dominant gene, with a 50 percent risk that an affected person will transmit the gene to each of his or her children. A rare recessive form has also been recognized. The mean age at onset of symptoms is 46 for the dominant form, but there is also a juvenile type due to a different gene. The sporadic form on average manifests at about 56 years of age. Mutation analysis for diagnostic purposes is available for the dominant gene type.

Our only child, a daughter, has Rett syndrome. Is this condition inherited, and if so, what are our risks of having another affected child?

This devastating disorder of the brain was first described by Dr. Andreas Rett in 1966, in Vienna. The frequency is between 1 in 10,000 to 1 in 15,000 births. Except for rare exceptions, only females are affected. Development appears to be normal for almost all, at least for the first six months of life. Thereafter, between six months and four years of age, normal development ceases. Typically there is deterioration in all areas of development, including behavior, personality, and communication skills. A particularly characteristic feature is hand wringing, flapping, or clapping, coupled with clumsiness, a lack of balance, and eventually a loss of the ability to walk. Autistic behavior and lack of emotional contact combine with developing mental retardation and seizures between 4 and 7 years of age. Regression is usually relentless. Between the ages of 5 and 20, most of those affected are confined to a wheelchair, being unable to use their arms or legs and having other severe, multiple handicaps due to retarded growth as well as the other effects of the disease. This syndrome is almost invariably lethal in males, pregnancy loss being the rule with only rare exceptions of live male births. Surviving males are likely to have severe to profound mental retardation and spasticity.

The Rett syndrome gene has been identified near the tip of the long arm of the X chromosome. Various mutations have been identified in affected females. Virtually all have been brand-new mutations—that is, not found in either mother or father. Well-described exceptions, however, have been noted in some mothers and their own sisters. There is also reason to believe that an atypical spectrum exists with severely affected females without a period of normal development on one end, and females with nonspecific developmental delays and learning deficits on the other end. For both such groups, absent a definite diagnosis, diagnostic sequencing of the Rett syndrome gene is now indicated. Fortunately, the risk of recurrence for this X-linked, dominant disorder is rare. Subsequent pregnancies could be monitored through prenatal diagnosis, but only if a specific mutation has been identified.

I have a son with autism. Is this a genetic disorder, and do I have an increased risk of having a second child with this disorder?

Autism is a serious disorder of brain function characterized by impaired social interactions, problems in communication, and repetitive and stereotyped behaviors. Most cases are probably evident at birth, but the hints are so subtle that they are frequently missed. Parents and doctor may fail to notice the newborn's lack of eye-to-eye contact with the mother when feeding, or the lack of anticipation of and responsiveness to being picked up. Other typical signs as the child grows older include extreme self-isolation, aloofness, withdrawal, apparent insensitivity to pain, treating people as objects, failure to develop lan-

guage, repetitive body movement, rocking or hand flapping, self-mutilation, head bang-
ing, and so on. Certainly the developmental difficulties are apparent in the first three years
of life. About 75 percent of those with autism have mental retardation, and about one-
third have seizures. There are more males than females affected, with a ratio of about 3:1.
The general prevalence in the population is about 1 in 2,000.

A significant genetic component has been recognized through studies of twins. In the
largest study thus far, concordance was noted in the vast majority of identical twins, compared with none noted among nonidentical twins. The risk of having a second autistic
child is between 2 and 6 percent.

The cause of this developmental brain disorder appears to be due to an interaction of
susceptibility genes and environmental factors (see Chapter 7). The precise locations of
the susceptibility genes are yet to be established with certainty, although certain chromosomes (for example, 7 and 13) are prime suspects. Meanwhile, twin and family studies
have pointed to a broader spectrum of manifestations. For example, even the nonautistic,
identical twins of affected children have frequently been noted to exhibit some mild social
and communication abnormalities. Other family members of affected children also have
been observed to have a marked increase in the frequency of such difficulties.

Care should be taken to separate out autistic behavior, which may occur in association
with specific genetic disorders such as the fragile X syndrome (see Chapter 9), and tuberous sclerosis. Recognizable medical disorders with associated autism are estimated to occur in only 10 to 25 percent of cases.

**My grandmother and her sister both developed cataracts around the age of 70 years.
They both say that it was due to excess sun exposure during their childhood. However, I
wondered if this was genetic and if I am at risk.**

A cataract is an opaque lens in the eye, resulting in distorted and blurred vision. Both
opinions are correct. A major study of twins in England with cataract essentially concluded that both genes and environmental effects contribute roughly equally to the development of the common form of cataracts seen in aging. Ultraviolet radiation from sun-
light is thought to play some role in causation. There are, therefore, likely susceptibility
genes that increase the risk of cataract for those genetically predisposed.

Age-related cataract remains the most common cause of blindness worldwide.

In addition, there are a large number of genetic disorders and birth defects in which
cataracts may be found. Environmental factors that may lead to cataract at birth include
injury, radiation exposure, diseases of the mother, exposure to certain drugs, and viral illness (such as German measles). Between 10 and 25 percent of cataracts at birth are inherited or are part of a genetic disorder.

**My grandmother and my father both developed glaucoma and both have become legally
blind. Am I at risk for developing this disorder?**

Glaucoma is one of the most common causes of blindness in the world. It represents a
group of diseases causing irreversible loss of the nerve cells that line the back of the eye
(retina) and that normally receive and transfer visual images to the brain. Various types of
glaucoma exist, and the onset is mostly in adulthood, although juvenile and infantile
forms also are recognized. Causes vary, and include genetic-environmental interactions
(multifactorial disorder; see Chapter 10), and complications of other diseases.
Unfortunately, there is no cure for this condition.

The occurrence of glaucoma within families is well recognized, but the mode of inheritance is mostly complex. Some studies show an increased likelihood of glaucoma, by as

much as 5 to 20 times, among family members of an affected person as compared with families without any affected individuals. In some families, a clearly dominant form of inheritance exists, with a resultant 50 percent risk for each of the offspring of an affected person. This pattern of inheritance is decidedly uncommon, and a risk somewhere below 5 percent is more likely to be the rule. The markedly increased frequency of glaucoma among blacks has not yet been explained.

My father and paternal grandmother both have a neuromuscular condition called Charcot-Marie-Tooth disease. I believe this is not a rare disorder, and I am concerned about my personal risk, as well as the risk for my children.

Charcot-Marie-Tooth (CMT) disease is also known as hereditary motor and sensory neuropathy. Characteristic symptoms and signs include slowly progressive weakness and wasting of muscles in the legs and much later in the arms. A lack of an ankle or knee jerk (reflex) is common, and in about 70 percent of cases there is impairment of sensation, mainly in the legs. High arches in the feet are present in at least 67 percent of affected individuals, and lack of balance and some tremor in the upper limbs occur in about 33 percent. Special tests of nerve conduction assist in the clinical diagnosis. However, precise diagnoses are usually achievable by DNA or gene analysis.

CMT may be inherited as a D, R, or X-L disorder. The form of inheritance in this family is either D or X-L, and gene analysis would be necessary to clinch the precise diagnosis. Different genes causing CMT have been recognized on chromosomes 17, 1, 8, and X. Only with a precise diagnosis can one reliably assess personal and family risks. Consultation with a clinical geneticist is important in this and all other complex genetic disorders.

APPENDIX B'

Genetics Resource Organizations

An almost automatic response to hearing about a genetic disorder or birth defect that affects an individual or a loved one is to seek more information. Disease-associated support groups, a selection of which are listed below, are frequently helpful in directing individuals to important sources of information or giving local referrals to a specialist. The support groups listed below can be contacted on the Internet via the Alliance of Genetic support groups—a coalition of voluntary organizations and professionals (http://www.geneticalliance.org). Through this important organization, these and other organizations can be identified. Additional helpful information of similar type can be found through the National Maternal and Child Health Clearinghouse (http://www.nmchc.org/). The Family Village provides information for parents of individuals with disabilities (http://www.familyvillage@waisman.wisc.edu).

I have repeatedly been consulted by patients who have obtained information through the Internet, become totally confused, and not infrequently misled. This course of events has occurred because of a combination of lack of medical knowledge and understanding, incorrect interpretations of the diagnosis, misunderstanding of how specific information applies to the individual, and, in some cases, incorrect or misleading factual data put out on the Internet. Readers are cautioned to develop a healthy skepticism for data gleaned from the Internet. For reliable guidance and counseling, they should consult a clinical geneticist.

Additional genetics/health-related selected Internet sites are listed in Appendix C.

A-T Children's Project, Deerfield Beach, FL

Aarskog Syndrome Parents Support Group, Levittown, PA

AboutFace International, Toronto, Ontario, Canada

AboutFace USA, Chicago, IL

Acid Maltase Deficiency Association, San Antonio, TX

Acoustic Neuroma Association, Atlanta, GA

Adoptive Families of America, Inc., St. Paul, MN

Agenesis of the Corpus Callosum Network, Orono, ME

AHEPA Cooley's Anemia Foundation, Inc., Washington, DC

Aicardi Syndrome Newsletter, Inc., Louisville, KY

Alagille Syndrome Alliance, Tigard, OR

Albino Fellowship Lancashire, UK

Alpha 1 Association, Minneapolis, MN

Alpha One Foundation, Miami, FL

Alveolar Capillary Dysplasia Association, Naperville, IL

Alzheimer's Association, Chicago, IL

Alzheimer's Disease Education and Referral Center, Silver Spring, MD

Ambiguous Genitalia Support Network, Clements, CA

American Association of Kidney Patients, Tampa, FL

American Autoimmune Related Diseases Association, Inc., East Detroit, MI

American Behçet's Disease Association, Memphis, TN

American Cancer Society, Atlanta, GA

American Diabetes Association, Alexandria, VA

American Foundation for the Blind, New York, NY

American Foundation of Thyroid Patients, Katy, TX

American Hemochromatosis Society, Inc., Delray Beach, FL

American Juvenile Arthritis Organization, Atlanta, GA

American Porphyria Foundation, Houston, TX

American Pseudo-Obstruction and Hirschsprung's Disease Society, Inc., North Andover, MA

American Self-Help Clearinghouse and New Jersey Self-Help Clearinghouse, Cedar Knolls, NJ

American Sickle Cell Anemia Association, Cleveland, OH

American Sleep Apnea Association, Washington, DC

American Syringomyelia Alliance Project, Inc., Longview, TX

American Thyroid Association, Bronx, NY

Americans With Disabilities Act Information Line, Washington, DC

Amyotrophic Lateral Sclerosis Association, Calabasas Hills, CA

Androgen Insensitivity Support Group, Burlington, MA

Angelman Syndrome Foundation, Inc., Westmont, IL

Anorchidism Support Group, Romford, UK

Any Baby Can, Inc., San Antonio, TX

Apert Syndrome Pen Pals, Providence, RI

Apoyo al Nino Down, Washington, DC

ARVD.com [Arrhythmogenic Right Ventricular Dysplasia], Washington, DC

Association for Children with Down Syndrome, Inc., Plainview, NY

Association for Glycogen Storage Disease, Durant, IA

Association for Macular Diseases, Inc., New York, NY

Association for Neuro-Metabolic Disorders, Ann Arbor, MI

Association for Spina Bifida & Hydrocephalus, Peterborough, UK

Association for the Bladder Extrophy Community, Wake Forest, NC

Association of Birth Defect Children, Inc., Orlando, FL

Association of Congenital Diaphragmatic Hernia Research, Advocacy & Support, Creedmoor, NC

Association of Genetic Support of Australasia, Inc., Surrey Hills, NSW, Australia

Autosomal Recessive Polycystic Kidney Disease Family Support Group, Kirkwood, PA

Barth Syndrome Family Support Network, Perry, FL

Basal Cell Carcinoma Nevus Syndrome/Gorlin Syndrome Home Page

Batten Disease Support and Research Association, Reynoldsburg, OH

Because I Love You, Winnetka, CA

Beckwith-Wiedemann Support Network, Ann Arbor, MI

Benign Essential Blepharospasm Research Foundation, Inc., Beaumont, TX

Blepharophimosis, Ptosis, Epicanthus Inversus Family Network, Pullman, WA

Canadian Multiple Endocrine Neoplasia Type 1 Society, Inc., Meota, Saskatchewan, Canada

Canadian Organization for Rare Disorders, Coaldale, Alberta, Canada

Canadian Porphyria Foundation, Neepawa, Manitoba, Canada

Canavan Foundation, Inc., New York, NY

Cardiac Arrhythmias Research & Education Foundation, Inc., Irvine, CA

CardioFacioCutaneous Syndrome Family Network, Inc., Vestal, NY

Cerebro-Costo-Mandibular Syndrome Support Group, Burlington, NJ

Charcot-Marie-Tooth Association, Chester, PA

Charcot-Marie-Tooth International, St. Catharines, Ontario, Canada

CHARGE Syndrome Foundation, Inc., Columbia, MO

Children & Adults with Attention Deficit Hyperactivity Disorder, Landover, MD

Children's Craniofacial Association, Dallas, TX

Children's Liver Alliance, Staten Island, NY

Children's Medical Center, The Genetics Center, Dayton, OH

Children's PKU Network, Del Mar, CA

5p- Society, Stanton, CA

8p Duplication Support Group, Dayton, OH

Chromosome 9p- Network, Las Vegas, NV

11q Research and Resource Group, Petersburg, VA

Disorders of Chromosome 16 Foundation, Vernon Hills, IL

22q and You Center, Philadelphia, PA

Chromosome 22 Central, Timmins, Ontario, Canada

Chromosome Deletion Outreach, Boca Raton, FL

Chronic Granulomatous Disease Association, Inc., San Marino, CA

Cleft Palate Foundation, Chapel Hill, NC

Coffin-Lowry Syndrome Foundation, Sammamish, WA

Coffin-Siris Syndrome Support Group, Antioch, CA

Colorectal Cancer Network, Kensington, MD

Congenital Lactic Acidosis Support Group, Denver, CO

Conjoined Twins International, Prescott, AZ

CONNAITRE LES SYNDROMES CERE-BELLEUX, Champs-sur-Marne, France

Conversations! The Newsletter for Women Who Are Fighting Ovarian Cancer, Amarillo, TX

Cooley's Anemia Foundation, Inc., Flushing, NY

Cornelia de Lange Syndrome Foundation, Inc., Avon, CT

Corporation for Menkes Disease, Fort Wayne, IN

Craniosynostosis and Parents Support, Inc., Beaufort, SC

Crohn's & Colitis Foundation of America, Inc., New York, NY

CSID Parent Support Group, Redmond, WA

Cushing Support & Research Foundation, Boston, MA

Cyclic Vomiting Syndrome Association, Canal Winchester, OH

Cystic Fibrosis Foundation, Bethesda, MD

Cystinosis Foundation, Inc., Oakland, CA

Cystinuria Support Network, Redmond, WA

Dandy-Walker Syndrome Network, Apple Valley, MN

Depression and Related Affective Disorders Association, Baltimore, MD

Diabetes Insipidus Foundation, Inc., Pottstown, PA

Dubowitz Syndrome Information and Parent Support

Dubowitz Syndrome Parent Support Network, Vincennes, IN

Dysautonomia Foundation, Inc., New York, NY

Dystonia Medical Research Foundation, Chicago, IL

Dystrophic Epidermolysis Bullosa Research Association of America, Inc., New York, NY

Ehlers-Danlos National Foundation, Los Angeles, CA

EITAN—Israeli Union of Rare Disorders, Tel Aviv, Israel

Ellis-Van Creveld Support Group, Honeoye Falls, NY

Engelhorn Foundation for Rare Diseases, Luxembourg

Epilepsy Foundation National Office, Landover, MD

FACES, The National Craniofacial Association, Chattanooga, TN

FacioScapuloHumeral Society, Inc., Lexington, MA

Families of Spinal Muscular Atrophy, Libertyville, IL

Families with Moyamoya Support Network, Cedar Rapids, IA

Family Empowerment Network, Supporting Families Affected with FAS/FAE, Madison, WI

Fanconi Anemia Research Fund, Inc., Eugene, OR

Floating Harbor Syndrome Support Group of North America, Grand Rapids, MI

Footsteps Institute for Rare Diseases, Tacoma, WA

Foundation for Ichthyosis and Related Skin Types, Lansdale, PA

Foundation for Nager & Miller Syndromes, Glenview, IL

FRAXA Research Foundation, Inc., Newburyport, MA

Freeman-Sheldon Parent Support Group, Salt Lake City, UT

Friedreich's Ataxia Research Alliance, Arlington, VA

Gluten Intolerance Group of North America, Seattle, WA

Goldenhar Syndrome Research and Information Fund, St. Petersburg, FL

Hallervorden-Spatz Syndrome Association, El Cajon, CA

Hemochromatosis Foundation, Inc., Albany, NY

Hereditary Colon Cancer Association, Boynton Beach, FL

Hereditary Disease Foundation, Los Angeles, CA

Hermansky-Pudlak Syndrome Network, Inc., Oyster Bay, NY

HHT Foundation International, Inc., New Haven, CT

Histiocytosis Association of America, Pitman, NJ

Holt-Oram SyndromeSupport, Cleveland, UK

Human Growth Foundation, Glen Head, NY

Huntington's Disease Society of America, New York, NY

Hydrocephalus Association, San Francisco, CA

Hydrocephalus Support Group, Inc., Chesterfield, MO

Hypertrophic Cardiomyopathy Association, Hibernia, NJ

Immune Deficiency Foundation, Towson, MD

Indiana Parent Information Network, Inc., Indianapolis, IN

Inherited High Cholesterol Foundation, Salt Lake City, UT

International Children's Anophthalmia Network, Philadelphia, PA

International Dyslexia Association, Baltimore, MD

International Fibrodysplasia Ossificans Progressiva Association, Winter Springs, FL

International Joseph Diseases Foundation, Inc., Livermore, CA

International Mitochondrial Disease Network

International Organization of Glutaric Acidemia, Blairsville, PA

International Rett Syndrome Association, Clinton, MD

International Society for Mannosidosis and Related Diseases, Inc., Baltimore, MD

International Tremor Foundation, Overland Park, KS

Intersex Society of North America, Ann Arbor, MI

Intestinal Multiple Polyposis and Colorectal Cancer Foundation, Conyngham, PA

Inverted Duplication Exchange and Advocacy, Thomasville, PA

IP Support Network, Wayne, MI

Iron Overload Diseases Association, Inc., N. Palm Beach, FL

Jeune Syndrome Information and Support Network, Lorain, OH

Joubert Syndrome Foundation, Rock, MI

Juvenile Diabetes Foundation International, New York, NY

Kabuki Syndrome Network, Regina, Saskatchewan, Canada

Kids with Heart National Association for Children's Heart Disorders, Green Bay, WI

Klippel-Trenaunay Support Group, Edina, MN

Kniest Syndrome Group, Faribault, MN

Langer-Giedion Syndrome Association, Toronto, Ontario, Canada

Late Onset Tay-Sachs Foundation, Erdenheim, PA

Laurence Moon Bardet Biedl Syndrome Network, Purchase, NY

Lissencephaly Network, Inc., Fort Wayne, IN

Little People of America, Inc., Lubbock, TX

Lowe Syndrome Association, Inc., West Lafayette, IN

Lupus Foundation of America, Rockville, MD

Macular Degeneration International, Tucson, AZ

Malignant Hyperthermia Association of the United States, Sherburne, NY

Maple Syrup Urine Disease Family Support Group, Goshen, IN

March of Dimes Birth Defects Foundation, White Plains, NY

Meniere's Network, Nashville, TN

Metabolic Information Network, Dallas, TX

ML4 Foundation, Brooklyn, NY

Moebius 3Warriors2 International, Neshkoro, WI

Mommies Enduring Neonatal Death (MEND), Coppell, TX

Multiple Hereditary Exostoses Coalition, Olmstead Falls, OH

Multiple Hereditary Exostoses Support Group of the Netherlands, Olderberkoop, The Netherlands

Multiple Sclerosis Association of America, Cherry Hill, NJ

MUMS National Parent-to-Parent Network, Green Bay, WI

Muscular Dystrophy Association, Tucson, AZ

Myasthenia Gravis Foundation of America, Inc., Chicago, IL

Myoclonus Research Foundation, Fort Lee, NJ

Myotubular Myopathy Resource Group, Texas City, TX

Nail-Patella Syndrome Support Group, Holland, PA

Narcolepsy and Cataplexy Foundation of America, New York, NY

Narcolepsy Network, Inc., Cincinnati, OH

National Adrenal Diseases Foundation, Great Neck, NY

National Alliance of Breast Cancer Organizations, New York, NY

National Alopecia Areata Foundation, San Rafael, CA

National Association for Continence, Spartanburg, SC

National Association for Parents of the Visually Impaired, Watertown, MA

National Association for Pseudoxanthoma Elasticum, Inc., Denver, CO

National Ataxia Foundation, Plymouth, MN

National Depressive and Manic-Depressive Association, Chicago, IL

National Down Syndrome Congress, Atlanta, GA

National Down Syndrome Society, New York, NY

National Easter Seal Society, Chicago, IL

National Fathers' Network, Bellevue, WA

National Foundation for Ectodermal Dysplasias, Mascoutah, IL

National Fragile X Foundation, Oakland, CA

National Gaucher Foundation, Rockville, MD

National Graves' Disease Foundation, Brevard, NC

National Hemophilia Foundation, New York, NY

National Incontinentia Pigmenti
Foundation, New York, NY

National Keratoconus Foundation, Los
Angeles, CA

National Kidney Foundation, New York,
NY

National Lymphedema Network, Oakland,
CA

National Marden Walker Organization,
New Haven, KY

National Marfan Foundation, Port
Washington, NY

National MPS Society, Inc., Downingtown,
PA

National Multiple Sclerosis Society, New
York, NY

National Neurofibromatosis Foundation,
New York, NY

National Niemann-Pick Disease
Foundation, Inc., Fort Atkinson, WI

National Organization for Albinism and
Hypopigmentation, East Hampstead,
NH

National Organization for Rare Disorders,
Inc., New Fairfield, CT

National Psoriasis Foundation, Portland,
OR

National Scoliosis Foundation, Stoughton,
MA

National Self-Help Clearinghouse, New
York, NY

National Sjogren's Syndrome Association,
Phoenix, AZ

National Support Group for
Arthrogryposis, Sonora, CA

National Tay-Sachs & Allied Diseases
Association, Inc., Boston, MA

National Tuberous Sclerosis Association,
Inc., Landover, MD

National Urea Cycle Disorders
Foundation, La Canada, CA

National Vitiligo Foundation, Tyler, TX

Nephrogenic Diabetes Insipidus
Foundation, Eastsound, WA

Nephrogenic Diabetes Insipidus Network,
Pottstown, PA

Neurofibromatosis, Inc., Lanham,
MD

Nevus Network, Woodbridge, VA

Noonan Syndrome Support Group, Inc.,
The Upperco, MD

Oculopharyngeal Muscular Dystrophy
Website

Ollier/Maffucci Self-Help Group, Sumter,
SC

Online Infantile Systemic Hyalinosis
Support Group

Online Myotonic & Congenital Dystrophy
Support Group, International Freedom,
PA

Online Support for Ankylosing Spondylitis

Opitz G/BB Family Network, Grand Lake,
CO

Organic Acidemia Association, Plymouth,
MN

Organization for Myelin Disorders
Research and Support, Cincinnati, OH

Osteogenesis Imperfecta Foundation, Inc.,
Gaithersburg, MD

Oxalosis and Hyperoxaluria Foundation,
St. Louis, MO

Paget Foundation, New York, NY

Pallister Hall Foundation, White River
Junction, VT

Pallister-Killian Family Support Group,
Fort Worth, TX

Parent Assistance Committee on Down
Syndrome, Briarcliff Manor, NY

Parents and Researchers Interested in
Smith-Magenis Syndrome (PRISMS),
Francestown, NH

Parents Helping Parents, Santa Clara, CA

Parents of Galactosemic Children, Inc.,
Sparks, NV

Parkinson's Disease Foundation, Inc., New
York, NY

Pediatric Adolescent Gastroesophageal
Reflux Association, Germantown, MD

Pediatric Neurotransmitter Disease
Association, Plainview, NY

Periodic Paralysis Association, Monrovia,
CA

Peter's Anomaly Online Support Group

Peutz Jeghers Syndrome Online Support
Group, Santa Rosa, CA

Pierre Robin Network, Quincy, IL

Polycystic Kidney Research Foundation,
Kansas City, MO

Potter's Syndrome Online Support
Group

Prader-Willi Syndrome Association USA,
Sarasota, FL

Proteus Syndrome Foundation, Colorado
Springs, CO

Prune Belly Syndrome Network,
Naugatuck, CT

Pulmonary Hypertension Association,
Silver Spring, MD

PXE International, Inc., Sharon, MA

Pyridoxine-Dependent Seizures Registry,
Seattle, WA

Rare Chromosome Disorder Support
Group, Surrey, UK

Reaching Out: WAGR/Aniridia Network,
Lincoln Park, MI

Restless Legs Syndrome Foundation, Inc.,
Rochester, MN

Retinoblastoma Support Group New
England, Bow, NH

Rubinstein-Taybi Parent Group USA,
Smith Center, KS

Scleroderma Foundation, Danvers, MA

Share and Care Cockayne Syndrome
Network, Inc., Stanleytown, VA

SHARE—Pregnancy and Infant Loss
Support, Inc., St. Charles, MO

Shwachman-Diamond Syndrome
International, St. Louis, MO

Shy-Drager/Multiple System Atrophy
Support Group, Inc., Austin, TX

Sibling Information Network, Storrs, CT

Sibling Support Project, Seattle, WA

Sickle Cell Disease Association of America,
Inc., Culver City, CA

Sjogren's Syndrome Foundation, Inc.,
Jericho, NY

Smith-Lemli-Opitz Advocacy and
Exchange, Glen Mills, PA

Society for Alström Syndrome Families,
Mount Desert, ME

Sotos Syndrome Support Association,
Wheaton, IL

Spina Bifida Association of America,
Washington, DC

Stickler Involved People, Augusta, KS

Sturge-Weber Foundation, Mt. Freedom,
NJ

Sudden Arrhythmia Death Syndromes
Foundation, Salt Lake City, UT

Support and Education Network for FAS
Parents and Caregivers, Paramus, NJ

Support Organization for Trisomy 18, 13,
and Related Disorders, Rochester, NY

TEF VATER International Support
Network, Upper Marlboro, MD

Thrombocytopenia Absent Radius
Syndrome Association, Egg Harbor
Twp., NJ

Tourette Syndrome Association, Inc.,
Bayside, NY

Treacher Collins Foundation, Norwich, VT

Trichorhinophalangeal Syndromes (TRPS-
I-II-III) Mailing List

Triple X Support Group, London, UK

Trisomy 9 International Parent Support,
Highland, CA

Turner's Syndrome Society of the United
States, Minneapolis, MN

United Cerebral Palsy Association, Inc.,
Washington, DC

United Leukodystrophy Foundation, Inc.,
Sycamore, IL

United Mitochondrial Disease
Foundation, Monroeville, PA

Usher Family Support, Minneapolis, MN

VHL Family Alliance, Brookline, MA

Williams Syndrome Association, Royal
Oak, MI

Wilson's Disease Association, Brookfield,
CT

Wolf-Hirshhorn (4p-) Parent Contact
Group, Woodbridge, VA

World Arnold Chiari Malformation
Association, Newtown Square, PA

Worldwide Education and Awareness for
Movement Disorders, New York, NY

www.myotonicdystrophy.com, Crystal
Lake, IL

Xeroderma Pigmentosum Society, Inc.,
Poughkeepsie, NY

XLH Network, Bowie, MD

XXYY Online Support Group, CO

Y-ME National Breast Cancer
Organization, Chicago, IL

APPENDIX C

Genetics- and Health-Related Internet Sites

A large number of Internet sites provide health information. The reader is cautioned not to uncritically accept all such information and is encouraged to seek specialist guidance. In searching for information, preferential choices should first be for professional organizations/societies, such as the American College of Medical Genetics; and for support groups or voluntary organizations, for example, the American Cancer Society and the American Heart Association. Some helpful Internet sources are listed here.

Internet Sources

Aging
American Geriatrics Society, http://www.americangeriatrics.org/
Brain Aging and Dementia (University of California, Irvine),
    http://www.alz.uci.edu/
GenerationA, people over age 50, http://www.generationa.com/950555805.shtml
National Institute on Aging, http://www.nih.gov/nia/

Alzheimer's Disease
Alzheimer Society of Canada, http://www.Alzheimer.ca
Alzheimer's Association, http://www.alz.org/
Alzheimer's Disease, http://www.alzheimers.com/
Alzheimer's Disease Education and Referral Center, http://www.alzheimers.org/
Alzheimer's Disease International (UK-based), http://www.alz.co.uk
Alzheimer's Society, http://www.alzheimers.org.uk/

Bioethics
Bioethics.net
Canadian Bioethics Society, http://chris.macdonald@dal.ca
Human Genome Project, http://www.ornl.gov/hgmis/
National Human Genome Research Institute, http://nhgri.nih.gov/ELSI/
National Institutes of Health, http://www.nih.gov/sigs/bioethics

Birth Defects
Autism Society of America, http://www.autism-society.org
March of Dimes, http://www.modimes.org/
National Association for Down Syndrome, http://www.nads.org/
National Down Syndrome Congress, http://www.ndsccenter.org/index3.htm
National Down Syndrome Society, http://www.ndss.org/

National Fragile X Foundation, http://www.nfxf.org/body_home.htm
Spina Bifida and Hydrocephalus Association of Canada, http://www.sbhac.ca/
Spina Bifida Association of America, http://www.sbaa.org/
Brain and Nervous System
ALS Association, http://www.alsa.org/
American Academy of Neurology, http://www.aan.com/
American Epilepsy Society, http://www.aesnet.org/
American Neurological Association, http://www.aneuroa.org/
Child Neurology Society, http://www1.umn.edu/cns/
Child Neurology Web Site, http://www.waisman.wisc.edu/child-neuro/
Dystonia Medical Research Foundation, http://www.dystonia-foundation.org/dmrf.html
Epilepsy Foundation of America, http://www.efa.org/
Huntington's Disease Society of America, http://www.hdsa.org/
International League Against Epilepsy, http://www.websciences.org/engel/
Movement Disorder Society, http://www.movementdisorders.org/
Muscular Dystrophy Association (USA), http://www.mdausa.org/
National Institute for Neurological Disorders and Stroke (NINDS), http://www.ninds.nih.gov/
National Multiple Sclerosis Society (USA), http://www.nmss.org/
National Parkinson Foundation, http://www.Parkinson.org/
Neurosciences on the Internet, http://www.neuroguide.com/
Stroke Information (NINDS), http://www.ninds.nih.gov/patients/Disorders/STROKE/strokehp.htm/
Whole Brain Atlas, http://www.med.harvard.edu/AANLIB/home.html
Cancer
American Academy of Dermatology, http://www.aad.org
American Cancer Society, http://www.cancer.org
Association of Cancer Online Resources, http://www.acor.org
Cancer Facts, http://www.cancerfacts.com
Estimating Breast Cancer Risks, CD ROM, http://www.cancertrials.nci.nih.gov/forms/
    CtRiskDisk.html
Meg Walsh, http://www.oncology.com
National Alliance of Breast Cancer Organizations, http://www.nabcoinfo
National Cancer Institute, http://cancernet.nci.nih.gov/
SHARE: Self-Help for Women with Breast Cancer, http://www.sharecancersupport.org
Susan Love, M.D., http://www.SusanLoveMD.com
U.S. Cancer Institute, Atlas of Cancer Mortality in the USA, http://www.nci.nih.gov/atlas.
Deafness
Deafness Research Foundation, http://www/hearinghealth.net
HEAR, http://www.hearnet.com
Hearing Concern, http://www.hearingconcern.com
NPC, http://www.nonnoise.org
US NIDCD, http://www.nidcd.nih.gov
Diabetes
American Diabetes Association, http://www.diabetes.org/
Canadian Diabetes Association, http://www.diabetes.ca/
International Diabetes Federation, http://www.idf.org/
Joslin Diabetes Center, http://www.Joslin.org/
For the Disabled
The Council for Responsible Genetics, http://www.gene-watch.org/org.html

Counterpart.org, http://www.counterpart.org/

Designing more usable websites, http://www.trace.wisc.edu/world/web/

Eye on Disability, http://www.disabilitynow.org.uk

Family Voices, http://www.familyvoices.org/

From the Window, http://www.atschool.eduweb.co.uk/hojoy/hojoy/cv.html

Health Field, http://www.coe.fr/soc-sp/

Rehabilitation International, http://www.rehab-international.org/

U.S. Government, Information Resource for People with Disabilities, http://www.disability.gov

User-Friendly URLs Bobby, http://www.cast.org/bobby/

Web Accessibility Initiative, http://www.w3.org/WAI/

WebABLE, http://www.webable.com

WeMedia, http://www.wemedia.com

Genetics

American College of Medical Genetics, http://www.faseb.org/genetics/acmg/acmenu.htm

Blazing a Genetic Trail, http://www.hhmi.org/GeneticTrail

Centers for Disease Control, http://www.cdc.gov/genetics

Contact a Family, http://www.cafamily.org.uk/index.html

Genes and disease, http://www.ncbi.nlm.nih.gov/disease/

Genetic Conditions/Rare Conditions Support Groups & Information, http://www.kumc.edu/gec/support/

Genetics Resource Center, http://www.pitt.edu/~edugene/resource/

Human Genome Project Information, http://www.ornl.gov/hgmis/

National Coalition for Health Professional Education in Genetics, http://www.nchpeg.org

National Organization for Rare Disorders, http://www.pcnet.com/~orphan/

Office of Rare Diseases, http://rarediseases.org/

Online Mendelian Inheritance In Man, http://www3.ncbi.nlm.nih.gov/omim

Progress Eucational Trust (PET), UK, BioNews, http://www.progress.org.uk/news

Rare Genetic Diseases in Children, http://mcrcr2.med.nyu.edu/murphp01/homenew.htm

Sickle Cell Information Center, http://www.emory.edu/PEDS/SICKLE/

U.S. National Institutes of Health, Talking Glossary of Genetics, http://www.nhgri.nih.gov/DIR/VIP/Glossary

Health and General

American Medical Association, http://www.ama-assn.org/

American Medical Infomatics Association, http://www.amia.org/

California Health Care Foundation: E-Reports, http://ehealth.chcf.org

Center for Disease Control Prevention guidelines, http://aepo-xdv-www.epo.cdc.gov/wonder/prevguid/prevguid.htm

Health on the Net Foundation, http://www.hon.ch

Health Privacy Project, http://www.healthprivacy.org

Internet Healthcare Coalition Draft Guidelines, http://www.ihealthcoalition.org/ethics/draftcode.html

NHS Direct, http://www.nhsdirect.nhs.uk

[U.S.] Federal Trade Commission, http://www.ftc.gov

U.S. National Academies, http://www.national-academies.org/headlines/

U.S. National Library of Medicine, http://www.medlineplus.gov

World Health Organization, http://www.who.int/

Heart Disease

American Heart Association, Congestive Heart Failure, http://www.americanheart.org/chf/

LQT European Information Center, http://www.bielnews.ch/cyberhouse/qt/engl/ drugs.html

National Heart, Lung, and Blood Institute, http://www.nhlbi.nih.gov/

Sudden Arrhythmia Death Syndrome Foundation, http://www.sads.org/

Infectious Disease

Centers for Disease Control and Prevention, http://www.cdc.gov/

Scottish Centre for Infection and Environmental Health, Fit for Travel, http://www.axl.co.uk/scieh

U.S. Food and Drug Administration, Foodborne Disease, http://vm.cfsan.fda.gov/%7Emow/intro.html

Mental Health

National Alliance for Research on Schizophrenia and Depression, http://www.mhsource.com/narsad/

National Depressive and Manic-Depressive Association, http://www.ndmda.org/

National Service Framework for Mental Health, http://www.doh.gov.uk/nst/mental-health.htm

World Fellowship for Schizophrenia and Allied Disorders, http://www.world-schizophrenia.org/

Nutrition and Diet

American Dietetic Association, http://www.eatright.org/index.html

American Society for Clinical Nutrition, http://www.faseb.org/ascn/

Arbor Nutrition Guide, http://www.arborcom.com

BBC Online Fighting Fat, http://www.bbc.co.uk/health/fightingfat/

Cyberdiet.com, http://www.cyberdiet.com/

FoodFitness, http://www.foodfitness.org.uk/

Nutrition Navigator, http://www.navigator.tufts.edu/

U.S. Food and Drug Administration, http://www.fda.gov/fdahomepage.html

UK Health Education Authority, Fast Food Information, http://www.thinkfast.co.uk/

USDA Food and Nutrition Information Center, http://www.nalusda.gov/fnic/

Pregnancy, Parenting, Population, and Obstetrics/Gynecology

6 Billion Human Beings, http://www.popexpo.net/eMain.html

American College of Obstetricians and Gynecologists, http://www.acog.com/

BabyCenter, http://www.babycenter.com

British Pregnancy Advisory Service, http://www.bpas.org

Gyne-Web, http://www.gyneweb.fr

International Planned Parenthood Federation, http://www.ippf.org

Labor of Love, The, http://www.thelaboroflove.com

OBGYN.net, http://www.parenthoodweb.com

U.S. National Academies, http://www.national-academies.org/headlines/

Zero to Three, Boston University and the Erikson Institute, Zeroing in on the early years, http://www.zerotothree.org

Sleep Disorders

National Sleep Foundation, http://www.sleepfoundation.org/

U.S. National Center on Sleep Disorders Research, http://www.nhlbi.nih.gov/about/nscdr

Smoking

Action on Smoking and Health (ASH), http://ash.org

Center for Disease Control, Tobacco Information and Prevention Source, http://www.cdc.gov/tobacco/index.htm

Nicotine Anonymous, http://rampages.onramp.net/~nica/

Quitnet, http://www.quitnet.org/qn_main.jtml

University of Geneva, Institute of Social and Preventive Medicine, http://www.stop-tabac.ch

Speech, Language, and Hearing

American Speech-Language-Hearing Association, http://www.asha.org/

Transplantation

CenterSpan, http://www.centerspan.org

Eurotransplant, http://www.eurotransplant.nl

International Society for Heart and Lung Transplantation, http://www.ishlt.org/

TransWeb, http://www.transweb.org

United Network for Organ Sharing, http://www.unos.org/

APPENDIX D

Reference Books

The enormous amount of information in this book is based upon published research studies over many years. Since this is not a medical text, the decision was made not to quote the thousands of research studies upon which this book is based, but rather to provide only a list of the books that were consulted. All facts quoted in this book have published reliable sources.

The books are listed under selected categorical headings rather than by chapter. This method obviates the need for multiple repetition of the same book in many chapters. By and large, all these references are medical or academic texts.

Genes, Chromosomes, and Genetic Disorders

Alberts, M. J., ed. *Genetics of Cerebrovascular Disease.* Armonk, NY: Future, 1999.

Baraitser, M. *The Genetics of Neurological Disorders*, 3rd ed. New York: Oxford University Press, 1997.

Barker, D. J. P. *Mothers, Babies, and Disease in Later Life.* London: BMJ Publishing Group, 1994.

Beighton, P., ed. *McKusick's Heritable Disorders of Connective Tissue*, 5th ed. St. Louis, MO: Mosby-Year Book, 1993.

Belldegrun, A., Kirby, R. S., and Oliver, R.T.D. *New Perspectives in Prostate Cancer.* Oxford, England: Isis Medical Media, 1998.

Cavalli-Sforza, L. L. *Genes, Peoples, and Languages.* New York: Farrar, Straus & Giroux, 2000.

Cavalli-Sforza, L. L., Menozzi, P, and Piazza, A. *The History and Geography of Human Genes.* NJ: Princeton University Press, 1994.

Cohen, M. M., Jr. *The Child with Multiple Birth Defects*, 2nd ed. Oxford Monographs on Medical Genetics, Oxford University Press, 1996.

Compston, Ebers G., Lassmann, H., McDonald, I., Matthews, B., and Wekerle, H. *McAlpine's Multiple Sclerosis.* New York: Churchill Livingstone, 1998.

Friedman, J. M., Gutmann, D. H., MacCollin, M., and Riccardi, VM. *Neurofibromatosis,* 3rd ed. Baltimore, MD: Johns Hopkins University Press, 1999.

Friedmann, Theodore. *The Development of Human Gene Therapy.* Cold Spring Harbor, NY: 1998.

Frieri, M., Kettelhut, B., ed. F*ood Hypersensitivity and Adverse Reactions: A Practical Guide for Diagnosis and Management.* New York: Marcel Dekker, 1999.

Gardner, R. J. M., Sutherland, G.R. *Chromosome Abnormalities and Genetic Counseling,* 2nd ed. New York: Oxford University Press, 1996.

Gehring, W. J. *Master Control Genes in Development and Evolution: The Homeobox Story.* New Haven, CT: Yale University Press, 1998.

Gelehrter, T. D., Collins, F. S., and Ginsburg, D. *Principles of Medical Genetics,* 2nd ed. Baltimore, MD: Williams & Wilkins, 1998.

Glover, T. D., Barratt, and C. L. R., eds. *Male Fertility & Infertility.* Cambridge, England: Cambridge University Press, 1999.

Hagerman, R. J., and Cronister, A. *Fragile X Syndrome: Diagnosis, Treatment and Research,* 2nd ed. Baltimore, MD: Johns Hopkins University Press, 1996.

Jones, K. L. *Recognizable Patterns of Human Malformation,* 5th ed. Philadelphia, PA: W. B. Saunders Company, 1997.

Jorde, L. B., Carey, J. C., Bamshad, M. J., and White, R. L. *Medical Genetics,* 2nd ed. St. Louis, MO: Mosby-Year Book, 1999.

Lechler, R., ed. *Handbook of HLA and Disease.* London, England: Academic Press, 1994.

McKusick, V. A. *Mendelian Inheritance in Man: A Catalog of Human Genes & Genetic Disorders,* 12th ed. Baltimore, MD: Johns Hopkins University Press, 1998.

Mueller, R. F., and Young, I. D., eds. *Emery's Elements of Medical Genetics,* 9th ed. New York: Churchill Livingstone, 1995.

Pinsky, L., Erickson, R. P., and Schimke, R. N. *Genetic Disorders of Human Sexual Development.* New York: Oxford University Press, 1999.

Rimoin, D. L., Connor J. M., and Pyeritz, R. E., eds. *Emery and Rimoin's Principles and Practice of Medical Genetics,* 3rd ed. New York: Churchill Livingstone, 1997.

Scriver, R., Beaudet, A. L., Sly, W. S., and Valle, D. *The Metabolic Basis of Inherited Disease,* 6th ed. New York: McGraw-Hill, 1989.

Segal, N. L. *Entwined Lives: Twins and What They Tell Us About Human Behavior.* New York: Dutton Books, 1999.

Stevenson, R. E., Hall, J. G., and Goodman R. M. *Human Malformations and Related Anomalies.* New York: Oxford University Press, 1993.

Stevenson, R. E., Schwartz, C. E., and Schroer, R. J. *X-Linked Mental Retardation.* New York: Oxford University Press, 2000.

Strachan, T., and Read, A. P. *Human Molecular Genetics,* 2nd ed. Oxford, England: BIOS Scientific Publishers Limited, 1999.

Thurman, E., and Susman, M. *Human Chromosomes: Structure, Behavior, Effects,* 3rd ed. New York: Springer-Verlag, 1993.

Cardiovascular Disease and Hypertension

Barter, P. J., and Rye, K-A, ed. "Plasma Lipids and Their Role in Disease." In *Biology,* 5. Amsterdam, The Netherlands: Harwood Academic, 1999.

Betteridge, D. J., Illingworth, D. R., and Shepherd, J., eds. *Lipoproteins in Health and Disease.* London, England: Arnold, 1999.

Charney, P., ed. *Coronary Artery Disease in Women: What All Physicians Need to Know.* Philadelphia, PA: American College of Physicians, 1999.

Goldman, L., and Braunwald, E., eds. *Primary Cardiology.* Philadelphia, PA: W.B. Saunders Company, 1998.

Harvey, R. P., and Rosenthal, N., eds. *Heart Development*. San Diego, CA: Academic Press, 1999.

Laragh, J. H., and Brenner, B. M., ed. *Hypertension: Pathophysiology, Diagnosis, and Management*, 2nd eds. New York: Raven Press, 1995.

Diabetes and Obesity

Bouchard, C. *The Genetics of Obesity*. Boca Raton, LA: CRC Press, 1994.

Bouchard, C., and Bray, G. A., eds. *Regulation of Body Weight: Biological and Behavioral Mechanisms*. New York: J. Wiley & Sons, 1996.

Bray, G. A., Ryan, and D. H., eds. *Nutrition, Genetics and Obesity*. Baton Rouge, LA: Louisiana State University Press, 1999.

Kahn, C. R., and Weir, C. G., eds. *Joslin's Diabetes Mellitus*, 13th ed. Philadelphia, PA: Lippincott, Williams & Wilkins, 1994.

Kettelkahn, L., and Bowman, B. A. *Obesity: A Major Global Public Health Problem*. Annual Review of Nutrition, 1999.

Pickup, J. C., and Williams, G., eds. *Textbook of Diabetes*. Boston, MA: Blackwell Science, 1997.

Cancer

Benson, A. B., III, ed. *Gastrointestinal Oncology*. Boston, MA: Kluwer Academic, 1999.

Bowcock, A. M., ed. *Breast Cancer: Molecular Genetics, Pathogenesis, and Therapeutics*. Totoway, NJ: Humana Press, 1999.

DeVita, Jr., V. T., Hellman, S., and Rosenberg, S. A., eds. *Cancer: Principles and Practice of Oncology*, 6th ed. Philadelphia, PA: Lippincott, William & Wilkins, 2000.

Ellisen L. W., and Haber, D. A. "Hereditary Breast Cancer." *Annual Review of Medicine* 425(1998).

Foulkes, W. D., and Hodgson S. V., eds. *Inherited Susceptibility to Cancer: Clinical, Predictive and Ethical Perspectives*. New York: Cambridge University Press, 1998.

Holland, J. F., and Frei, E. *Cancer Medicine*, 3rd ed. Baltimore, MD: Williams & Wilkins, 1993.

Kaisary, A. V., Murphy, G. P., Denis, L., and Griffiths, K., eds. *Textbook of Prostate Cancer: Pathology, Diagnosis, and Treatment*. London, England: Martin Dunitz, 1999.

Levin, V. A., ed. *Cancer in the Nervous System*. New York: Churchill Livingstone, 1996.

Morgan, M. W. E., Warren, R., and della Rovere, G. Q., eds. *Early Breast Cancer: From Screening to Multidisciplinary Management*. London, England: Harwood Academic, 1998.

Offit, Kenneth. *Clinical Cancer Genetics: Risk Counseling & Management*. New York: John Wiley, 1998.

Roses, D. F., ed. *Breast Cancer*. New York: Churchill Livingstone, 1999.

Shaw, G. L., ed. *Cancer Genetics for the Clinician*. Boston, MA: Kluwer Academic Publishers, 1999.

Vogelstein, B., and Kinzler, K. W. *The Genetic Basis of Human Cancer*. New York: McGraw-Hill HPD, 1997.

Young, G. P., Rozen, P., and Levin, B., eds. *Prevention and Early Detection of Colorectal Cancer*. Philadelphia, PA: W.B. Saunders, 1996.

Mental Illness

McGue, M., and Bouchard, Jr., T. J. "Genetic and Environmental Influences on Human Behavioral Differences." *Annual Review of Neuroscience* 21, no.1(1998).

Alzheimer's Disease and Aging

Cassel, C. K., et al. *Geriatric Medicine*, 3rd ed. New York: Springer-Verlag, 1997.

Hales, R. E., Yudofsky, S. C., and Talbott, J. A. *Textbook of Psychiatry*, 3rd ed. Washington, DC: American Psychiatric Press, 1999.

LeWitt, P. A., and Oertel, W. H., eds. *Parkinson's Disease: The Treatment Options*. London, England: Martin Dunitz, 1999.

Post, S. G., and Whitehouse, P. J., eds. *Genetic Testing for Alzheimer Disease: Ethical and Clinical Issues*. Baltimore, MD: Johns Hopkins University Press, 1998.

Tallis R., Fillit, H., and Brocklehurst, J. C. *Brocklehurst's Textbook of Geriatric Medicine and Gerontology*, 5th ed. Edinburgh: Churchill-Livingstone, 1998.

Terry, R. D., Katzman, R., Bick, K. L., and Sisodia, S. S., eds. *Alzheimer Disease*, 2nd ed. Philadelphia, PA: Lippincott, Williams & Wilkins, 1999.

Prenatal Genetic Diagnosis

Milunsky, A., ed. *Genetic Disorders and The Fetus: Diagnosis, Prevention, and Treatment*, 4th ed. Baltimore, MD: Johns Hopkins University Press, 1998.

Genetic Counseling, Ethics and Law

Andrews, L. B., Fullarton, J. E., Holtzman, N. A., and Motulsky, A. G. *Assessing Genetic Risks: Implications for Health and Social Policy*. Committee on Assessing Genetic Risks, Institute of Medicine, National Academy Press, 1994.

Chadwick, R., Schickle, D., Have, H., and Wiesing, U., eds. *The Ethics of Genetic Screening*. Boston, MA: Kluwer Academic, 1999.

Kluger-Bell, K. *Unspeakable Losses: Understanding the Experience of Pregnancy Loss, Miscarriage, and Abortion*. New York: W.W. Norton, 1998.

Knoppers, B. M., Caulfied, T., and Kinsella, T. D., eds. *Legal Rights and Human Genetic Material*. Toronto, ONT: Emond Montgomery Publications, 1996.

Rothstein, M. A., ed. *Genetic Secrets: Protecting Privacy and Confidentiality in the Genetic Era*. New Haven, CT: Yale University Press, 1997.

Woods, J. R., Jr., and Esposito Woods, J. L., eds. *Loss During Pregnancy or in the Newborn Period: Principles of Care with Clinical Cases and Analyses*. NJ: Jannetti, Pitman, 1997.

Treatment of Genetic Disorders

Carella, A. M., ed. *Autologous Stem Cell Transplantation: Biological and Clinical Results in Malignancies*. Amsterdam, The Netherlands: Harwood Academic, 1997.

Reiffers, J., Goldman, J. M., Armitage, and J. O., eds. *Blood Stem Cell Transplantation*. St. Louis, MO: Mosby, 1998.

INDEX